Photosynthesis and Plant Development

Photosynthesis and Plant Development

Proceedings of a conference
held at the 'Limburgs Universitair Centrum',
Diepenbeek, Belgium, 23–29 July 1978

Edited by

R. Marcelle
Laboratory of Plant Physiology, Research Station of Gorsem,
B–3800 Sint-Truiden, Belgium

H. Clijsters & M. Van Poucke
Department SBM, Limburgs Universitair Centrum,
B–3610 Diepenbeek, Belgium

Springer-Science+Business Media, B.V. 1979

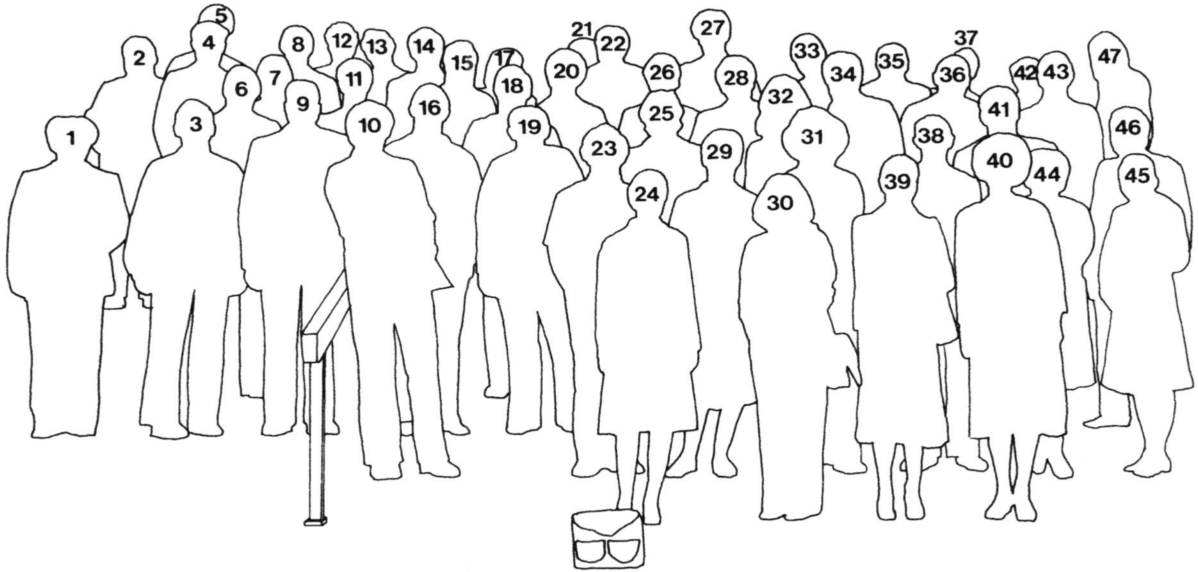

1. CARLIER 2. DUBOIS 3. OHKI 4. KANDELER 5. BUSCHMANN 6. BRUGMANS
7. PERSIJN 8. VAN POUCKE 9. FERREE 10. LANDSBERG 11. DELMOTTE 12. BUGGELN
13. VENDRIG 14. LEDENT 15. STEER 16. CHARLES-EDWARDS 17. BAERT 18. PEARSON
19. ŠESTÁK 20. BROUGHTON 21. VAN LAER 22. TREHARNE 23. TAMAS 24. LURIE
25. FARQUHAR 26. AYRES 27. LENZ 28. TERRY 29. MOUSSEAU 30. BODSON
31. VAN ASSCHE 32. BOSSELAERS 33. SACHS 34. FRIEND 35. GILLES 36. MARCELLE
37. HONNAY 38. HO 39. VAN GENECHTEN 40. CAERS 41. HABESHAW 42. BOON
43. CLIJSTERS 44. LIS 45. STARCK 46. ANTOSZEWSKI 47. VALCKE

Sponsored by

Het Ministerie voor Nationale Opvoeding en Nederlandse Cultuur, Brussels

Het Nationaal Fonds voor Wetenschappelijk Onderzoek, Brussels

De Vlaamse Leergangen te Leuven

Het Instituut tot Aanmoediging van het Wetenschappelijk Onderzoek in Nijverheid en Landbouw, Brussels

Het Limburgs Universitair Centrum, Diepenbeek

Het Opzoekingsstation van Gorsem, Sint-Truiden

ISBN 978-94-009-9627-4 ISBN 978-94-009-9625-0 (eBook)
DOI 10.1007/978-94-009-9625-0

Library of Congress Cataloging in Publication Data CIP
Main entry under title:

Photosynthesis and plant development.

Sponsored by het Ministerie voor Nationale Opvoeding en Nederlandse Cultuur, Brussels, and others.
Includes indexes.
1. Photosynthesis—Congresses. 2. Plants—Development—Congresses. I. Marcelle, R. II. Clijsters, H. III. Poucke, M. van. IV. Belgium. Ministerie van Nationale Opvoeding en Nederlandse Cultuur.
QK882.P553 518.1'3342 79–14173
ISBN 978-94-009-9627-4

Cover design: Max Velthuijs

Originally published by Dr W. Junk bv Publishers in 1979
Softcover reprint of the hardcover 1st edition 1979

CONTENTS

Session IV: Photosynthesis and source-sink relations

Session V: Photosynthesis and Nitrogen Metabolism

Session VI: Photosynthesis and Pesticides, Photosynthesis and Pathogen Infection

PREFACE

This book contains the proceedings of a symposium, held in the Limburgs Universitair Centrum, Diepenbeek, Belgium, from July 23 to 29, 1978. It can be considered as a continuation of the publication 'Environmental and Biological Control of Photosynthesis' (Dr. W. Junk b.v. Publishers, The Hague, 1975). In the last meeting, however, emphasis was much more on the biological control of photosynthesis. The sequence of the different topic sessions and papers on the symposium programme is maintained in the publication; the editors are aware of the fact that different contributions might figure equally well in more than one session.

A limited number of speakers was invited; time for discussion therefore could generously be provided to the audience. We thank all the participants for their active contribution to the success of this conception of organizing a meeting. The quality of a discussion session also depends on the performance of the discussion leader; thanks are due to Drs. G. Bernier, J.J. Landsberg, C.J. Pearson, R. Sachs, I.A. Tamas and K.J. Treharne who took the chair with scientific authority and enthusiasm. In order to keep in memory the flavor of this aspect of the meeting, an account of the discussion on one particular topic, the relationship between photosynthesis and flowering, was made by Dr. R. Sachs (reporter) and Dr. G. Bernier (discussion leader) and is published here after the reports on this topic.

The sponsorship of all organizations listed on page vi is gratefully acknowledged. We express our gratitude to Prof. Dr. L. Verhaegen, rector of the Limburgs Universitair Centrum, and to Prof. Dr. A. Soenen, director of the Research Station of Gorsem, for their encouraging support in organizing this meeting. Particular thanks are due to Dr. A. Grijpdonck, Miss T. de Backere and their staff for taking the burden of most of the administrative work. The skilful collaboration of Miss G. Vliegen, Symposium secretary, was appreciated by all the participants. We would like to thank all our assistants and technicians for their help during the meeting.

R. Marcelle, H. Clijsters and M. Van Poucke

INAUGURAL ADDRESS: PLANT DEVELOPMENT AND CROP YIELD

P.F. WAREING
(Aberystwyth, U.K.)

ABSTRACT

The influence of plant development on dry matter production and crop yield is considered under the following main headings: (1) direct effects upon photosynthesis, (2) effects on the efficiency of light interception, (3) effects on 'source/sink' relations, (4) effects on the partition and distribution of assimilates.

Developmental influences on photosynthesis include changes in activity during leaf ontogeny and plastid development, and the modulation of photosynthetic rate by growth substances.

Some of the most important effects of development on crop yield are mediated through leaf development and hence on the efficiency of light interception. Yield in both arable and grass crops can be increased by higher rates of leaf growth in the early spring. Recent work has demonstrated considerable genetic variation within species in the temperature minima and temperature coefficients for growth, and such variation offers opportunities for improvement of yield by breeding. Yield in certain cereal crops is affected by the length of life of the 'flag' leaf, which in turn is affected by temperature.

The influence of developmental processes on 'source/sink' relations is considered, especially in relation to the partition and distribution of assimilates. The possible role of growth substances in determining sink strength and in the transport of assimilates is discussed.

INTRODUCTION

The term development is used in various senses in biology, but I prefer to use it in the broadest sense as applying to the complex changes which an individual plant undergoes during its life-cycle from fertilization to death. Development normally involves both quantitative changes, *i.e.*, growth and qualitative changes in the various cells, tissues and organs of the plant body normally included in the term differentiation. A comprehensive discussion of photosynthesis in relation to development used in its broadest sense would be very wide ranging and beyond the scope of this short paper. I propose, therefore, to limit my discussion to certain aspects of the subject and to deal particularly with development in relation to crop yield. However, it may be useful, as a general introduction to this symposium, first to outline some of the broader aspects of photosynthesis in relation to development.

Developmental changes may be internally 'programmed', as illustrated, for example, by the morphological and metabolic changes occurring during the development of a leaf, but these processes may be modified by external factors, such as temperature or photoperiod. Effects of external factors fall within the scope of this symposium to the extent that they appear to evoke 'programmed' or adaptive responses affecting photosynthesis either directly or indirectly through developmental effects, as in the effect of photoperiod on leaf senescence or on the onset of flowering and fruiting.

In considering development in relation to crop yield, we are concerned not only with the photosynthetic rate of individual leaves, but also with the overall rate of dry matter production by the whole plant and this will depend not only on the photosynthetic rate but also upon the leaf area available for intercepting solar radiation. It is useful, for many purposes, to distinguish between the total capacity of the plant to produce assimilates ('source strength') and the capacity of the growth centres and storage organs to utilize assimilates ('sink strength').

Now,

$$\text{source strength} = \text{source size} \times \text{source activity}$$

i.e. rate of assimilation per plant = leaf area × rate of assimilation per unit leaf area.

By analogy,

$$\text{sink strength} = \text{sink size} \times \text{sink activity}$$

where sink activity may be defined as the rate of uptake or incorporation of assimilates per unit weight of sink tissue (Wilson, 1972; Wareing & Patrick, 1975).

Developmental factors will influence source strength by effects on the photosynthetic rate and upon the leaf area, and sink strength by effects upon the sink size and sink activity, as will be considered below.

Considering factors affecting photosynthesis directly, it is clear that the photosynthetic rate may be modified by various aspects of development, including changes occurring during the development and subsequent ageing of the individual leaf. Developmental changes within the whole plant may also affect the photosynthetic rate of individual leaves indirectly as occurs with the onset of reproduction, senescence and dormancy.

Leaf development plays a central role in determining source strength but young leaves also act as sinks. The rate of leaf development will affect the efficiency of light interception and hence dry matter production, especially during the early growth of arable crops. On the other hand, developing leaves constitute a major sink for assimilates until they commence to export assimiliates themselves. Hence the partition of assimilates as between photosynthetic and non-photosynthetic tissues is very important, and will be influenced greatly by developmental factors.

In most crop plants, only part of the plant is utilized and hence the proportion of the total dry matter which is accumulated in the useful part of the plant ('harvest index') is important, and will depend upon the sink strength of that part. As we have seen, sink strength depends both upon sink size and sink activity; both parameters are primarily determined by developmental factors, such as the number of fruits and their rates of growth (Biscoe & Gallagher, 1977).

The possible role of growth substances in the regulation of photosynthesis has been reviewed recently (Treharne, 1978) and the subject will therefore be summarized only briefly here. Some of the effects of applied growth substances on photosynthetic rate are attributable to effects upon stomatal aperture, but I shall confine my account to more direct effects upon photosynthesis.

There have been a number of reports that application of exogenous auxins, gibberellins and cytokinins promote photosynthesis in a wide range of species (Coulomb & Paquin, 1959; Lester *et al.*, 1972; Marcelle & Oben, 1973; Poskuta *et al.*, 1975; Treharne & Stoddart, 1968; Wareing *et al.*, 1968) whereas application of abscisic acid and synthetic gibberellin antagonists reduce the photosynthetic rate (Boyer, 1971; Cummins *et al.*, 1971; Kriedemann *et al.*, 1972; Kriedemann & Loveys, 1975; Marcelle & Oben, 1973). On the other hand, correlations have been demonstrated between changes in photosynthetic rate and variations in levels of endogenous growth substances in response to treatments such as water stress (Loveys & Kriedemann, 1974) and partial defoliation (Neales *et al.*, 1971; Treharne *et al.*, 1970).

A distinction has to be made between the effects of growth substances on etiolated or developing leaves, and effects upon fully mature green leaves. In etiolated tissue, benzyladenine (BA) reduced the lag phase for the production of grana and apparently promoted the synthesis of ribulose 1.5 biphosphate carboxlyase (RuBPc) (Harvey *et al.*, 1974). The synthesis of RuBPc is stimulated by the application of kinetin to etiolated leaves exposed to light (Feierabend, 1966).

Ultrastructural morphogenesis of plastids of greening cereal leaves was promoted by gibberellic acid (GA_3), but with isolated etioplasts it was not effective (Wellburn *et al.*, 1973). In cereals, RNA metabolism is stimulated by GA_3 and inhibited by ABA (Poulson & Beevers, 1970). Activity of RuBPc was enhanced by GA_3 in etiolated cereal leaves, but after greening, activity was unaffected (Treharne *et al.*, 1971).

Considering now the effects of applied growth substances on expanding and mature green leaves, treatment with GA_3 and cytokinins (kinetin and BA) were found to increase RuBPc activity and photosynthesis of fully expanded leaves, but not of young expanding leaves containing high endogenous gibberellin levels (Treharne & Stoddart, 1968; Wareing *et al.*, 1968). Since the synthesis of Fraction I protein, which includes RuBPc, in the chloroplast occurs primarily during the phase of leaf expansion with only a low turnover rate in mature leaves (Smillie *et al.*, 1967; Treharne *et al.*, 1970), it would appear that the increase in RuBPc activity observed in mature leaves was due to enzyme activation rather than to enzyme synthesis. However, RuBPc consists of a major subunit formed in the chloroplast during development, and a minor subunit formed in the cytoplasm (Graham *et al.*, 1970). Synthesis of the minor subunit in the cytoplasm may not be limited to the phase of leaf expansion and hence growth substances may increase RuBPc activity by stimulating synthesis of the minor subunit or possibly by altering the permeability of the chloroplast membrane so promoting entry of the subunit into the chloroplast.

The significance of these various observations for normal chloroplast development is obscure, but since a considerable proportion of the endogenous gibberellins of the leaf appear to be centered in the chloroplast it seems likely that they play some role in regulating photosynthesis. Moreover, since exposure of etiolated cereal leaves to red light results in rapid increases in the levels of extractable gibberellins (reviewed by Wareing & Thompson, 1976) it is possible that such changes in endogenous growth substances are involved in the phytochrome control of plastid development.

LIGHT INTERCEPTION AND LEAF DEVELOPMENT

One of the most important aspects of development affecting dry matter production and crop yield concerns the development of the leaf canopy and its effects on the efficiency of light interception. A number of investigators have shown that during the vegetative phase, the rate of dry matter production by various types of crop is proportional to the amount of radiation intercepted by the leaf canopy (Biscoe & Gallagher, 1977). This question has been further discussed recently by Monteith (1977) who has shown that the relation between the quantity of intercepted radiation and the annual production of dry matter is closely similar for crops as different as apples, barley, potatoes and sugar beet and that a composite figure of about 1.4 g/MJ, corresponding to an efficiency of about 2.4%, can be derived by pooling the data for all these crops.

Monteith derives a relation between dry matter production (C) and the maximum photosynthetic rate (Pm) for British conditions and shows that dry matter production is not very sensitive to the value assumed for Pm, so that it may be difficult to increase crop yields much by breeding for small increases in maximum photosynthetic rate. The rate of photosynthesis of many temperate crop plants is relatively insensitive to temperature over the optimum range 20–30°C, so that the yield of temperate C3 plants in Britain is limited mainly by the availability of light, although the yields of C4 crops (with their higher optimum temperatures of 30–40°C for Pm) are limited by temperature as well as light under British conditions.

Now the amount of light intercepted by a crop during its growing season is a function of the seasonal distribution of its leaf area index. Ever since the pioneering work of Watson (1952), it has been recognized that, in temperate climates, one of the major factors limiting annual dry matter production in arable crops, such as sugar beet, is the slow development of a leaf canopy in the spring so that, at a time when daily solar radiation is increasingly rapidly, only a low proportion of this radiation is intercepted by the leaves or utilized in photosynthesis. Thus the rate of leaf development in the spring is a major factor limiting overall dry matter production in temperate crops, and the main prospect of achieving more efficient light interception by the crop lies in the possibility of more rapid leaf initiation and leaf growth in the spring.

The rise in mean temperatures in the spring lags behind the improvement in light conditions by several weeks. Growth processes, such as leaf initiation and leaf growth, generally show higher temperature minima and higher temperature coefficients than does photosynthesis (Monteith & Elston, 1971) so that crop rate is frequently limited by the rate at which assimilates can be utilized in growth, rather than by the rate of assimilate production in photosynthesis.

Consequently, low temperature generally constitutes the major factor limiting plant growth, even during late spring.

It should be noted that this latter conclusion applies not only to spring-sown arable crops, but also to grassland where the problem is not inefficient light interception due to low leaf area index, but to the slow rate of growth of the leaves, which constitute the harvested part of the plant. That is to say, even in grassland, dry matter production in the spring is 'sink limited' rather than 'source limited'.

The effect of warmer spring temperatures on leaf growth and the consequent increases in crop yield are well-illustrated by the results obtained with the potato crop at Tenby in South-West Wales, where the influence of the Gulf Stream creates milder winter and early spring conditions than prevail in inland areas of Britain. The growth of potatoes at Tenby is characterized by early emergence and rapid growth of leaves so that an effective leaf surface (leaf area index 3.0) can regularly be established by late May (Allen, 1977). A high leaf area index is thereby maintained for a longer period than in other areas and very high yields of over 60 tha^{-1} can be achieved. An experiment was carried out in which seed tubers of the c.v. Maris Piper were planted on the same date at Tenby and at Sutton Bonington in the Midland region of England and subjected to the same agronomic treatments at the two sites; it was found that a more rapid increase and longer maintenance of leaf surface was achieved at Tenby than at Sutton Bonington, and tuber yields were approximately twice as great at Tenby as at the other site.

Although temperature is probably the most important factor limiting early growth in most temperate regions, in some species, such as sugar beet, it appears that daylength may also be a limiting factor in the spring (Milford & Lenton, 1976). Thus, in sugar beet plants exposed to 16 h photoperiods, the leaf area and dry weights were markedly higher than with plants grown under 12 h photoperiods, despite the fact that both groups received similar amounts of visible radiation.

An adequate leaf canopy can be developed earlier in the spring by autumn sowing or by raising seedlings in a glasshouse for subsequent transplanting into the field. The advantages of autumn sowing were demonstrated by comparing the growth and yields of autumn and spring sown wheat (Watson *et al.*, 1963). Autumn-sown wheat achieved a maximum leaf area index (L) markedly earlier than spring sown plants and yield was 20 per cent greater from the autumn sown plants. Similarly autumn-sown kale achieved an adequate value of L earlier than spring-sown plants and this was reflected in a much greater yield of dry matter at harvest (Warne, 1961). In experiments with sugar beet, seedlings raised in a glasshouse for 5 weeks and subsequently planted out of doors markedly increased the yield of roots compared with that of plants sown directly in the field (Scott & Bremner, 1966).

It is evident that great benefits to crop yield would accrue from the development of cultivars which show higher growth rates at cool temperatures. It is common knowledge that there are marked differences between species with respect to their ability to grow at cool temperatures, but there have been relatively few studies on the extent of genetic variation within a single species. However, it has been shown that there is considerable ecotypic variation with respect to growth/temperature responses within various grass species, including *Dactylis glomerata, Lolium perenne* and *Festuca arundinacea* (Cooper 1964; Eagles, 1967). Mediterranean races of these species show higher leaf growth rates during the winter months than do Scandinavian races, and studies in controlled environments have shown that there are marked differences between populations of different origins in their ability to grow at cool temperatures (Cooper, 1964; Eagles, 1967). Similar genetical variation in growth responses to temperature have recently been shown for *Festuca rubra* (Ollerenshaw, *et al*., 1976).

However, more precise information is required not only on the temperature minima for growth, but on the growth rates over the whole temperature range to which the species is normally exposed. Hitherto it has been very costly and time-consuming to obtain precise, detailed information on growth/temperature responses, because of the difficulty of securing a sufficient number of controlled environment cabinets to cover a complete range of temperatures. However, using young seedlings in a controlled environment cabinet providing a linear temperature gradient over the range 5–25°C (Mason, 1976), it is possible to obtain quite rapidly, detailed and accurate growth/temperature response curves for a wide range of genotypes (Elias, Mason & Wareing, unpubl.) Detailed studies of this type on a range of grass species has fully confirmed the results of earlier studies and has shown that Mediterranean ecotypes of *D. glomerata* have higher growth rates at lower temperatures than to Scandinavian ecotypes, but the latter have much higher temperature coefficients (Fig. 1). Between these two extremes there are various intermediate types of response curves. Similar variations in temperature responses have been shown for several other species of grass.

The differences in temperature responses demonstrated in this way, using young seedlings at the first leaf stage, were shown to be well correlated with data obtained for older plants of the same populations grown under field conditions during the winter. Thus, it is possible to obtain accurate growth temperature data for different populations rapidly and at a very young stage, so that it should be possible to use early 'screening' of large numbers of young seedlings and thereby to accelerate the breeding of grass varieties for ability to grow at low temperatures. However, the ability for early growth in the spring carries with it the danger of frost damage, and indeed the Mediterranean forms of *D. glomerata* and *F. arundinacea* are significantly less frost resistant than the Scandinavian forms. Hence, in breeding new cultivars it will be necessary to combine the capacity for early growth with adequate cold resistance. The problem of chilling sensitivity in sub-tropical species, such as maize and *Phaseolus vulgaris*, seems to be of a different nature, and apparently chilling injury is due to a phase transition of the lipids of the cell membranes from a liquid to a semi-crystalline state (Lyons, 1973).

It is clearly important to extend this type of study to arable crops, such as sugar beet and potatoes, to determine the extent of genetic variation in temperature responses within these species.

THE METABOLIC AND PHYSIOLOGICAL BASIS OF ADAPTATION TO LOW TEMPERATURE

Very little is known as to the special aspects of metabolism which confer the

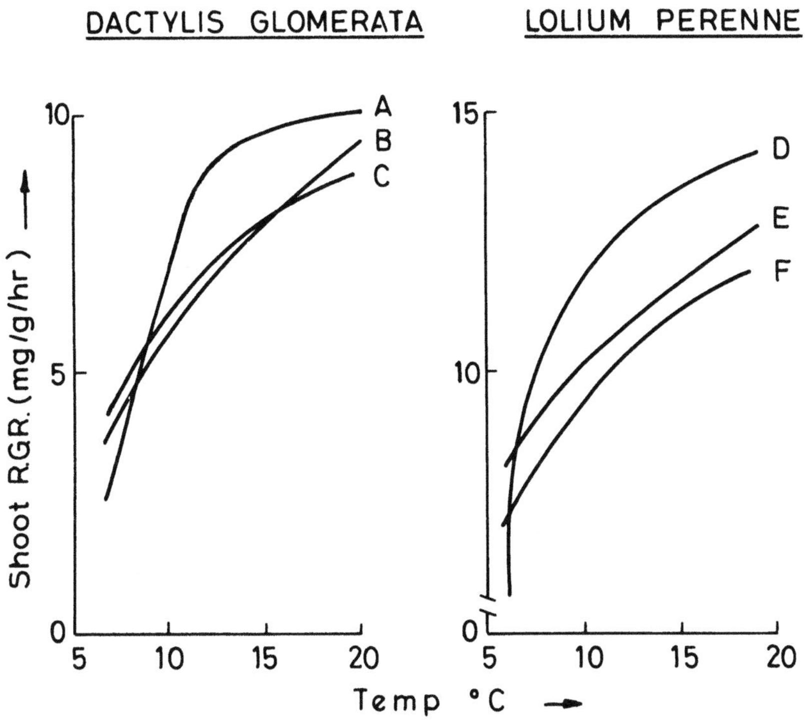

Fig. 1. Effects of temperature on relative growth rates of seedlings of *Dactylis glomerata* and *Lolium perenne*. Origin of seed parents:
a. *D. glomerata*: A – Holt, Northern Norway; B – Banat, Romania; C – Vila Viscosa, Portugal.
b. *Lolium perenne*: D – Hank, Zurich, Switzerland (1470 m altitude); E – Liguria, Italy; C-Cultivar s.23.

ability to grow at cool temperatures. There is evidence that plants adapted to grow at low temperatures have a higher respiration rate than those requiring warmer conditions (Mooney & Billings, 1961; Pisek, 1973; Ollerenshaw *et al.*, 1976; Steward and Ollerenshaw, 1977). The higher dark respiration rates, reflecting higher metabolic rates, presumably compensate for the effects of lower temperatures on metabolism and growth.

There have been a number of reports that the temperature to which the roots of plants are exposed has a marked effect on the growth of the shoots (Brouwer, 1965; Watts, 1972). Since the roots appear to be centres of cytokinin production and leaf growth is stimulated by cytokinins it was suggested that the reduced growth of leaves of grasses and other plants when the roots are maintained at cool temperatures (5°C or less) may be mediated through a reduction in the supply of root-produced cytokinins to the shoot. This possibility was investigated by Atkin *et al.* (1973) who reported that when the roots of maize were maintained at cool temperatures there was a marked reduction in the levels of endogenous cytokinins in the root sap, compared with those of plants maintained at higher temperatures.

In similar experiments with tomato plants (Menhenett & Wareing, 1975), it was found that the growth of the leaves and stems were markedly reduced in plants where the roots were maintained at 70°C compared with those of plants with a root temperature of 22°C, the air temperature being the same (22–24°C) for both groups of plants. Associated with these growth changes it was found that the flux of cytokinin and gibberellin activity in the root exudate was reduced by about 50% in the plants with reduced root temperature. Application of exogenous gibberellic acid increased the growth of stems and leaves of plants grown with low root temperature, but benzyladenine (BA) was not effective with whole plants; on the other hand, both GA_3 and BA increased the growth of leaf discs taken from plants grown at low temperatures.

In further experiments with two contrasting races of cocksfoot grass (*Dactylis glomerata*) (Menhenett & Wareing, 1976), low soil temperature resulted in markedly lower endogenous gibberellin levels in the shoots, but cytokinin levels were higher in plants grown at low root temperatures. The inhibitory effects of low root temperatures on leaf growth could be overcome by application of GA_3, but not by BA, since the shoot apical meristems and region of cell extension in grasses is close to the soil surface, it was not certain in these latter experiments that the effect of low soil temperature affected only the roots. However, the results were, so far as gibberellins were concerned, similar to those obtained with tomato, where this difficulty did not arise.

These various results suggest that the effects of low root temperatures on shoot growth may result from reduced production of gibberellins rather than cytokinins, by the roots. It is known that gibberellins may be produced in both shoots and roots (Jones & Phillips 1966; Reid & Carr, 1967), but there is no clear evidence that the shoots are dependent upon the roots for the supply of certain types of gibberellin. This is clearly a problem which merits further investigation.

LEAF PERSISTENCE AND YIELD

In grain crops, such as cereals, peas and beans, the production of new leaves is effectively terminated with the onset of the reproductive phase. Hence the assimilates accumulated in the grain are derived mainly from the existing leaves and the grain filling process can continue only so long as these leaves remain functional.

Both 'internal' and 'external' factors affect the longevity of the leaves. It is well-known that in certain species, leaf senescence is associated with the development of the fruits and that fruit removal delays senescence. There is good reason to think that the accelerating effect of developing fruits on leaf senescence is partly due to the export of carbohydrates, amino acids and other metabolites from the leaves and their accumulation in the fruits, resulting in degradation and ultimately in senescence and death of the leaf tissues. However, experiments with soybeans suggest that senescence in this species may involve a 'signal' of hormonal nature produced by the maturing seeds which induces senescence in the leaves (Lindoo & Nooden, 1977; Thimann, 1978). Thus, the grain filling process, which is dependent upon the functioning of the leaves, ultimately causes its own termination. Indeed, the ripening of monocarpic crops such as cereals and soybeans depends upon 'programmed death'. Such 'fruiting-induced' senescence is characteristic of many annual, monocarpic species, but clearly fruiting does not necessarily result in leaf senescence, as is illustrated by many of our common fruit crops such as apples and pears.

It has frequently been suggested that yield in cereals, such as wheat, in which the assimilates accumulated in the grain are largely derived from photosynthesis in the flag leaf and the ear itself, might be increased by delaying the onset of senescence and hence prolonging the grain-filling period, either by breeding or by cultural techniques. Indeed, it has been suggested that the high yields of certain barley varieties may be partly attributable to the relatively long life of the flag leaf. Moreover, it has been observed that high summer temperatures tend to reduce the life of the flag leaf so that wheat yields are higher in cool summers in Britain than in warmer ones.

Thus, whereas higher temperatures in the spring result in higher yields by promoting earlier development of an adequate leaf canopy, in cereals, at least, yield is increased by cooler temperatures, which prolong the life of the leaves. The possibility of prolonging the life of the flag leaves in cereals by chemical treatment at the appropriate stage is an attractive one, but I am not aware of any success in this direction, as yet.

SINK STRENGTH AND DEVELOPMENT

Another major impact of development on crop yield is exerted *via* effects upon sink strength. As we have seen, sink strength is determined both by 'sink size' and by 'sink activity', and developmental factors affect both parameters. It is possible for the whole plant to be either (1) source limited, when the actual rate of assimilate production is less than the potential maximum rate of consumption, or (2) sink limited, when the potential rate of production is greater than the actual rate of consumption. However, even where the accumulation of dry matter within the plant as a whole is not sinked limited, individual parts such as fruits or storage organs, may show sink limitation, so that in addition to total dry matter accumulation, it is important to consider the partition of assimilates within the plant, both (1) as between shoot and root and (2) as between the individual parts, such as leaves, stems, fruits, etc., within the shoot itself. There is much evidence, indeed, that increases in yield, both by unconscious selection in the past and by more recent breeding programmes, are due primarily to the partition of a greater proportion of total assimilates to the useful parts of the plant (*i.e.,* by a higher 'harvest index'), rather than to increased total dry matter production (Hayes, 1971; Table 1). Hence developmental factors affecting the partition of assimilates are of paramount importance in determining yield in many types of crop (Watson, 1971; Wareing & Patrick, 1975).

1. *Sink strength in the whole plant*

The total sink strength of whole plants exhibits considerable elasticity, since the growth and branching patterns of the shoots and roots are usually indeterminate, thus offering an array of competing sinks which may be expected to absorb readily any increase in assimilate production. At first sight, therefore, it would seem unlikely that overall dry matter accumulation is limited by sink strength. However, as we have already seen (p. 4), under cool temperature conditions in the spring, there is frequently a phase during which the overall growth rate is sink limited. With further rise in the mean temperature, the overall growth rate may no longer be sink limited, provided that other external factors are not limiting. However, under conditions of drought or mineral nutrient deficiency, overall growth rate may still be sink limited.

Table 1. The change in height and harvest index of barley varieties in the United Kingdom.

Variety	Date of Release	Height (cm)	Harvest Index
Spratt	Pre-1900	101	0.40
Plumage Archer	1914	93	0.49
Kenia	1933	83	0.55
Proctor	1953	81	0.54
Zephyr	1966	76	0.57

From Hayes (1971) and personal communication

There is considerable evidence that, under certain conditions, the photosynthetic rate can be modified according to the demand for assimilates (Wareing & Patrick, 1975). When source-sink ratios of whole plants are reduced (*e.g.*, by partial defoliation), the net photosynthetic rate of the remaining leaves increases, while increasing the source-sink ratios (*e.g.*, by removing fruits or shoot apices) tends to reduce the rate of assimilate accumulation. Under conditions where the potential rate of production exceeds the rate of consumption, there must be some 'feed-back' mechanism whereby the rate of assimilation is regulated to meet demand. The nature of such a feed-back control is not known but may, in part, depend upon the attainment of threshold levels of sucrose and other soluble assimilates in the source leaves (Neales & Incoll, 1968).

2. *Partition between shoot and root*

The partition of assimilates between shoot and root profoundly affects the overall growth rate of the plant. For maximum rate of dry matter production by the plant it is important that as high a proportion of assimilates as possible should be 'ploughed back' into leaf tissue and that 'expenditure of dry matter on the rest of the plant (stems, petioles, roots) should be no more than is required to support the leaves in an efficient arrangement and supply sufficient mineral nutrients and water' (Watson, 1971).

As is well-known, environmental factors and mineral nutrient conditions markedly affect the root-shoot ratio. Thus high nitrogen levels promote shoot growth as against root growth, as do also low light levels. Apparently, when there is severe competition between shoot and root for assimilates, the shoot system has the advantage.

Under any given set of environmental conditions, the root-shoot balance of a species is remarkably stable and if it is disturbed, for example by partial defoliation, the partition of assimilates for the next one to two weeks changes so that a high proportion are incorporated into new leaf tissue at the expense of root growth and so the original root-shoot balance is rapidly restored (Wareing, 1970). Evidently, there is some homeostatic mechanism maintaining a given root-shoot balance under any given set of conditions and this, in turn, implies an interplay between growth in root and shoot. Since roots supply water and mineral nutrients to the shoot and the latter supply carbohydrates to the roots, the possibility of such an interplay is immediately apparent. However, it is known that there is an interdependence between shoot and root in respect of other metabolites also, the roots supplying hormonal factors (*e.g.*, cytokinins) and certain species of amino acids to the shoot, while the latter supply vitamins, including thiamin, to the roots and these other metabolites may also be involved in the regulation of shoot-root balance.

3. *Partition within the shoot system*

Where the harvested part of the crop is a specific organ, such as a fruit, tuber or root stock, it becomes important to consider whether the growth rate of this organ is source-limited or sink-limited.

Although it is commonly assumed that assimilates move freely throughout the plant, there is frequently a characteristic pattern of assimilate movement within the shoot system and individual sinks may derive assimilates from fairly specific groups of leaves. For example, the developing wheat ear derives most of its assimilates from the uppermost leaf. Thus, source-sink relations may vary considerably within a single shoot system, so that some parts may be source limited and others sink limited. Moreover, the situation may well vary during development. For example, when developing embryos or fruits are very small their rate of accumulation of assimilates may be sink limited, but as they enlarge their sink strength will increase so that ultimately growth may become source limited.

The movements of assimilates to and from each leaf changes as it and the plant develop. A very young leaf obtains assimilates required for growth from older leaves; as it expands it enters a phase when it may be exporting assimilates although still a net importer, and then it becomes a net exporting organ with virtually no import. Initially, most of the assimilate exported from a young leaf is transported to the adjacent shoot apex. However, with the development of leaves above it, an increasingly greater proportion of its assimilates move to the roots.

Apart from the partitioning of assimilates into acropetal and basipetal moving streams, the pattern of distribution of assimilates from any one leaf is affected by competition between the various sinks within the plant. The existence of such competition is self-evident, and it is common knowledge that pruning, disbudding and fruit thinning normally lead to increased growth of the remaining sinks. There are many observations indicating that developing flowers and fruits have a high competitive ability and that the transition from vegetative to reproductive development is frequently marked by a change in pattern of assimilate distribution (Wardlaw, 1968). A developing fruit has a priority of demand on assimilates from neighbouring leaves and frequently can draw on assimilates from more distant sources. The ability of developing fruits to monopolize available assimilates is shown by the fact that the cessation of vegetative growth of the shoot and root frequently observed with the onset of reproduction can be reversed by fruit removal. Storage organs, such as potato tubers and rootstocks, also appear to dominate vegetative growth in a similar manner to fruits (Wardlaw, 1968).

Little is known as to what determines the competitive ability of a growth centre, but evidently each centre has a certain 'mobilizing ability' whereby it can 'pull' or 'attract' assimilates, against the competing abilities of other centres. The problem cannot be solved fully until the mechanism of phloem transport is better understood.

The most generally accepted theory of phloem transport is the Pressure Flow Theory, according to which the flow of dissolved organic substances depends upon the turgor pressure gradient along the path of transport. The maintenance of this turgor pressure difference between the source and sink will depend upon continual loading of sucrose into the sieve tubes at the source, and its removal at the sink, either by metabolism or by conversion into insoluble reserve substances. However, apart from the commonly accepted concept of sink activity in relation to the Pressure Flow Hypothesis, there is increasing evidence that hormones may be involved in determining sink activity and indeed in the transport process itself.

HORMONES AND THE TRANSPORT AND PARTITION OF ASSIMILATES

There is increasing evidence for the involvement of growth substances in both the transport and the partitioning of assimilates.

Recent evidence points to the possible involvement of auxins in the 'loading' of assimilates into the phloem in the leaf. Following the earlier demonstration that the uptake of sugars by bacteria, fungi and algae involves proton co-transport (Tanner *et al.*, 1977), it has been suggested (Giaquinta, 1977a, b) that the loading of sugars into phloem involves the establishment of a pH gradient between the apoplast (pH 5.6) and the sieve tube sap (pH 8–8.5) across the plasmalemma of the sieve tube. It is postulated that the proton gradient across the plasmalemma provides the driving force for sucrose uptake and is maintained by an ATP-driven proton-extrusion system. A charged complex (sucrose-H+ carrier) is held to be driven across the membrane by an electrical potential difference generated by the proton extrusion mechanism.

A very similar model for phloem loading has also been suggested by Malek and Baker (1977) who showed that loading of ^{14}C sugars into the phloem of *Ricinus communis* is stimulated by low pH and by K+ ions. A model was proposed for a proton co-transport of sugars from the apoplast driven by a linked (proton efflux/K^+ influx) pump which is ATP energized.

Further work by Malek and Baker (1978) has shown that loading into the phloem of petioles of *Ricinus* is greatly stimulated by fusicoccin, which is known to stimulate proton efflux in segments of *Avena* coleoptiles and internodes of *Pisum* (Cleland & Rayle, 1978); moreover, the auxin, indole-3-acetic acid (IAA), which also stimulates proton efflux in these latter tissues, was found to promote phloem loading in *Ricinus*, although the growth substance was less active than fusicoccin. Further work is required to determine whether endogenous IAA plays a role in normal phloem loading.

There is also evidence that auxins may be important in sink activity. The supply of assimilates to a sink involves their transfer from the phloem to the sink tissue, and this appears to occur via the apoplast, so that it will involve transfer across two membrane systems, *viz.* (1) transfer across the plasmalemma of the sieve tube into the apoplast ('unloading') and (2) uptake from the apoplast into the symplast of the sink tissue, the rate of which will be measured by its sink activity. It is probable that both unloading from the phloem and uptake by the sink tissues are active processes. The rate at which assimilates are accumulated by a sink can presumably be limited either by the rate of unloading or the rate of uptake by the apoplast.

It is now well established that endogenous growth substances are essential for all aspects of growth and hence must be required for sink activity whenever this involves the utilization of assimilates in growth, and it would seem significant that sites of active growth, such as young leaves and fruits, are normally found to contain high levels of endogenous growth substances. Thus the importance of growth substances in sink activity would be clear, even if they played no other role.

However, there is now a well-established body of evidence that application of exogenous growth substances, especially auxins and cytokinins, to non-growing tissues, such as stem internodes, can lead to increased movement of sucrose to the point of application (Davies & Wareing, 1965; Patrick & Wareing, 1973). These results can be obtained over relatively short periods (12 h) during which neither growth, nor increased metabolism of the tissues at the point of hormone application could be detected (Patrick & Wareing, 1976). Hence, it would appear that the accumulation of sucrose in response to applied growth substances is not due to metabolic demand in the ground tissue at the site of application. It would appear, rather, that the growth substances stimulate the active transfer of sucrose from the phloem to the ground tissue in the region of hormone application, by stimulating either unloading of sucrose from the phloem and/or its active uptake by the cells of the adjacent ground tissue.

Hitherto, it has been difficult to envisage how growth substances might promote the active transport of sucrose across membranes. However, the evidence that the loading of sucrose into the phloem in the leaf may involve proton co-transport (p. 13), suggests that a similar mechanism, operating in reverse, may be involved in the transfer of sucrose from the phloem into the sink tissue (Wareing, 1978). It is possible to envisage that either the unloading of sucrose from the sieve tube into the apoplast involves ATP-driven proton extrusion with active transport of sucrose in the opposite direction to that which occurs in the loading process, or alternatively, that there is proton co-transport of sucrose from the apoplast into the cells of the ground tissue. If either of these processes occur, then a possible mechanism for the promotion of sucrose accumulation at the site of application of auxins is provided, since there is considerable evidence that the mode of action of IAA in promoting cell extension involves the stimulation of an electrogenic proton pump coupled with passive K^+ uptake (Cleland & Rayle, 1978). The nature of this IAA-activated proton pump remains to be elucidated, but there are good reasons for thinking that it may involve an ATP-ase.

It remains to be seen whether the postulated involvement of growth substances in the regulation of sink activity is valid, but it is clear that a better understanding of the processes determining sink activity is required. Moreover, if the unloading of assimilates and their uptake by sink issues is under hormonal control it would seem likely that the hormonal status of a given sink may be an important factor in determining its competitive ability in relation to other sinks.

REFERENCES

Allen, E.J. Effects of date of planting on growth and yield of contrasting potato varieties in Pembrokeshire. *J. agric. Sci. Camb.* 89: *711—735* (1977).

Atkin, R.K., G.E. Barton & D.K. Robinson. Effect of root-growing temperature on growth substances in xylem exudate of *Zea mays. J. exp. Bot.* 24: *475–487* (1973).

Biscoe, P.V. & J.N. Gallagher. Weather, dry matter production and yield, in: Landsberg, J.J. & C.V. Cutting (eds.), Environmental Effects on Crop Physiology. Academic Press, London (1977).

Boyer, J.S. Nonstomatal inhibition of photosynthesis in sunflower at low leaf water potentials and high light intensities. *Plant Physiol.* 48: *532–536* (1971).

Brouwer, R. Root growth of grasses and cereals pp. 153–166, in: Milthorpe, F.L. & J.D. Ivins (eds.), The Growth of Cereals and Grasses, Butterworths, London (1965).

Cleland, R.E. & D.L. Rayle. Auxin H^+-excretion and cell elongation. *Bot. Mag. Tokyo,* Special Issue 1: *125—139* (1978).

Cooper, J.P. Climate variation in forage grasses. I. Leaf development in climatic races of *Lolium* and *Dactylis. J. app. Ecol.*, 1: *45–61* (1964).

Coulombe, L.J. & R. Paquin. Effet de l'acide gibberellique sur le metabolisme des plantes. *Can. J. Bot.*, 37: *897–902* (1959).

Cummins, W.R., H. Kende & K. Rashke. Specificity and reversibility of the rapid stomatal response to abscisic acid. *Planta* (Berl.) 99: *347–351* (1971).

Davies, C.R. & P.F. Wareing. Auxin-induced transport of radio-phosphorus in stems. *Planta* (Berl.) 65: *139–156* (1965).

Eagles, C.F. The effect of temperature in vegetative growth in climatic races of *Dactylis glomerata* in controlled environments. *Ann. Bot. N.S.* 31: *31–39* (1967).

Feierabend, J. Der Einfluss von Cytokininen auf die Bildung von Photosynthese enzymen in Roggenkeimlingen. *Planta* (Berl.) 84: *11–15* (1966).

Giaquinta, R. Possible role of pH gradient and membrane ATPase in the loading of sucrose into the sieve tubes. *Nature* (Lond.) 267: *369–370* (1977).

Giaquinta, R. Phloem loading of sucrose, pH dependence and selectivity. *Plant Physiol.*, 59: *750–755* (1977).

Graham, D., M.D. Hatch, C.R. Slack & R.N. Smillie. Light induced formation of enzymes of the C4-dicarboxylic acid pathway of the photosynthesis of detached leaves. *Phytochem.*, 9: *521–532* (1970).

Harvey, B.M.R., B.C. Lu & R.A. Fletcher. Benzyladenine accelerates chloroplast differentiation and stimulates photosynthetic enzyme activity in cucumber cotyledons. *Can. J. Bot.*, 52: *2581–2586* (1974).

Hayes, J.D. Ann. Report Welsh Plant Breeding Station for 1970, pp. 32–33 (1971).

Jones, R.L. & I.D.J. Phillips. Organs of gibberellin synthesis in light-grown sunflower plants. *Plant Physiol.*, 41: *1381–1386* (1966).

Kriedemann, P.E. & B.R. Loveys. Hormonal influences on stomatal physiology & photosynthesis pp. 227–236, in: Marcelle, R. (ed.), Environmental and Biological Control of photosynthesis. Junk, The Hague (1975).

Kriedemann, P.E., B.R. Loveys, G.C. Fuller & A.C. Leopold. Abscisic acid and stomatal regulation. *Plant Physiol.*, 49: *842–847* (1972).
Lester, D.C., O.G. Carter, F.M. Kelleher & D.R. Laing. The effect of gibberellic acid on apparent photosynthesis and dark respiration of stimulated swards of *Pennisetum clandestinum* Hochst. *J. agr. Res.*, 23: *205–213* (1972).
Lindoo, S.J. & L.D. Nooden. Studies on the behaviour of the senescence signal in Anoka soybeans. *Plant Physiol.*, 59: *1136–1140* (1977).
Loveys, B.R. & P.E. Kriedemann. Internal control of stomatal physiology and photosynthesis. I. Stomatal regulation and associated changes in levels of abscisic and phaseic acids. *Austr. J. Plant Physiol.*, 1: *407–415* (1974).
Lyons, J.M. Chilling injury in plants. *Ann. Rev. Plant Physiol.*, 24: *445–466* (1973).
Malek, F. & D.A. Baker. Proton co-transport of sugars in phloem loading. *Planta* (Berl.) 99: *275–282* (1977).
Malek, F. & D.A. Baker. The effect of fusicoccin on proton co-transport of sugars in the phloem loading of *Ricinus communis* L. *Plant Sci. Lett.,* 11, *233–239* (1978).
Marcelle, R. & G. Oben. Effects of some growth regulators on the CO_2 exchanges of leaves. *Acta Hort.*, 34: *55–58* (1973).
Mason, G. An improved temperature-gradient tunnel for use at low temperature. *Ann. Bot.* 40: *381–384* (1976).
Menhenett, R. & P.F. Wareing. Possible involvement of growth substances in the response of tomato plants (*Lycopersicon esculentum* Mill.) to different soil temperatures. *J. hort. Sci.* 50: *381–397* (1975).
Menhenett, R. & P.F. Wareing. Effects of soil temperature on the growth and hormone content of *Dactylis glomerata* L. (cocksfoot) in controlled environments. *J. exp. Bot.*, 27: *1259–1267* (1976).
Milford, G.F.J. & Lenton. J.R. Effect of photoperiod on growth of sugar beet. *Ann. Bot.*, 40: *1309–1315* (1976).
Monteith, J.L. Climate and the efficiency of crop production in Britain. *Phil. Trans. R. Soc. Lond. B.* 281: *277–294* (1977).
Monteith, J.L. & J.F. Elston. Micro-climatology and crop production. pp 23–42, in: Wareing, P.F. & J.P. Cooper (eds.), Potential Crop Production. Heinemann Educational Books, London (1971).
Mooney, H.A. & W.D. Billings. Comparative physiological ecology of arctic and alpine populations of *Oxyria digyna. Ecol. Mongr.* 31: *1–29* (1961).
Neales, T.F. & L.D. Incoll. The control of leaf photosynthesis rate by the level of assimilate concentration in the leaf; a review of the hypothesis. *Bot. Rev.* 34: *107–125* (1968).
Neales, T.F., K.H. Treharne & P.F. Wareing. A relationship between net photosynthesis, diffusive resistance and carboxylating enzyme activity in bean leaves. pp. 89–96, in: Hatch, M.D., C.B. Osmond & R.O. Slatyer (eds.), Photosynthesis and Photorespiration. Wiley-Interscience, New York (1971).
Ollerenshaw, J.H., W.S. Stewart, J. Gallimore & R.H. Baker. Low temperature growth in grasses from Northern latitudes. *J. Agric. Sci. Camb.* 87: *237–239* (1976).
Patrick, J.W. & P.F. Wareing. Auxin-promoted transport of metabolites in stems of *Phaseolus vulgaris* L. *J. exp. Bot.* 24: *1158–1171* (1973).
Patrick, J.W. & P.F. Wareing. Auxin-promoted transport of metabolites in stems of *Phaseolus vulgaris* L. *J. exp. Bot.* 27: *969–982* (1976).
Pisek, A. Effect of temperature on metabolic processes 1. Photosynthesis; 2. Respiration. pp. 102–133, in: Precht, H. *et al.* (eds), Temperature and Life. Springer-Verlag, Berlin (1973).
Poskuta, J., E. Parys, E. Ostrowska & E. Wolkowa. Photosynthesis, photorespiration, respiration and growth of pea seedlings treated with gibberellic acid (GA_3). pp. 201–209, in: Marcelle, R. (ed.), Environmental and biological control of photosynthesis. Junk, The Hague (1975).
Poulson, R. & L. Beevers. Effects of growth regulators on ribonucleic acid metabolism of barley leaf segments. *Plant Physiol.* 46: *782–785* (1970).

Reid, D.M. & D.J. Carr. Effects of a dwarfing compound, CCC, on the production and export of gibberellin-like substances by root systems. *Planta* (Berl.) 73: *1–11* (1967).
Scott, R.K. & P.M. Bremner. The effects of growth, development and yield of sugar beet of extension of the growth period by transplantation. *J. agric. Sci. Camb.* 66: *379–388* (1966).
Smillie, R.M., D. Graham, M.R. Dwyer, A. Grieve & N.F. Tobin. Evidence for the synthesis *in vivo* of proteins of the calcium cycle and the of the photosynthetic electron-transfer pathway of chloroplast ribosomes. *Biochem. Biophys. Res. Comm.* 28: 604 (1967).
Stewart, W.S. & J.H. Ollerenshaw. Intra-specific variation in rates of dark respiration and of photosynthesis at low positive temperatures in *Festuca rubra* L. pp. 419–426, in: Proc. XIII International Grassland Congress, Leipzig (1977).
Tanner, W., E. Komor, F. Fenzl & M. Decker. Sugar-proton co-transport systems. pp. 79–90, in: Marrè, E. & O. Ciferri (eds.), Regulation of Cell Membrane Activities in Plants. North-Holland Publ. Co. Amsterdam (1977).
Thimann, K.V. Senescence. *Bot. Mag. Tokyo,* Special Issue 1: *19–43* (1978).
Treharne, K.J. Photosynthesis and its hormonal control. pp. 153–158. in: Opportunities for Chemical Plant Growth Regulation 1978 (1978).
Treharne, K,J. & J.L. Stoddart. Effects of gibberellin on photosynthesis in red clover (*Trifolium pratense* L.). *Nature* (Lond.), 220: *457–458* (1968).
Treharne, K.J., J.L. Stoddart, J. Pughe, K. Paranjothy & P.F. Wareing. Effects of gibberellin and cytokinins on the activity of photosynthetic enzymes and plastid ribosomal RNA synthesis in *Phaseolus vulgaris* L. *Nature* (Lond.) 228: *129–131* (1970).
Treharne, K.J., J.L. Stoddart & C.L. Hedley. Effects of growth regulants GA_3 and CCC on the formation and activity of photosynthetic enzymes in Gramineae. pp. 2479–2509, in: Proceedings IInd International Congress on Photosynthesis (1971).
Wardlaw, I.F. The control and pattern of movement of carbohydrates in plants. *Bot. Rev.* 34: *79–105* (1968).
Wareing, P.F. Growth and its co-ordination in trees. pp. 1–21 in: Luckwill, L.C. & Cutting, C.V. (eds.) Physiology of Tree Crops. Academic Press, London & New York (1970).
Wareing, P.F. Hormonal regulation of assimilate movement. pp. 105–111, in: Opportunities for Chemical Plant Growth Regulation, Procs. Joint BCPC and BPGRG Symposium (1978).
Wareing, P.F. & J. Patrick. Source-sink relations and the partition of assimilates in the plant. pp. 481–499, in: Cooper, J.P. (ed.) Photosynthesis and Productivity in Different Environments, Cambridge University Press, London (1975).
Wareing, P.F. & A.G. Thompson. Rapid effects of red light on hormone levels. pp. 285–294, in: Smith, H. (ed.), Light and Plant Development, Butterworths, London & Boston (1976).
Wareing, P.F., M.M. Khalifa & K.J. Treharne. Rate-limiting processes in photosynthesis at saturating light intensities. *Nature* (Lond.) 220: *453–457* (1968).
Warne, L.G.G. Potential productivity of marrow stem and thousand-headed kale. *Nature* (Lond.) 192: *579* (1961).
Watson, D.J. The physiological basis of variation yield. *Adv. Agron.* 4: *101–145* (1952).
Watson, D.J. Size, structure and activity of the productive system of crops. pp. 76–88, in: Wareing, P.F. & Cooper, J.P. (eds.) Potential Crop Production, Heinemann Education Books, London (1971).
Watson, D.J., G.N. Thorne & S.A.W. French. Analysis of growth and yield of winter and spring wheats. *Ann. Bot.* 27: *1–22* (1963).
Watts, W.R. Leaf extension in *Zea mays* II. Leaf extension in response to independent variation of the temperature of the apical meristem, of the air around the leaves and of the root zone. *J. exp. Bot.* 23: *713–721* (1972).
Wellburn, F.A.M., A.R. Wellburn, T.L. Stoddart & K.J. Treharne. Influence of gibberellic and abscisic acids and the growth retardant, CCC, upon plastid development. *Planta* (Berl.) 111: *337–346* (1973).

Wilson, J. Warren. Control of crop processes. pp. 7–30, in: Rees, A.R., K.E. Cockshull, D.W. Hurd & R.G. Hard (eds.), Crop Processes in Controlled Environments. Academic Press, London & New York (1972).

Author's address:
Department of Botany and Microbiology
University College of Wales
Penglais, Aberystwyth, Dyfed SY23 3DA
United Kingdom.

SESSION I

PHOTOSYNTHESIS AND LEAF DEVELOPMENT

CHANGES IN ACTIVITIES OF PHOTOSYSTEMS DURING LEAF ONTOGENESIS

Z. ŠESTÁK
(Praha, Czechoslovakia)

ABSTRACT

The literature on changes in activities of Photosystems 1 and 2 (PS 1 and PS 2), photophosphorylation and amounts of components of the electron transport chain in the course of leaf ontogenesis is briefly reviewed. During greening of etiolated leaves the formation of PS 1 precedes that of PS 2. The amounts of cytochromes and plastoquinones usually increase simultaneously with the relative amount of P700. The leaf development in light from unfolding to the end of photosynthetic activity is characterized by two to three simultaneous maxima in activities of PS 1 and PS 2 and non-cyclic photophosphorylation of isolated chloroplasts. These peaks are reflected as shoulders in the ontogenetic course of photosynthetic gas exchange of the leaf. The ratio of activities of PS 1/PS 2 declines in the course of leaf development. The above findings are in agreement with the changes in chloroplast ultrastructure and chlorophyll *a* forms *in vivo*.

INTRODUCTION

Almost 70 years have elapsed from the time Lyubimenko (1910) published the first observations of a change in photosynthetic characteristics with leaf age. His observations of pigment changes were followed by similar ones including also photosynthetic rate in the well-known studies of Willstätter & Stoll (1918) on the 'assimilation number'. Further studies of ontogenetic changes in pigment contents and photosynthetic gas exchange were published in the following years (for their review see Šesták & Čatský, 1967) and new ones continue to appear (for reviews of ontogenetic changes in chlorophylls and carotenoids see Šesták, 1977a, 1978).

The discovery of Hill (1937) opened up a new epoch in the study of photosynthesis. The initial delight of working with a rather simple system of chloroplasts isolated from the complicated plant body was after several years replaced by sceptic observations of the various effects on the rate of the Hill reaction. In 1950, Clendenning & Gorham found that the age of leaves from which the chloroplasts have been isolated was one of important factors controlling the Hill activity.

The formulation of the electron transport chain in the 'Z-scheme' with two photosystems and coupled photophosphorylations has lately introduced the possibility that not only Photosystem 2 (PS 2) (represented by the Hill reaction) but also Photosystem 1 (PS 1) and other components of the electron transport chain may be modified during leaf ontogenesis. Having reviewed the respective literature, I found over 100 papers more or less dealing with the changes in photosystem activity and electron transport chain composition with leaf age (Šesták, 1977b). I would like to review here briefly these literature data and the results obtained with my co-workers in the Department of Physiology of Photosynthesis and Water Relations of our Institute.

Generally, there are three types of experiments studying leaf ontogenetic changes: (a) Measurements in the course of greening of etiolated leaves grown for various time periods in darkness. (b) Measurements in the course of normal development of a leaf, starting usually with leaf unfolding. (c) Comparisons of leaves of various ages in the insertion gradient at the given time.

GREENING OF ETIOLATED LEAVES

The performance of these experiments is rather easy: The plants are grown under controlled conditions in the dark and when they reach the required age, they are irradiated and the course of development of photosynthetic activities is analysed along with the formation of chlorophylls. Nevertheless, the development of leaves in the dark is different from that in the light, the area, anatomy and chemical composition of leaves at the beginning of irradiation are different from those of leaves of any age developing continuously in a normal irradiation regime. The comparison of leaves etiolated for various time periods shows not only the effect of leaf development, but also of artificial senescence introduced by the exhaustion of basic organic substrates. Hence, the results of these experiments depict only model situations, the application of which to the normal leaf development must be done with extreme care.

The chlorophyll synthesizing ability usually declines with the age of the etiolated leaf: the ability of chloroplasts to reorganize the prolamellar body membranes into primary lamellar layers decreases (Henningsen & Boynton, 1974), the content of soluble protein is lower (Obendorf & Huffaker, 1970), the regeneration of protochlorophyllide following leaf illumination is slower, the activity of enzymes of the chlorophyll biosynthesis pathway declines, etc. (for further references see Šesták, 1977a). These differences may affect also the observed courses of formation of the photochemical and biochemical apparatus of photosynthesis, as its activities and amounts are mostly related to the chlorophyll ($a + b$) content.

Detectable activity of PS 2 was observed after 2 to 6 hr illumination of etiolated leaves. In the following hours the activity mostly sharply increased, the appearance of the peak value depended on the type of PS 2 reaction (Henningsen & Boardman, 1973, and others). The start of PS 2 activity lagged ca. 2 hr after the start of chlorophyll formation, but both these characteristics developed in parallel in relation to leaf fresh weight (Plesničar & Bendall, 1973). The shape of the 'light curve' of PS 2 changed during leaf greening: its rate saturated with radiant energy was higher after 70 than 2 hr greening (Henningsen & Boardman, 1973).

The measurable activity of PS 1 was usually found sooner than that of PS 2 and peaked also sooner (Więckowski, 1971; Henningsen & Boardman, 1973; Plesničar & Bendall, 1973). The prior activity of PS 1 may be connected with the dark formation of cytochromes (cyt) f, b_6 and b-559$_{LP}$ supposed to be associated with PS 1, while the PS 2 cyt b-559$_{HP}$ is formed only after several hours of irradiation (Boardman *et al*., 1972; Plesničar & Bendall, 1972, 1973; Henningsen & Boardman, 1973). The contents of cyt b_{559}, C-550 and plastoquinones (PQ) A and C per dry matter or leaf area units increased during leaf greening in parallel with P700 (Barr & Crane, 1970; Baker & Butler, 1976).

Cyclic photophosphorylation bound to PS 1 starts usually at the same time as PS 1 (Oelze-Karow & Butler, 1971; Plesničar & Bendall, 1973), preceding thus the onset of non-cyclic photophosphorylation connected with the start of PS 2 activity (Gyldenholm & Whatley, 1968; Raïtsina, Lebedeva & Vecher, 1968).

A special model case is represented by experiments in which etiolated leaves are irradiated with flashes: they are not discussed here.

LEAF DEVELOPMENT IN LIGHT

The studies of normal leaf development are valid only when the leaf is cultivated under steady irradiance, temperature, humidity and nutrition. Even in controlled environment chambers the complications with irradiance remain, as the lately formed leaves shade the studied leaf, mostly irregularly. Therefore, the newly formed leaves are sometimes pinched off immediately after their appearance ('decapitated plants'): this procedure shifts the appearance of some cardinal points of leaf characteristics, and it is difficult to distinguish the effects of undisturbed leaf irradiance from those of disturbed growth correlations of the plant. The bending of the stem part with younger leaves is certainly a less interfering procedure.

The activity of Hill reaction as a characteristic of PS 2 has been measured in the course of leaf ontogenesis more often than the PS 1 activity (for review see Šesták, 1977b). The most popular experimental material has been the primary leaves of *Phaseolus vulgaris*. The general course found by several authors was an increase in the PS 2 rate to a maximum followed by a decline. Unfortunately, only a few papers depict the whole leaf ontogenesis, from leaf unfolding to the end of its photosynthetic activity.

More detailed studies with chloroplasts from French bean primary leaves (Strnadová & Šesták, 1974; Šesták *et al*., 1975; Šesták, Zima & Strnadová, 1977; Šesták, Zima & Wilhelmová, 1978) and pumpkin cotyledons (Harnischfeger, 1974) showed a first maximum of Hill activity after leaf unfolding or at the beginning of cotyledon development, a second peak mostly after the end of the main leaf blade growth and chlorophyll accumulation and, sometimes, a third peak preceding the end of positive photosynthetic balance of the leaf (Fig. 1). The observations of Vecher & Lebedeva (1967) and Vecher, Lebedeva & Raïtsina (1967) with pea and French bean chloroplasts and Andersen *et al*. (1972) with bundle sheath chloroplasts of the second maize leaf are in agreement with this ontogenetic course.

The course of changes in PS 1 activity (Mehler reaction) (Šesták, Zima & Strnadová, 1977) and non-cyclic photophosphorylation (Šesták *et al*., 1975; Šesták, Zima & Strnadová, 1977) in chloroplasts of primary bean leaves is similar to that of PS 2 (Figs. 1 and 3). The rather sharp second peak of all photochemical activities measured appears on the same day of leaf

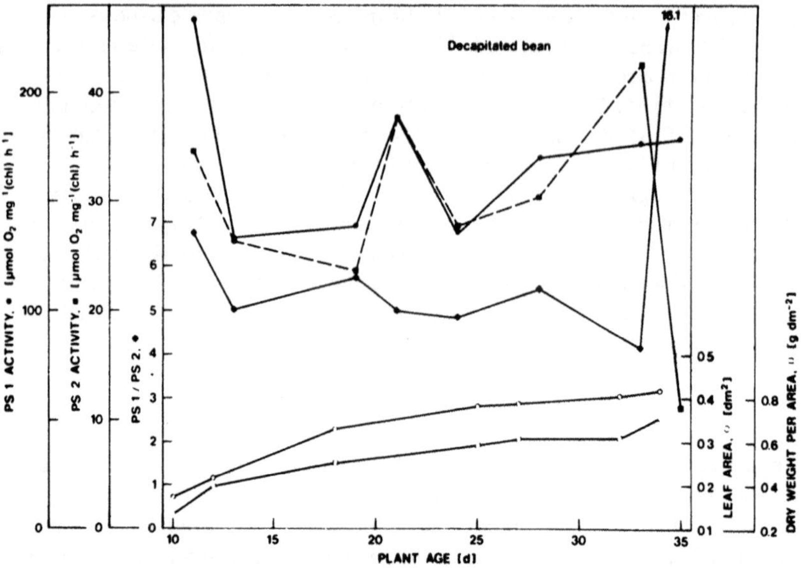

Fig. 1. Changes in the polarographically (Pt/Ag(AgOH) electrode of the Clark type) measured activities of Photosystem 1 (DCPIP red.→MV; means of 3 measurements) and Photosystem 2 ($H_2O \rightarrow K_3[Fe(CN)_6]$; means of 6 measurements) and their ratio during the ontogenesis of primary leaves of decapitated plants of *Phaseolus vulgaris* cv. Harzgruss. The leaves from which the chloroplasts were isolated (Tricine buffer, pH 7.5) are characterized by leaf area and thickness expressed as dry weight per area. (For details see Šesták, Zima & Strnadová 1977.)

development. Similar shapes of ontogenetic courses were found by Andersen *et al.* (1972) for PS 1 in mesophyll and bundle sheath chloroplasts from the second maize leaf, and by Heyes & Dale (1971) for cyclic and non-cyclic photophosphorylation in primary leaves of *Phaseolus.*

Harnischfeger (1974) supposed that this complicated ontogenetic course of PS 2 activity is induced by the changing response to pH of the reaction medium. This possibility was not confirmed in our experiments using isolation and/or reaction media of different pH (Šesták, Zima & Wilhelmová, 1978). The observed ontogenetic course was also not induced by bean cultivar, methods of determination of photosystem activities (Strnadová & Šesták, 1974) or measurement irradiance (Šesták, Zima & Strnadová, 1977). In the last third of leaf development, PS 2 activity is limited mainly by its dark reactions, the chloroplasts acquire a shade character (Fig. 2 from Šesták, Zima & Strnadová, 1977). Factors affecting the length of life span of the leaf, *e.g.*, decapitation, and the general leaf metabolism, *e.g.*, water deficit, affect the day of appearance of the second photochemical maximum or the rates of photochemical reactions, but not the general character of the ontogenetic changes (Pospíšilová, Zima & Šesták, 1976; Šesták, Zima & Strnadová, 1977).

The ratio of activities of both photosystems (PS 1/PS 2) declines during leaf ontogenesis (Pospíšilová, Zima & Šesták, 1976; Šesták, Zima & Strnadová, 1977). This decline (Fig. 1) is in agreement with the changes in the ratio of fragments enriched with PS 1 or PS 2 (Šesták, 1969): from chloroplasts of young leaves relatively more French press particles enriched with PS 1 are isolated, while in preparations from old leaves PS 2 enriched fragments prevail. Also the ontogenetic changes in the P700 content are in agreement with the changes in the PS 1/PS 2 ratio (Šesták & Demeter, 1976). Raafat, Száz & Horváth (1970)

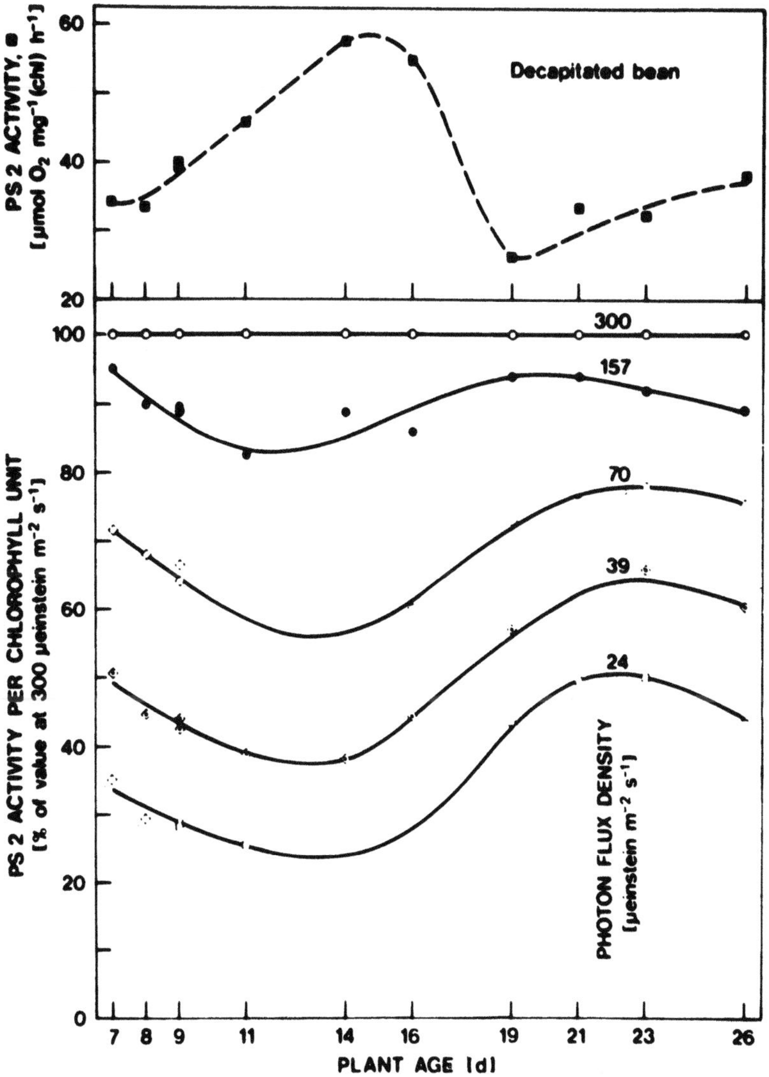

Fig. 2. Ontogenetic course of Photosystem 2 ($H_2O \rightarrow K_3[Fe(CN)_6]$) rate measured polarographically at 300 μeinstein $m^{-2}s^{-1}$ (400–700 nm) in chloroplasts from primary leaves of decapitated plants of *Phaseolus vulgaris* cv. Jantar (top) and its relative changes by measuring at lower photon flux densities (24 to 157 μeinstein m^{-2} s^{-1}) (bottom). Life span of the leaf was shortened in this experiment and hence the 1st peak of activity was not recorded. Electrode and buffer as in Fig. 1; means of 3 determinations. (From Šesták, Zima & Strnadová 1977.)

connected these changes with the ratio of 'bound' and 'not bound' chlorophyll (Chl) forms *in vivo*.

The capacity of the electron transfer sequence between PS 2 and PS 1 is sometimes characterized by the amounts of cyt and PQ: Harnischfeger (1973) found that the amount of cyt f relative to chl was steady between 10 and 38 days of pumpkin cotyledons life, it declined at the end of the cotyledon expansion, and then increased again. Also Vecher & Predkel' (1974) found an increase in cyt f and b_6 contents from 3 – 5 d to 8 – 10 d barley leaves followed by a decline. Kaval'chuk & Vechar (1973) found a steady increase in PQ amount per dry matter unit of 4 to 19 d primary leaves of rye. Hence, the changes in photosystem activities during leaf ontogenesis are connected with the changes in photosynthetic unit size and composition. Keresztés and Faludi-Dániel (1973) found in young maize leaves photosynthetic units containing 290 chl molecules, while in old leaves 863 molecules.

The ontogenetic changes in photochemical activities are not parallel to those of net photosynthetic CO_2 uptake, but the second and third photochemical maxima are reflected as shoulders in the part of the gas exchange curve (Fig. 3) declining after reaching the main maximum (Šesták *et al.*, 1975). The reflection of these changes in chloroplast and thylakoid ultrastructure remains to be found in future experiments.

LEAF INSERTION GRADIENT

The evaluation of published data on differences in photochemical activities of photosynthesis in leaves of various position on the stem shows two drawbacks of many studies: (1) leaves of only two to three age groups are compared (old – mature – young, or lower – middle – upper); (2) only rarely the measurements in ontogenetic gradient are repeated in several periods of plant development.

The data are rather variable: most often an increase in photosystem and photophosphorylation activities to a maximum near to the 'photosynthetic maturity' of the leaf is observed followed by a decline, but contradictory results (steady increase or decline) have also been published (for review see Šesták 1977b). Per dry matter unit, maximum amounts of cyt f and b_6 in the middle leaf tier of potato were found by Vecher & Predkel' (1974), while the amounts of PQ A and C increased from upper to lower tobacco leaves (Barr & Arntzen, 1969).

The maximum photochemical activities reached by a leaf generally increase from leaves formed at the beginning of plant life to leaves formed in ca. one third of vegetative phase of plant ontogenesis, and then decline.

CONCLUSIONS

Changes in the activities of photosystems occurring during leaf ontogenesis have been thoroughly studied mainly in the last ten years. The well-known character of increase to a maximum of activities followed by a decline must be complemented with secondary maxima per chlorophyll amount just after leaf unfolding and prior to the end of the positive photosynthetic balance of the leaf. The activities of PS 1, PS 2 and non-cyclic photophosphorylation develop in parallel, nevertheless the ratio of PS 1/PS 2 rates declines in the course of leaf ontogenesis. The above mentioned changes are accompanied by the variation in individual components of the photosynthetic unit. Nevertheless, exact simultaneous studies of photochemical activities, components of electron transport chain, enzymes of photochemical reactions and carboxylation, chlorophylls *in vivo*, chloroplast and thylakoid ultrastructure, and photosynthetic gas exchange during the development of individual leaves and in sequentially measured leaf insertion gradients are badly needed.

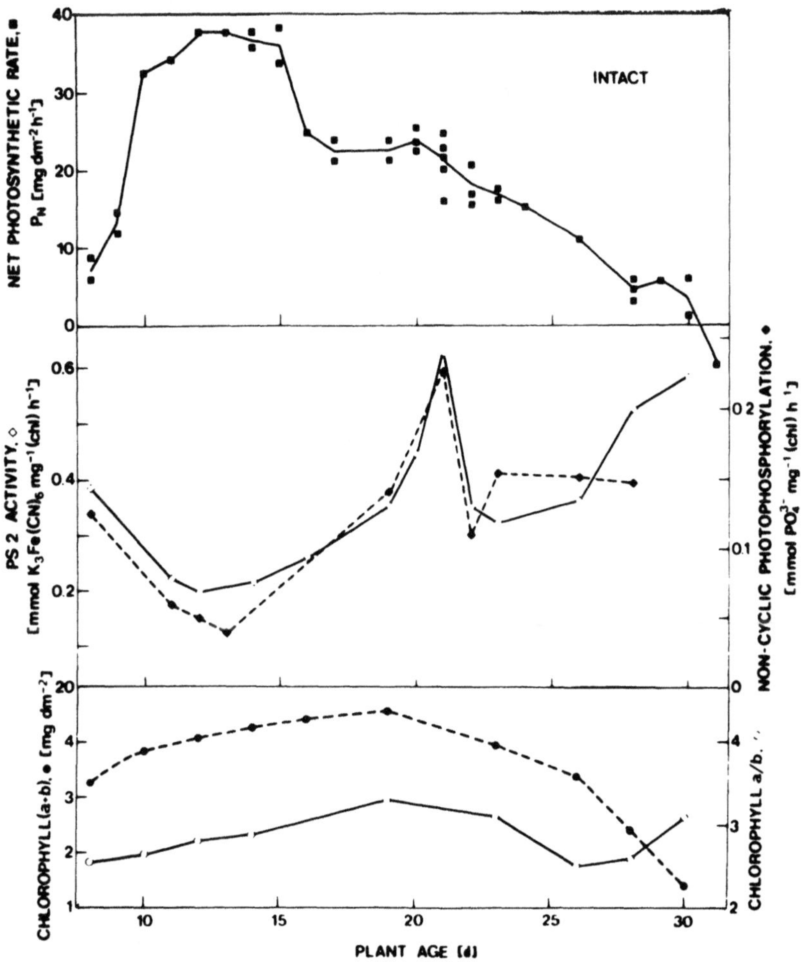

Fig. 3. The activity of non-cyclic photophosphorylation follows the course of the Photosystem 2 ($H_2O \rightarrow K_3$ [$Fe(CN)_6$]) activity (both measured spectrophotometrically) during development of primary leaves of intact plants of *Phaseolus vulgaris* L. cv. Jantar (bottom). The appearance of the second maximum of photochemical activities and an eventual increase in relative photochemical activities at the end of leaf life span are reflected in the ontogenetic course of net photosynthetic CO_2 influx (measured with infra-red CO_2 analyzers) as shoulders (top). For comparison, chlorophyll ($a + b$) amounts and a/b ratios are supplemented. (From Šesták *et al.* 1975.)

REFERENCES

Andersen, K.S., J.M. Bain, D.G. Bishop & R.M. Smillie. Photosystem II activity in agranal bundle sheath chloroplasts from Zea mays. *Plant Physiol.* 49: *461–466* (1972).

Baker, N.R. & W.L. Butler. Development of the primary photochemical apparatus of photosynthesis during greening of etiolated bean leaves. *Plant Physiol.* 58: *526–528* (1969).

Barr, R. & C.J. Arntzen. The occurrence of δ-tocopherylquinone in higher plants and its relation to senescence. *Plant Physiol.* 44: *591–598* (1969).

Barr, R. & F.L. Crane. Comparative studies on plastoquinones. V. Changes in lipophilic chloroplast quinones during development. *Plant Physiol.* 45: *53–55* (1970).

Boardman, N.K., J.M. Anderson, R.G. Hiller, A. Kahn, P.G. Roughan, T.E. Treffry & S.W. Thorne. Biosynthesis of the photosynthetic apparatus during chloroplast development in higher plants. pp. 2265–2287, in: Forti, G., M. Avron & A. Melandri (eds.), Photosynthesis, two centuries after its discovery by Joseph Priestley. Junk, The Hague (1972).

Clendenning, K.A. & P.R. Gorham. Photochemical activity of isolated chloroplasts in relation to their source and previous history. *Can. J. Res.* 28 C: *114–139* (1950).

Gyldenholm, A.O. & F.R. Whatley. The onset of photophosphorylation in chloroplasts isolated from developing bean leaves. *New Phytol.* 67: *461–468* (1968).

Harnischfeger, G. Chloroplast degradation in ageing cotyledons of pumpkin. *J. exp. Bot.* 24: *1236–1246* (1973).

Harnischfeger, G. Studies on chloroplast degradation *in vivo* II. Effects of aging on Hill activity of plastids from *Cucurbita* cotyledons. *Z. Pflanzenphysiol.* 71: *301–312* (1974).

Henningsen, K.W. & N.K. Boardman. Development of photochemical activity and the appearance of the high potential form of cytochrome b-599 in greening barley seedlings. *Plant Physiol.* 51: *1117–1126* (1973).

Henningsen, K.W. & J.E. Boynton. Macromolecular physiology of plastids. IX. Development of plastid membranes during greening of dark-grown barley seedlings. *J. Cell Sci.* 15: *31–55* (1974).

Heyes, J.K. & J.E. Dale. A virescent mutant of *Phaseolus vulgaris*: photosynthesis and metabolic changes during leaf development. *New Phytol.* 70: *415–426* (1971).

Hill, R. Oxygen evolved by isolated chloroplasts. *Nature* 139: *881–882* (1937).

Kaval'chuk, R.A. & A.S. Vechar. Lipidny sastaŭ khlaraplastaŭ prarostkaŭ zhyta. [Lipid composition of chloroplasts of rye seedlings.] *Vestsi Akad. Navuk belarus. SSR. Ser. biyal. Navuk* 1973 (4): *20–28* (1973).

Keresztes, A. & A. Faludi-Dániel. Ultrastructure, pigment content and photosynthetic activity of the normal and mutant chloroplasts in developing *Tradescantia* leaves. *Acta biol. Acad. Sci. hung.* 24: *175–189* (1973).

Lyubimenko, V.N. Soderzhanie khlorofilla v khlorofillnom zerne i energiya fotosinteza. [Chlorophyll content in a chloroplast and energy of photosynthesis.] *Tr. SPb. Obshch. Estestvozn.*, Ser. 3 – bot. 41: *1–266* (1910).

Obendorf, R.L. & R.C. Huffaker. Influence of age and illumination on distribution of several Calvin cycle enzymes in greening barley leaves. *Plant Physiol.* 45: *579–582* (1970).

Oelze-Karow, H. & W.L. Butler. The development of photophosphorylation and photosynthesis in greening bean leaves. *Plant Physiol.* 48: *621–625* (1971).

Plesničar, M. & D.S. Bendall. The development of photochemical activities during greening of etiolated barley. vol. 3. pp. 2367–2374, in: Forti, G., M. Avron & A. Melandri (eds.), Photosynthesis, two centuries after its discovery by Joseph Priestley. Junk, The Hague (1972).

Plesničar, M. & D.S. Bendall. The photochemical activities and electron carriers of developing barley leaves. *Biochem. J.* 136: *803–812* (1973).

Pospíšilová, J., J. Zima & Z. Šesták. Effect of hydration level in primary bean leaves on the activity of photosystems 1 and 2 in isolated chloroplasts. *Biol. Plant.* 18: *473–497* (1976).

Raafat, A., K. Szász & I. Horváth. Effect of leaf age on the chlorophyll fractions extracted with two different acetone concentrations. *Agrochimica* 14: *392–397* (1970).

Raïtsina, G.I., T.D. Lebedeva & A.S. Vecher. Fotokhimicheskaya aktivnost' khloroplastov izolirovannykh iz list'ev gorokha v protsesse zeleneniya. [Photochemical activity of chloroplasts isolated during greening of pea leaves.] *Dokl. Akad. Nauk belorus. SSR* 12: *731–734* (1968).

Šesták, Z. Ratio of photosystem 1 and 2 particles in young and old leaves of spinach and radish. *Photosynthetica* 3: *285–287* (1969).

Šesták, Z. Photosynthetic characteristics during ontogenesis of leaves. 1. Chlorophylls. *Photosynthetica* 11: *367–448* (1977a).

Šesták, Z. Photosynthetic characteristics during ontogenesis of leaves. 2. Photosystems, components of electron transport chain, and photophosphorylations. *Photosynthetica* 11: *449–474* (1977b).

Šesták, Z. Photosynthetic characteristics during ontogenesis of leaves. 3. Carotenoids. *Photosynthetica* 12: *89–109* (1978).

Šesták, Z. & J. Čatský. Sur les relations entre le contenu en chlorophylle et l'activité photosynthétique pendant la croissance et le vieillissement des feuilles. pp. 213–262, in: Sironval, C. (ed.), Le chloroplaste, croissance et vieillissement. Masson et Cie, Paris (1967).

Šesták, Z., J. Čatský, J. Solárová, H. Strnadová & I. Tichá. Carbon dioxide transfer and photochemical activities as factors of photosynthesis during ontogenesis of primary bean leaves. pp. 159–166, in: Nasyrov, Yu.S. & Z. Šesták (eds.), Genetic aspects of photosynthesis. Junk, The Hague (1975).

Šesták, Z. & S. Demeter. Changes in circular dichroism and P700 content of chloroplasts during leaf ontogenesis. *Photosynthetica* 10: *182–187* (1976).

Šesták, Z., J. Zima & H. Strnadová. Ontogenetic changes in the internal limitations to bean-leaf photosynthesis. 2. Activities of photosystems 1 and 2 and non-cyclic photophosphorylation and their dependence on photon flux density. *Photosynthetica* 11: *282–290* (1977).

Šesták, Z., J. Zima & N. Wilhelmová. Ontogenetic changes in the internal limitations to bean-leaf photosynthesis. 4. Effect of pH of the isolation and/or reaction medium on the activities of photosystems 1 and 2. *Photosynthetica* 12: *1–6* (1978).

Strnadová, H. & Z. Šesták. Reliability of methods used for determining ontogenetic changes in Hill reaction rate. *Photosynthetica* 8: *130–133* (1974).

Vecher, A.S. & T.D. Lebedeva. Aktivnost' reaktsii Khilla s raznymi okislitelyami v protsesse razvitiya list'ev. [Rate of Hill reaction with various oxidants during leaf ontogenesis.] *Dokl. Akad. Nauk belorus. SSR* 11: *727–730* (1967).

Vecher, A.S., T.I. Lebedeva & G.I. Raïtsina. Aktivnost' reaktsii Khilla i fotosinteticheskogo fosforilirovaniya v izolirovannykh khloroplastakh list'ev gorokha. [Rates of Hill reaction and photosynthetic phosphorylation in isolated chloroplasts of pea leaves.] *Dokl. Akad. Nauk belorus. SSR* 11: *451–454* (1967).

Vecher, A.S. & K.I. Predkel'. Porfirinsoderzhayushchie pigmenty i aktivnost' nekotorykh gemoproteidnykh oksidoreduktaz v list'yakh i plastidakh. [Porphyrin containing pigments and activity of some haemoproteid oxidoreductases in leaves and plastids.] pp. 313–319, 412, in: Shlyk, A.A. (ed.), Khlorofill. Nauka i Tekhnika, Minsk (1974).

Więckowski, S. Photosynthetic activity of bean leaves and the relative content of chlorophyll in particles isolated from chloroplasts after irradiation of etiolated seedlings. *Photosynthetica* 5: *44–49* (1971).

Willstätter, R. & A. Stoll. Untersuchungen über die Assimilation der Kohlensäure. Berlin (1918).

Author's address:
Institute of Experimental Botany
Czechoslovak Academy of Sciences
160 00 Praha 6
Flemingovo nám. 2
Czechoslovakia.

EFFECT OF LEAF AGE ON PHOTOSYNTHESIS AND PHOTORESPIRATION*

S. LURIE, N. PAZ, N. STRUCH AND B.A. BRAVDO
(Rehovot, Israel)

ABSTRACT

Four month old tobacco plants (*Nicotiana rustica*) containing seven leaves were examined leaf by leaf for net photosynthesis and CO_2 compensation point. The stomatal and mesophyll resistances were calculated. Each leaf was then ground separately and the homogenate assayed for ribulose bisphosphate carboxylase and oxygenase activity. It was found that the highest net photosynthetic activity was in the middle leaves, although the highest RuBPC activity occurred in the youngest leaves. However, the youngest leaves also had higher stomatal resistance which would lower the net photosynthesis of these leaves.

The RuBPO activity did not parallel that of RuBPC. It was lowest in the youngest and oldest leaves and maximal in the middle leaves. The ratio of RuBPC/RuBPO increased steadily from the oldest to youngest leaves and showed a good inverse correlation to the compensation point measured in whole leaves.

INTRODUCTION

There have been a number of studies on the changes which occur in the biochemical composition and metabolic activity of senescing leaves. Measurement of gas exchange by mature leaves have revealed an extended phase of declining photosynthesis as the leaves senesce (Freeland, 1952; Nixon and Wedding, 1956; Hopkinson, 1964; Šesták & Catsky, 1962; Richardson, 1957; Hardwick *et al.*, 1968). There is a similar pattern of decline in respiration with, in some species, a transient increase in respiration at later stages of foliar senescence (Hover & Gustafson, 1926). Perhaps the most striking changes in senescence are losses of chlorophyll, nucleic acids and protein (Wollgiehn, 1967; Spenser & Titus, 1973; Hardwick *et al.*, 1968) accompanied by a general increase in proteolytic activity as measured *in vitro* (Peterson & Huffaker, 1975).

Converse to the decline in photosynthesis during leaf senescence, studies of CO_2 uptake in young leaves show that photosynthesis increases as the leaves expand (Jewiss & Woledge, 1967; Rawson & Woodward, 1976; Woodward & Rawson, 1976). In effect during its life a plant is composed of a changing number of leaves whose photosynthetic capabilities do not remain constant. The

*The authors would like to thank Shoshana Shalom for invaluable technical assistance. This research was partially funded by a grant from the Central Fund of the Hebrew University of Jerusalem and from the Israeli American binational foundation.

changes in photosynthesis may be due to changes in stomatal resistance, increase or decrease in photosynthetic enzyme activity, a change in the activity of the photorespiratory pathway, or a combination of these factors. To see which exerts the major control on photosynthesis at different leaf age, tobacco (*Nicotiana rustica*) plants were examined leaf by leaf along the stem.

METHODS AND MATERIALS

Four month old *Nicotiana rustica* plants were used containing seven leaves from the base to the inflorescence. Leaves were counted from the base up, 1 being the oldest and 7 the youngest. For each plant the leaves were measured individually for CO_2 uptake and transpiration. The apparatus used was an open system with a flow rate of 600 cc/min which minimized boundary layer interference. A 10 dm^2 section of leaf was inserted into a small chamber and gas of a known concentration of CO_2 and fixed humidity was forced over the upper and lower sections of the leaf. Illumination was with HPLR Philips 500 W lamps which gave 800 μEins. $m^{-2}s^{-1}$. The CO_2 and H_2O concentrations at the inlet and outlet of the chamber were simultaneously measured using an IR CO_2 analyzer (UNOR, Maihak, W. Ger) and dew point hygrometer (E.G. & G., Calif., U.S.A.).

After gas exchange measurements were performed, the leaves were removed from the plant, weighed and measured for leaf area. Each leaf then had three one centimeter diameter discs cut, one at the tip and one on either side of the midrib. These were extracted in 80% acetone and chlorophyll measured according to Arnon (1949).

Each leaf was then ground in 10 ml of 0.1 M Tris-HC1 pH 8.2, 0.01 M $MgC1_2$, 0.02 M $NaHCO_3$, 0.001 M dithiothreotol, filtered through 8 layers of cheesecloth, and centrifuged for 30 minutes at 10,000 × g. The supernatant was assayed for RuBPC and RuBPO activity after the procedure of Lorimer *et al.* (1976). An aliquot of the supernatant was precipitated in 10% TCA, and analyzed for protein according to Lowry (1951).

RESULTS

The leaf weight and leaf area of the leaves increased from the youngest leaves to the third from the oldest leaf, at which point they plateaued (Fig. 1 A). This may mean that leaves 7 through 4 are still expanding, or it may be that the first three leaves produced are the largest and the later ones never reach the size of the first three. In that case only leaves 7 and 6 are still growing. These are the leaves where the protein and chlorophyll content are still increasing. In leaves 5 and 4 protein and chlorophyll content are at a plateau and from leaf 3 senescence has begun and both chlorophyll and protein are decreasing (Fig. 1 A).

Figure 1 B expresses this data per unit area also showing the standard deviations of the measurements. Leaf weight per cm^2 is constant along the plant stem. Chlorophyll is also almost constant until leaf 4. After that in the older leaves there is a fall in the chlorophyll per cm^2. Protein shows two phases. In the three youngest leaves it is stable per unit area and after that it shows a decrease per area in the older leaves.

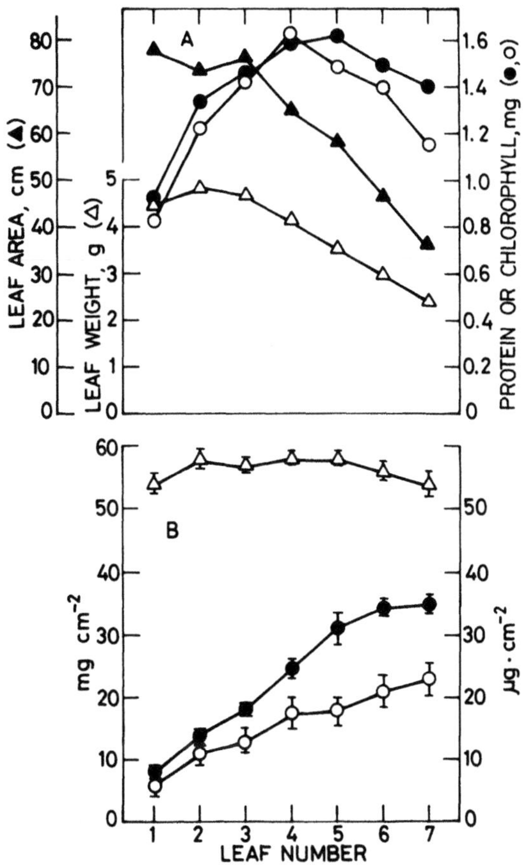

Fig. 1. A. The leaf area (▲), leaf weight (△) and leaf protein (●) and chlorophyll (○) in tobacco leaves of different ages from oldest (leaf number 1) to youngest (leaf number 7).
B. Leaf weight (△) per cm² expressed in mg and leaf protein (●) and chlorophyll (○) per cm² expressed in μg. B shows the standard deviation of the data.

When the CO_2 uptake of these leaves is compared to the activity of RuBPC and RuBPO from extracts of the leaves, the following picture emerges (Fig. 2 A). Three plants, each with 7 leaves were measured leaf by leaf for CO_2 uptake and then the enzymatic activity of the leaves was examined. The photosynthetic capability of the whole leaf increased from the youngest leaf to leaf 5 and 4 and from them decreased again to the oldest leaf 1. The carboxylase activity on the other hand was more or less constant from leaf 7 to leaf 4 and then decreased to leaf 1. The oxygenase activity showed a gradual increase from leaf 7 to leaf 4 and then decreased to leaf 1. Thus the ratio of RuBPC/RuBPO (the 'photosynthetic potential') increases steadily from the oldest leaf (1) to the youngest leaf (7). If the same plants are examined on the basis of activity per cm², a similar picture is seen in CO_2 uptake, but the RuBPC and RuBPO activities continue to increase in the youngest leaves (Fig. 2 B). Figure 2 B shows the standard deviations of the measurements. The same tendency is seen if the enzymatic activity is expressed on a per mg protein basis (Table 1). The activity of RuBPC is lowest in the two oldest leaves and highest in the youngest leaf. In the intervening leaves it is about the same value for all leaves. Because of changes in the RuBPO rate, however, the ratio of RuBPC/RuBPO steadily increases from the oldest to the youngest leaves.

Net photosynthesis as measured by CO_2 exchange is more or less paralleled by the RuBPC activity measured *in vitro* from leaves 1 to 5. It is in the two youngest leaves that the discrepancy appears. From the enzyme assays they would appear to have very efficient photosynthesis while the gas exchange shows a decrease in net photosynthesis. Therefore the stomatal resistance of these leaves was measured to see if that was the limiting factor in gas exchange in the young leaves. Figure 3 shows that the stomatal resistance is higher both in the two oldest leaves and in the youngest leaf.

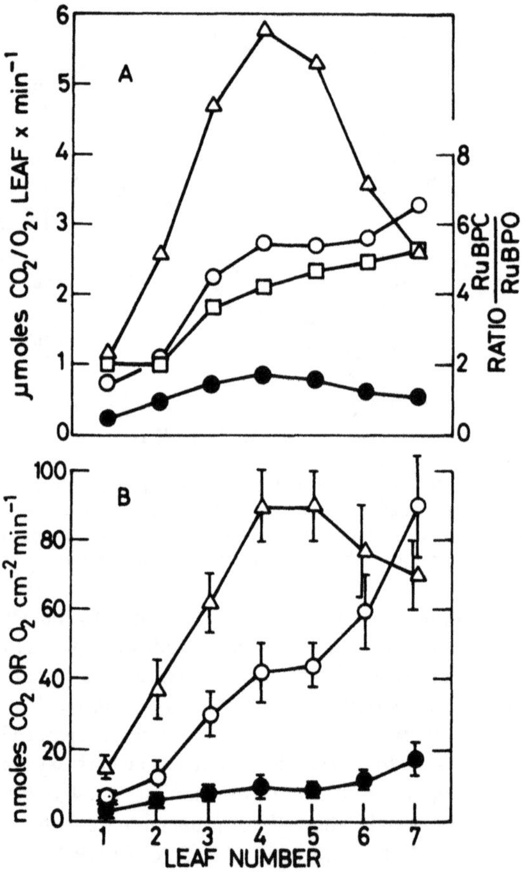

Fig. 2. A. CO_2 uptake of whole leaves (△) compared with CO_2 fixation (○) and O_2 fixation (●) of enzyme extracts of the leaves. (□) shows the ratio of the Co_2 fixation/O_2 fixation.

B. The same data shown in A expressed on a per cm^2 basis with the standard deviations of the data.

Table 1. The activity of RuBPC/O in leaves of different positions on the stem

Leaf number	RuBPC (μmoles mg protein^{-1}min^{-1})	RuBPO (μmoles mg protein^{-1}min^{-1})	Ratio $\frac{\text{RuBPC}}{\text{RuBPO}}$
1	0.422	0.231	1.8
2	0.851	0.638	1.4
3	1.394	0.369	3.8
4	1.590	0.363	4.4
5	1.133	0.255	4.4
6	1.340	0.250	5.2
7	2.165	0.405	5.3

Since it was seen that the *in vitro* measurements showed a change in the ratio of carboxylase to oxygenase activity from leaf to leaf, it was of interest to compare this with measurements in the leaves themselves. Accordingly the compensation point and mesophyll resistance of the leaves were measured. The compensation point is a measure of the photorespiration *in vivo* and is juxtaposed in Table II with the RuBPO/RuBPC ratio which we term the 'photorespiration potential'. The mesophyll resistance is a measure among other things of the carboxylation efficiency and is compared with the RuBPC/RuBPO ratio or 'photosynthesis potential'. A correlation analysis shows significance at P = 0.05 for the compensation point and the ratio of

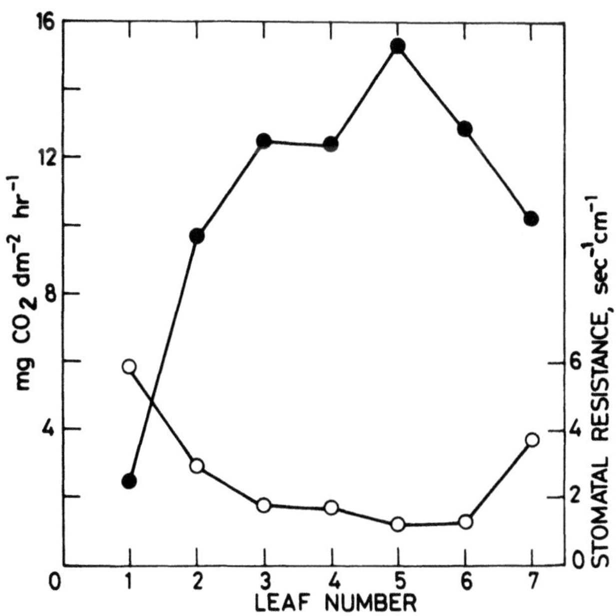

Fig. 3. CO_2 uptake (●) of whole leaves from the oldest (leaf 1) to the youngest (leaf 7) compared to stomatal resistance (○) of the same leaves.

Table 2. A comparison of the compensation point and mesophyll resistance as measured in whole leaves to the ratio of carboxylase and oxygenase from enzyme extracts of these leaves.

Leaf Number	Compensation point (ppm)	RuBPO / RuBPC	Mesophyll resistance (sec cm^{-1})	RuBPC / RuBPO
1	46.2	0.47	94.1	2.1
2	47.4	0.50	20.5	2.0
3	38.7	0.27	16.6	3.7
4	24.3	0.23	19.6	4.3
5	18.7	0.20	15.3	4.9
6	19.9	0.20	18.1	4.9
7	19.2	0.18	20.5	5.3

oxygenase to carboxylase. The mesophyll resistance does not correlate well with the ratio of carboxylase to oxygenase.

It appears, therefore, that the enzymatic measurements reflect in most cases what is the actual situation in the leaves. The net photosynthesis is dependent not just on the relative activities of the carboxylase and oxygenase, but also on the stomatal resistance. In mature leaves (leaves 5, 4 and 3) the stomatal resistance is very low and the determining factor is the carboxylation activity. In senescing leaves (leaves 1 and 2) both the stomatal resistance and the enzymatic activity combine to lower the net photosynthesis of the leaves. In young leaves (leaves 6 and 7) the enzymatic activity is high and the limiting factor for net photosynthesis appears to be the mesophyll resistance in leaf 6 and both the mesophyll and stomatal resistance in leaf 7.

DISCUSSION

We have found, as have others, that as the leaves age there is a decrease in chlorophyll, protein and net photosynthesis. On a per protein basis the photosynthetic activity is declining faster than the other factors, indicating that much of the protein loss is RuBPC. This was found also by Hardwick *et al.* (1968) in *Perilla frutescens* and by Peterson and Huffacker (1975) in barley. We have also shown that photorespiration as measured by RuBPO activity decreases as leaves age, although at a different rate than the decrease of RuBPC. In the oldest leaves there is an increase in stomatal and mesophyll resistance which would also decrease the photosynthesis as measured in whole leaves.

Mature tobacco leaves have the highest levels of chlorophyll, protein and net photosynthesis. However, no more than one third of the total leaves are at this stage of development at any one time. Moving up or down the stem there is a decrease in these parameters. Frey and Moss (1976) found in barley that the RuBPC activity was greatest in young, fully expanded leaves and decreases as they went down the stem. Our findings are somewhat different in that we find that in still expanding leaves the RuBPC activity is higher than in fully expanded leaves, but that activity decreases as measurements are taken further down the stem.

Plants differ as to when in their leaf development they exhibit maximal photosynthetic activity. Cucumbers (Hopkinson, 1964), peas (Smillie, 1962) and beans (Saeki, 1959) show maximal photosynthesis before the leaves are fully expanded; while soybean (Woodward & Rawson, 1976) and tall fescue (Jewiss & Woledge, 1967) show maximal photosynthesis at time of complete expansion. In tobacco we find that maximal photosynthesis as measured in whole leaves occurs in fully expanded leaves, but that potential photosynthesis as seen from the RuBPC/RuBPO ratios is highest in young leaves.

Rawson & Woodward (1976) examined photosynthesis and transpiration in expanding leaves of tobacco. They found high stomatal and mesophyll resistances in expanding leaves which decreased inversely to the increase in photosynthesis. In their case peak photosynthesis was reached when the leaves were 60–80% expanded. After the peak the photosynthesis declined which they felt was due to lack of sinks in the young plants. In our case there is a sink present in the form of flowers and developing seeds. In the tobacco of these experiments there was no increase in mesophyll resistance in young leaves, but the stomatal resistance was higher, which could account for the discrepancy between the rates of photosynthesis measured *in vivo* and *in vitro*.

The activity of RuBPO was maximal in mature leaves and lower in either old or young leaves. The ratio of RuBPO/RuBPC and the compensation point of the leaves increased as one went from young to old leaves. This agrees with the findings of Salin and Homann (1971) that young tobacco leaves show less photorespiration than mature leaves.

Comparing the measurements of whole leaves and enzyme extracts, it appears that there is a good correlation between the ratio of the carboxylase/oxygenase activities and both photosynthesis and photorespiration of whole leaves, except when stomatal resistance is the limiting factor. It appears that this initial reaction of the pathways of either photosynthetic carbon reduction or photorespiration is usually a limiting factor for the total process.

REFERENCES

Arnon, D.I. Copper enzymes in isolated chloroplasts. *Plant Physiol.* 24: *1–15* (1949).

Freeland, R.O. Effect of age of leaves upon the rate of photosynthesis in some conifers. *Plant Physiol.* 27: *685–690* (1952).

Frey, N.M. & D.N. Moss. Variation of RuDP Case activity in barley. *Crop Sci.* 16: *209–213* (1976).

Hardwick, K., M. Wood & H.W. Woolhouse. Photosynthesis and respiration in relation to leaf age in *Perilla frutescens* (I.) *Britt. New Phytol.* 67: *79–86* (1968).

Hopkinson, J.M. Studies on the expansion of the leaf surface. *J. exp. Bot.* 15: *125–137* (1964).

Hover, J.M. & F.G. Gustafson. Rate of respiration as related to age. *J. gen. Physiol.* 10: *33*ff. (1926).

Jewiss, O.R. & J. Woledge. The effect of age on the rate of apparent photosynthesis in leaves of tall fescue (*Festuca arundinacea* Schreb.). *Ann. of Bot.* 31: *661–671* (1967).

Lorimer, G.H., M.R. Badger, & T.J. Andrews. Activation of ribulose 1,5-bisphosphate carboxylase. *Biochemistry* 15: *529–536* (1976).

Lowry, O.H., N.J. Rosebrough, A.L. Farr & R.J. Ranall. Protein measurement with the phenol reagent. *J. Biol. Chem.* 193: *265–275* (1951).

Nixon, R.W. & R.T. Wedding. Age of date leaves in relation to efficiency of photosynthesis. *Proc. Am. Soc. hort. Sci.* 67: *265*ff. (1956).

Peterson, L.W. & R.C. Huffacker. Loss of ribulose 1,5-diphosphate carboxylase and

increase in proteolytic activity during senescence of detached primary barley leaves. *Plant Physiol.* 55: *1009–1015* (1975).
Rawson, H.M. & R.G. Woodward. Photosynthesis and transpiration in dicotyledonous plants. I. Expanding leaves of tobacco and sunflower. *Aust. J. Plant Physiol.* 3: *247–256* (1976).
Richardson, S.D. The effect of leaf age on the rate of photosynthesis in detached leaves of tree seedlings. *Acta Bot. Neerl.* 6: *445* (1957).
Saeki, T. Variation of photosynthetic activity with ageing of leaves and total photosynthesis in a plant community. *Bot. Mag., Tokyo* 72: *404–408* (1959).
Salin, M.L. & P.H. Homann. Changes in photorespiratory activity with leaf age. *Plant Physiol.* 48: *193–196* (1971).
Šesták, Z. & J. Catsky. Intensity of photosynthesis and chlorophyll content as related to leaf age in *Nicotiana sanderae. Biologia Pl.* 4: *131* (1962).
Smillie, R.M. Photosynthetic and respiratory activities of growing pea leaves. *Plant Physiol.* 27: *716–721* (1962).
Spenser, P.W. & J.S. Titus. Apple leaf senescence: leaf disc compared to attached leaf. *Plant Physiol.* 51: *89–92* (1973).
Wollgiehn, R. Nucleic acid and protein metabolism of excised leaves. *Soc. Exp. Biol. Symp.* 21: *231–246* (1967).
Woodward, R.G. & H.M. Rawson. Photosynthesis and transpiration in dicotyledonous plants. II. Expanding and senescing leaves of soybean. *Aust. J. Plant Phsyiol.* 3: *257–267* (1976).

Authors' address:
Faculty of Agriculture
Hebrew University
Rehovot
Israel.

THE PHOTOSYNTHETIC CHARACTERISTICS OF APPLE LEAVES (CV. GOLDEN DELICIOUS) DURING THEIR EARLY GROWTH*

R.L. WATSON AND J.J. LANDSBERG
(Bristol, U.K.)

ABSTRACT

Measurements of the rate of net photosynthesis by the spur leaves of Golden Delicious apple trees were made at intervals between leaf emergence and maturity. The same leaves were used for each series of measurements, which were made by gas exchange at a range of photon flux densities. Leaf area was determined non-destructively and the number of stomata per unit leaf area assessed from photographs taken with a scanning electron microscope using leaves similar to those under study. Stomata were classified visually as mature or immature. Dry weight per unit leaf area was determined on these leaves.

The proportion of mature stomata increased as leaf area increased, as did dry weight per unit leaf area (from ~ 50 to 80 g m^{-2} between early May and the end of June). There was no evidence that the initial slope of the P_n – photon flux density curve (a) changed with time – it remained at about 2.2×10^{-3} mg μE^{-1}. Mesophyll conductance (g_m) increased as the leaves expanded, reaching 1.5 to 2×10^{-3} s m^{-1} in mid-May; leaf expansion was complete by early June and subsequent changes in g_m showed no consistent trend. The maximum values of stomatal conductance (g_s (max), s m^{-1}) were recorded at about the time g_m reaches its maximum; g_s (max) then declined. The most marked and consistent trend was shown by dark respiration (R_d), which decreased from about 0.08 mg m^{-2} s^{-1} in early May to about 0.04 mg m^{-2} s^{-1} in mid-June.

Using the measured leaf photosynthesis parameters, and typical orchard light and temperature values, hourly values of net photosynthesis were calculated, day by day from the time of leaf emergence and compared to measured leaf growth rates (dry matter increments) to estimate the point at which the leaves were likely to become net exporters of assimilate. The calculations indicated that leaves become net exporters of assimilate about 10 days after emergence. The leaf would have produced its own weight in assimilate 30 days after emergence. In the first 90 days of its life, a leaf would produce about 770 mg of assimilate, maximum rate being 14 mg day^{-1} after 60 days.

INTRODUCTION

Where apples are grown near their northern limits the leaves appear in early April and maximum leaf area is reached in July–August. The first leaves which emerge are cluster, or spur leaves, derived from buds formed during the

*This work was partly supported by NATO research grant No. 765.

previous autumn. A spur may carry up to eight leaves, although usually only three to five develop. When shoot growth begins in late May–early June new leaves are produced successively at the nodes and this process continues until the end of shoot extension in July–August. Maximum leaf area is attained when the last of the shoot leaves is fully expanded.

This study was concerned with young spur leaves on a four-year-old Golden Delicious tree in the spring of 1977. The objective was to examine the variation in the parameters of an analytical model of leaf net photosynthesis during the growth of young leaves and to evaluate the consequences of these changes in terms of leaf carbon balance.

MODEL AND MEASUREMENTS

The net photosynthesis model used was a simplified form of a model of leaf photosynthesis by Charles-Edwards & Ludwig (1974) (see Thornley, 1976). Net photosynthesis (P_n, mg m^{-2} s^{-1}) is given by the equation

$$P_n = \frac{\alpha Q_p g_m C_i}{\alpha Q_p - g_m C_i} R_d \qquad (1)$$

where

α (mg $CO_2 \mu E^{-1}$) is the initial slope, Q_P (μE m^{-2} s^{-1}) is photon flux density, g_m (m s^{-1} or mm s^{-1}) is mesophyll conductance, C_i (mg m^{-3}) is the concentration of CO_2 in the intracellular spaces and R_d (mg m^{-2} s^{-1}) is the rate of dark respiration. At saturating photon flux density photosynthesis is given by

$$P_{max} = \frac{g_m C_a - R_d}{1 + g_m/G} \qquad (2)$$

where

C_a is ambient CO_2 concentration (mg m^{-3}) and $G = (1.37r_a + 1.605\,r_s)^{-1}$; r_a and r_s are boundary layer and stomatal conductance values respectively.

This model has been used by Watson *et al.* (1978) in a study of the photosynthetic characteristics of mature leaves on similar trees over two seasons; its properties and the underlying assumptions are discussed in that paper. We note here that the mesophyll conductance values are in some part a function of the rate of photorespiration, for which it was not possible to obtain acceptable estimates.

To solve the Model C_i is eliminated, leading to a quadratic involving α, Q_p, g_m, R_d and G. Parameter values were obtained using a maximum likelihood procedure (MLP; Ross, 1975), available on both the Rothamsted and Bristol computers, applied to between 10 and 30 measurements of P_n at a range of photon flux densities and two C_a levels (approx. 300 and 600 ppm). Each set of measurements provided a complete light response curve. Input data were net photosynthesis rates and corresponding photon flux densities, r_s, r_a and C_a. Each set of measurements provided a set of parameter values for which the program provided standard errors.

Measurements were made on four spur leaves on a four-year-old Golden Delicious tree growing in a large plastic container (dustbin) in 1:3 peat-loam mix with adequate water and fertilizer. They were started in early May, as soon as the leaves were large enough to insert in an assimilation chamber. Each leaf was on a different spur. The intention was to use the same leaves throughout, but one was damaged after the first set of measurements and a new one had to be chosen.

The work was done in a glasshouse, the tree being brought in from outside for the photosynthetic measurements, which were made by infra-red gas analysis in an open-system leaf chamber. The chamber could be manoeuvred into position so that the leaf could be inserted without stress or damage. This was achieved by using a horizontally split circular chamber through which a radial 8 mm hole was bored at the join. The leaf with its petiole wrapped in 'blu-tack' was put into the lower slot; clamping the lid formed an air tight seal so that the low flow rates ($< 1 l\,min^{-1}$) could be used, allowing accurate measurement of CO_2 fluxes from the small leaves. Chamber temperature was controlled at about 22°C by Peltier blocks, but otherwise the system and procedures used were as described by Watson *et al.* (1978). Maximum attainable photon flux density was about 1500 $\mu E\ m^{-2}$.

Light response curves for each leaf were measured at two ambient CO_2 concentrations. The measurements took about a day to complete.

When photosynthesis measurements were made the area of each leaf was determined by tracing carefully round it and determining the area of the tracing. Tracings were replicated three times. When each series of photosynthesis measurements were made ten leaves similar to those in the experiment were removed from the tree and their area, fresh weight and dry weight determined. Using these leaves, and also some collected in late April, stomatal frequency (number per unit area), and type (mature or immature) were examined under a scanning electron microscope. Stomata were defined as mature when the stomatal aperture was clearly visible. Immature stomata appeared to be covered by a membrane or layer of epidermal cells. After the last series of photosynthesis measurements in mid-June the leaves which had been used in the experiment were themselves removed and their dry weight determined.

RESULTS

The growth (in area) of the leaves used for measurements is shown in Fig. 1; leaf 5 was the replacement leaf. Leaf 4 hardly grew at all and this (not unusual in early spur leaves) is reflected in the inconsistent parameter values obtained from this leaf (Fig. 2).

The changes with time in α, g_m, R_d and the maximum stomatal conductance (g_s(max), $mm\ s^{-1}$) measured in any set are shown in Fig. 2. There is no evidence (Fig. 2a) that α is not constant; an average value of about $2.2 \times 10^{-3}\ mg\mu E^{-1}$ is indicated by these data. The mesophyll conductance (g_m, Fig. 2b) values for leaves 1 and 3 appeared to reach a maximum in mid to late May; there is some suggestion that they were lower in June. Dark respiration (R_d, Fig. 2c) showed the clearest time trend, the values for all leaves dropping steadily as the leaves matured. The time trend in dry matter per unit leaf area is also plotted in Fig. 2c; R_d appears to be inversely related to this parameter. The maximum values of stomatal conductance (g_s(max), Fig. 2d) were low in early May, reached a peak in mid-May (with the exception of leaf 4) and then declined. The initial rise in g_s (max) corresponded to the increase in the proportion of mature stomata (Fig. 2d).

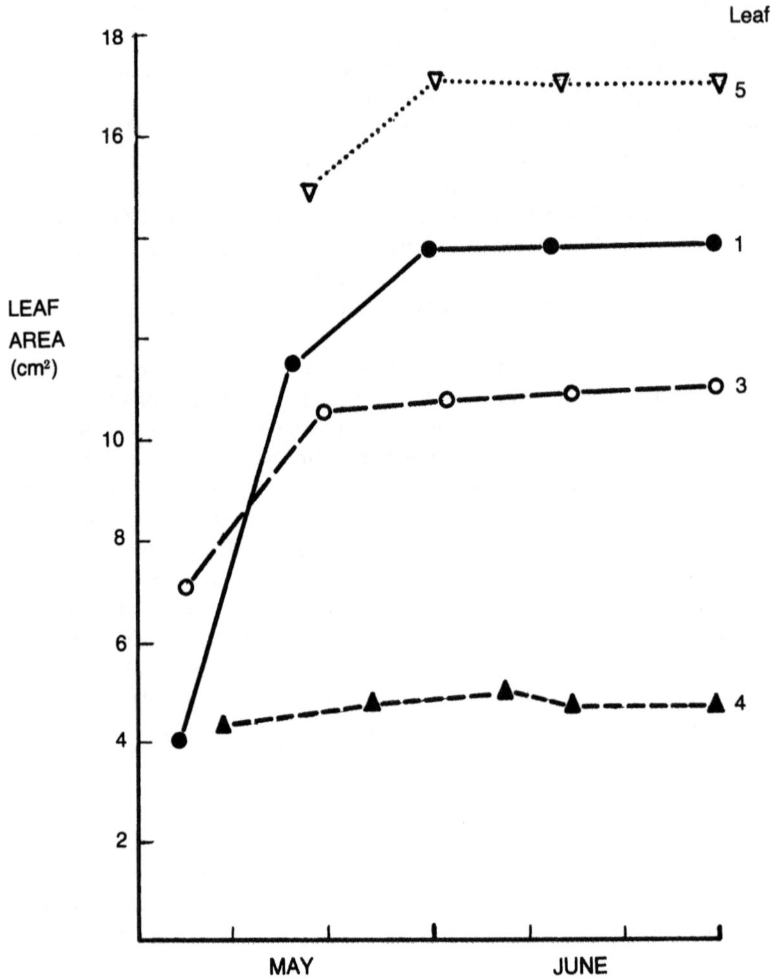

Fig. 1. Areas of the young spur leaves on which photosynthesis measurements were made. Leaf 5 was a replacement for leaf 2 (damaged).

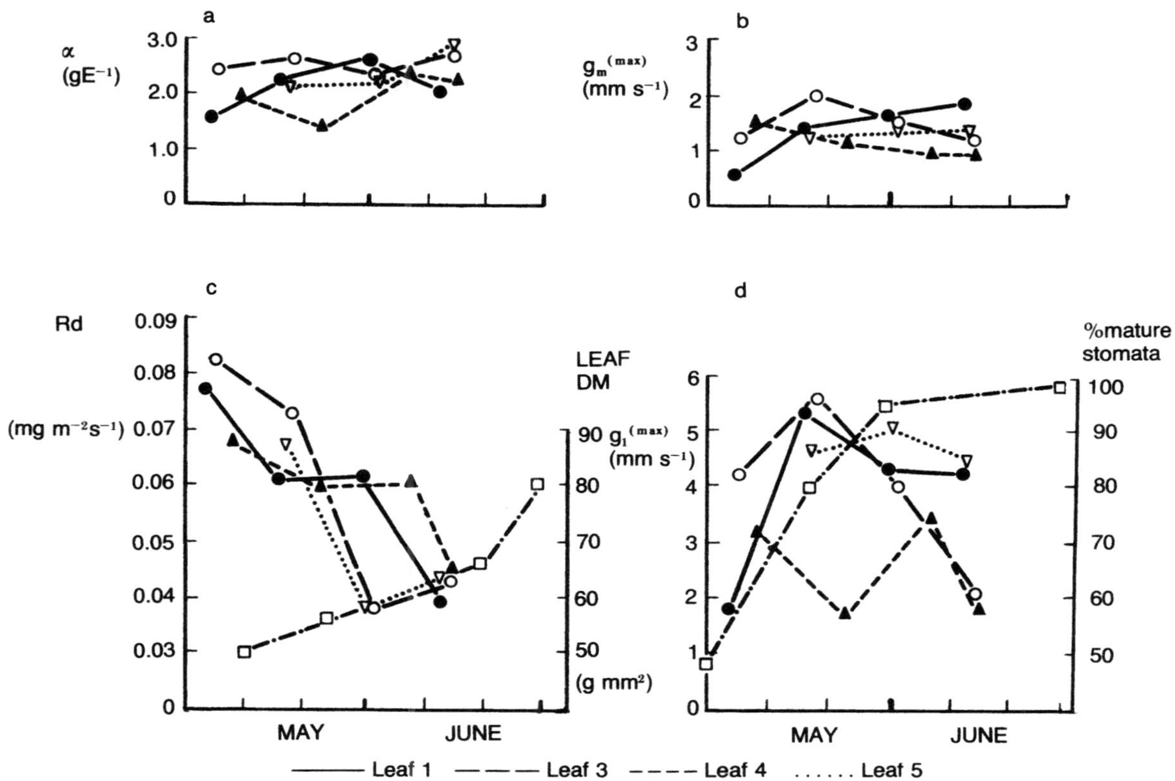

Fig. 2. Changes in time in a) the initial slope (α) of the light response curve; b) the mesophyll conductance (g_m); c) dark respiration (R_d); and dry matter per unit leaf area d) maximum stomatal conductance (g_s(max)) and proportions of mature stomata.

To evaluate the implications of the observed changes in the parameters of equation (1) we calculated light response curves using mean model parameter values from leaves 1 and 3 from each series of measurements. These are given in Table 1 and the light response curves calculated with them are shown in Fig. 3a.

Table 1. Mean values of the parameters of equation 1 obtained from leaves 1 and 3.

Date	α (mg μE^{-1})	g_m (mm s^{-1})	R_d at 22°C (mg m^{-2} s^{-1})	g_s (mm s^{-1})
4–5 May	2.04×10^{-3}	0.93	0.08	3.0
17–18 May	2.44×10^{-3}	1.74	0.07	5.4
end May	2.48×10^{-3}	1.60	0.05	4.2
13–14 June	2.41×10^{-3}	1.55	0.04	3.1

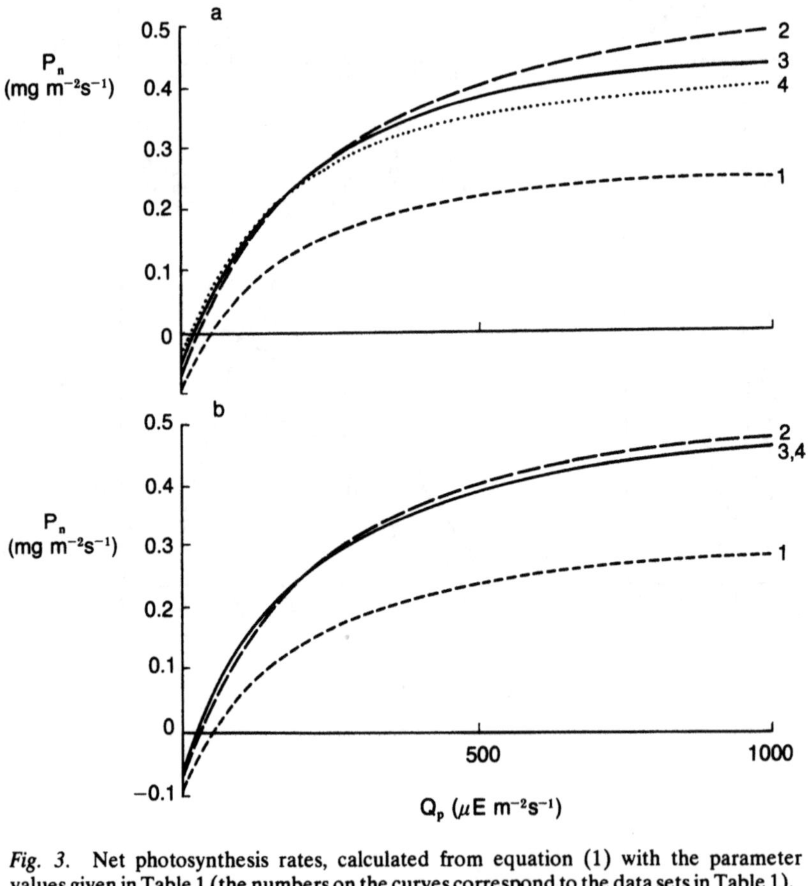

Fig. 3. Net photosynthesis rates, calculated from equation (1) with the parameter values given in Table 1 (the numbers on the curves correspond to the data sets in Table 1). a) using mean measured values of g_s (max); b) with g_s (max) set to 5 mm s^{-1}

The curves are slightly artificial since stomatal conductance was taken to be constant (= g_s(max)) at all values of photon flux density, although it is known that g_s is reduced at photon flux densities of less than about 500 μE m^{-2} (Warrit, 1977). However the difference in P_n caused by this reduction at low energy levels is negligible.

To eliminate the effects of g_s another set of curves was calculated with the same parameter values but with g_s set at 5 mm s^{-1} in every case. The results (Fig. 3b) are very similar except that the slight reduction in P_n at high photon flux densities (Fig. 3a) caused by the reduction in g_s(max), is not apparent. The increase in g_m from 0.3×10^{-3} (curve 1) to 1.74×10^{-3} m s^{-1} (curve2) has a considerable effect on P_{max}, but the slight decrease in g_m with time from mid-May does not greatly affect the position of the curves.

The effects of g_s(max) and g_m can be evaluated by considering the partial derivatives of equation (2), *viz*.

$$\frac{\delta P_{max}}{\delta g_m} = \frac{C_a + R_d/G}{(1 + g_m/G)^2} \tag{3}$$

and

$$\frac{\delta P_{max}}{\delta G} \quad \frac{(g_m C_a - R_d)g_m}{(1 + g_m/G)^2 G^2} \tag{4}$$

Equation (3) shows that for any value of G the effect of a change in g_m decreases as g_m increases; with representative values of the other parameters $\delta P_{max}/\delta g_m = 428$ when $g_m = 0.5 \times 10^{-3}$ and 164 when $g_m = 2.5 \times 10^{-3}$m s^{-1}. Similarly, from equation (4), the effects of changes in G decrease as the value of the parameter increases; $\delta P_{max}/\delta G = 197$ when $G = 1 \times 10^{-3}$, 100 when $G = 2 \times 10^{-3}$ and 41 when $G = 4 \times 10^{-3}$ m s^{-1}. These values of G represent changes in g_s (max) from about 2 to 3.5 to 8 mm s^{-1} ($r_s \triangleq 500$, 285 and 125 s m^{-1}). It appears that increasing stomatal conductance above about 3 to 3.5 mm s^{-1} confers little advantage in terms of photosynthesis but increases in g_m beyond the highest value observed (1.74×10^{-3} m s^{-1}, Table 1) would lead to significant increases in P_n.

Leaf carbon balance

Starting from 1 April the carbon balance of a 'typical leaf' was calculated by hourly simulation of net photosynthesis and dark respiration. Values for the parameters of the leaf photosynthesis model were estimated from empirical curves (Fig. 4) based on the data given in Fig. 2. Hourly values of stomatal conductance were calculated from the model given by Warrit (1977), who showed that changes in g_s of Golden Delicious apple leaves, in relation to light flux density, could be described by a simple rectangular hyperbola.

Actual leaf growth rate (LGR, mg dry matter day^{-1}) was calculated from curves describing time trends in leaf area (see Figs 1 and 5) and dry matter per unit leaf area (Fig. 2c). Average daily short-wave radiant energy income and daylength were calculated from empirical curves, the energy income curve being based on climatological records for the south of England. Maximum energy flux density (on a horizontal surface), and hence hourly values of short-wave energy flux density were then calculated for each day, assuming that the time course of radiation could be approximated by a truncated cosine curve. No corrections were made, in the photosynthesis calculations, for angle of incidence to the leaves. Daily maximum and minimum temperatures were calculated from empirical relationships describing the seasonal changes in these parameters at Long Ashton; the diurnal course of temperature was assumed to be sinusoidal, with the minimum temperature at 0200 and the maximum at 1400 h. Leaf respiration rate was taken to be a function of temperature, as given by Watson *et al*. (1978).

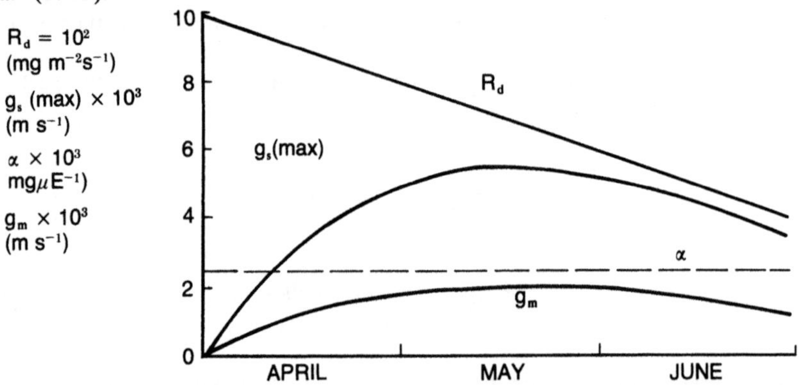

Fig. 4. Time course of the parameter values used in the simulation of leaf carbon balance. The curves are described by equations in the program.

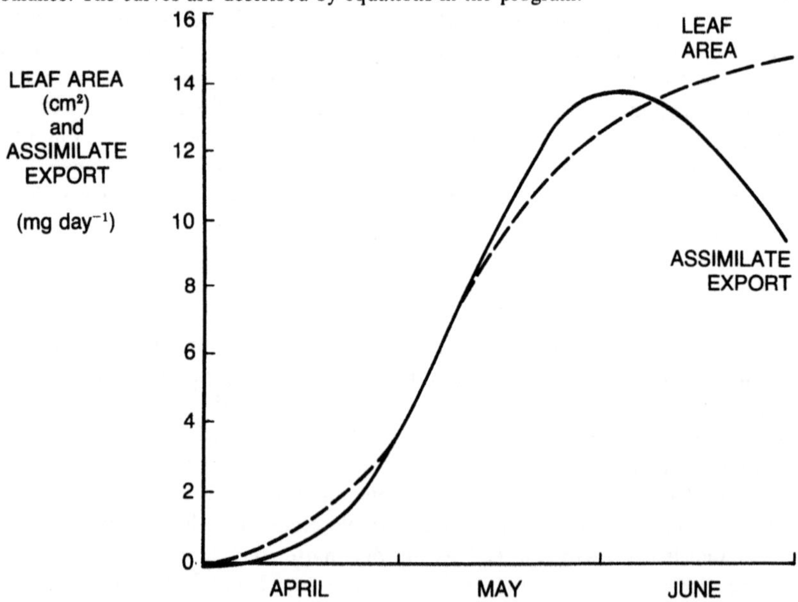

Fig. 5. Time course of assimilate export (mg day^{-1}) by the simulated leaf. The leaf area is also shown.

Daily carbon production (A, mg) by the leaf was calculated as total (24 h) CO_2 uptake, less dark respiration, multiplied by 0.65 to convert to carbohydrate (Šesták, Catsky & Jarvis, 1971). The difference between A and LGR gives an estimate of the amount of assimilate exported by the leaf.

The leaf area curve used in the simulation, and the calculated daily carbon balance of the leaf, are shown in Fig. 5. A leaf with the given photosynthetic characteristics, under the radiant energy and temperature conditions used in the calculations, would produce enough assimilate daily to meet its own growth requirements about 10 days after emergence, after which it would become a net exporter. After a further 20 days it will have equalled its own weight in total dry matter production. Total dry matter production by the leaf during the first 90 days of its life would be about 770 mg, with a maximum production rate of about 14 mg day^{-1}, reached after about 60 days.

DISCUSSION

Watson *et al.* (1978) found considerable leaf to leaf variation in the parameter values of the net photosynthesis model. This is also apparent in these results. They suggested that, for the leaves of Golden Delicious apple trees, the initial slope (α) could be taken as 2.50×10^{-3} mgμE^{-1}, that g_m is about 3.50 mm s^{-1} for extension and 2.50 mm s^{-1} for spur leaves, and $R_d = 0.006 \exp(0.09\,T)$ mg m^{-2} s^{-1} (where T = leaf temperature, °C) We obtained a similar value for α, with no evidence of variation with leaf age. The highest values ($\sim$ 2 mm s^{-1}) for the mesophyll conductance of the young spur leaves used here were lower than those given by Watson *et al.* Disregarding the values from leaf 4, which did not grow, the data suggested that g_m increased up to about the time of full bloom (mid-May) and then decreased slightly; for the simulation of leaf carbon balance this was taken to be the pattern (see Fig. 4).

Dark respiration per unit leaf area at 22°C decreased steadily with age and this was incorporated in the simulation by making the constant in the equation describing the temperature dependence of R_d time-dependent. The lowest R_d values measured by Watson *et al.* were in early August, after which there was a tendency for them to increase. The highest values they measured were in senescing leaves. Because of the upward trend in dry matter per unit leaf area (Fig. 2c), the reduction in respiration with leaf age would be greater if respiration were expressed on a dry matter basis. Between early May and late June R_d (22°C) decreased from approximately 1.5×10^{-6} mg m^{-2} s^{-1} to about 0.6×10^{-6} mg m^{-2} s^{-1}; a 2.5-fold decrease compared to the change from 0.07 mg m^{-2} s^{-1} to 0.04 mg m^{-2} s^{-1} (1.75-fold) on a leaf area basis.

If we again ignore leaf 4 the rise in the value of g_s (max) between the first two sets of measurements corresponds to the rise in the proportion of mature stomata (Fig. 2d) but, since there are strong indications of a subsequent decrease in g_s (max) this cannot constitute a complete explanation for its variability. Warrit (1977) found that late in the year (September, October) stomata became less responsive to changes in light intensity and it may be that the reduction in g_s (max) reflects a decrease in stomatal responsiveness in these spur leaves.

Calculations with equations (3) and (4) indicate that when g_s (max) reaches 3 to 3.5 mm s^{-1} the limitation to P_{max} lies in the mesophyll conductance – a conclusion also indicated by Fig. 3. It therefore appears that stomatal conductance becomes non-limiting for photosynthesis at an early stage of leaf development but mesophyll conductance imposes severe limitations to P_{max} (cf curve 1, Fig. 3). The factors affecting mesophyll conductance, as defined here, warrant investigation.

The results obtained from the simulation of the leaf carbon balance reflect the assumptions and simplifications made, especially with regard to parameter values in the first month (April) of the leaf growth period. The conclusion that the leaf becomes a net exporter of assimilate less than 10 days after emergence, when its area is only about 5% of final area, must therefore be treated with caution. The decline in the rate of assimilate export after the end of May is due to the slight reduction in both g_m and g_s(max) from late May onwards, combined with increasing respiration losses, caused by increasing temperatures. The calculated assimilate production rates and total assimilate production by the leaf reflect its potential rather than any actual situation, since no account was taken of leaf and sun angles and mutual shading. Nevertheless the results of the simulation are of interest and we consider that they constitute a reasonable generalization of the experimental data.

REFERENCES

Charles-Edwards, D.A. & J. Ludwig. A model for leaf photosynthesis by C_3 plant species. *Ann. Bot.* 38: *921–930* (1974).

Ross, G.J.S. Simple non-linear modelling for the general user. Proceedings of the 40th Session of the International Statistical Institute (Warsaw) vol. 2, 585–593 (1975).

Šesták, Z., J. Catsky & P.G. Jarvis. Plant Photosynthesis Production. Manual of Methods. Dr. W. Junk N.V. Publishers, The Hague (1971).

Thornley, J.H.M. Mathematical models in plant physiology. Academic Press, London (1976).

Warrit, B. Studies on the stomatal behaviour of apple leaves. Ph. D. Thesis, University of Bristol (1977).

Watson, R.L., J.J. Landsberg & M.R. Thorpe. Photosynthetic Characteristics of the leaves of 'Golden Delicious' apple trees. *Plant, Cell and Environment 1, 51–58* (1978).

Authors' address:
Microclimatology Section
Long Ashton Research Station
Long Ashton
Bristol
United Kingdom.

COMPARATIVE STUDY OF PHOTOSYNTHETIC EFFICIENCY AND LEAF ARCHITECTURE DURING LEAF DEVELOPMENT*

J. DE GREEF, J.P. VERBELEN, R. CAUBERGS, E. MOEREELS AND E. SPRUYT.

(Wilrijk, Belgium)

ABSTRACT

Primary leaves of dark-grown bean seedlings are fully developed after about 10 days of etiolation. When 8-day-old etiolated leaves are selectively illuminated with continuous white light, they turn green but they stay as small as the dark controls, while the leaves of plants fully exposed to white light become green and expand very fast (10-fold increase in 6 days). Per unit area photosynthesis is much higher in expanding leaves for 3 days. Thereafter the difference becomes negligible in spite of the increasing leaf expansion. The overall photosynthetic efficiency of fully mature leaves is thus entirely related to expansion.

When 3-day-old and 9-day-old etiolated seedlings are subjected to continuous white light for 9 days they both develop primary leaves of the same size, but the former contain much more cells per leaf. Leaves having a smaller number of much larger cells show a higher photosynthetic activity per gram fresh weight and per μg chlorophyll. Chloroplast development of both leaf types looks qualitatively and quantitatively the same. On the contrary, the small, green leaves obtained by selective illumination of etiolated leaves, have much smaller, but structurally normal chloroplasts.

During leaf development photosynthetic efficiency seems to be dependent on a proper relationship between chloroplast and cell size, on the one hand, and cell volume/free space ratio, on the other.

INTRODUCTION

In developmental biology photosynthesis is generally studied in relation to plastid development. Very little attention is paid to the influence of leaf architecture and cell size on the photosynthetic efficiency of the plastids. Leaf development is light dependent and is controlled by the phytochrome system (Parker *et al*., 1949; Downs, 1955).

We could demonstrate that the development of etiolated bean seedlings can strongly be influenced when different plants parts are selectively illuminated. The fully developed primary leaves of intact etiolated plants turn completely green but do not expand when they are exposed to continuous white light (WL) while the remainder of the plant is shielded (De Greef *et al*., 1976). The stem parts kept in darkness show further etiolation growth. Giving 5 min of red light (RL) on the stem separately while the leaves are kept under continuous WL, leaf

*Support from the Belgian Science Foundation (F.K.F.O., grant no. 9009) is gratefully acknowledged.

expansion occurs. RL followed by far red light (FR) reverses the effect. From these data we concluded that leaf expansion is controlled by a phytochrome-mediated inter-organ co-operation.

When 3-day-old and 9-day-old etiolated seedlings were subjected to continuous WL for 9 days, they both developed primary leaves of the same size, although the leaf size of the former was initially 10-fold smaller than that of the 9-day-old seedlings. In the younger plants leaf growth was due to both cell division and cell expansion, while the increase of leaf size in 9-day-old seedlings was only due to cell expansion so that cell division must have stopped (Verbelen, 1976).

Based on these observations we were interested to know whether the architecture of the leaf during its development had any influence upon its photosynthetic efficiency.

MATERIAL & METHODS

In all our experiments we used etiolated bean seedlings (*Phaseolus vulgaris* L. cv. Limburg, French beans). The plants were raised from their seeds as described previously (De Greef & Caubergs, 1973). Great care was taken of selecting uniform experimental material.

The averages of leaf area and weight were made by sampling a large number of plants. The leaf area was determined by making photocopies of fresh leaves. The leaf shades were cut out and after weighing their surface area was calculated. In other experiments they were measured directly using a portable area meter Model LI 3000, Lambda Instruments Corp. The fresh weight of the leaves was measured to the nearest 0.001 g.

Cell size and leaf architecture were studied by light microscopy on pieces of the leaf blade after clearing them in a sodium hypochloride solution and in chlorallactophenol. Epidermal diffusion resistance was recorded with an automatic porometer MK II, Delta-T Devices.

Light shielding was performed by a black body effect. The primary leaves were pulled through light-tight slits to the outside of a black box, while the remainder of the seedling was kept inside the box.

White light (12.000 Lux) was administered from a luminaire of 20 fluorescent tubes (Power Groove, Cool White, 215 W). Pigment estimations were made as previously outlined (De Greef & Caubergs, 1972).

0_2-evolution was measured by the manometric technique with a photochemical Warburg apparatus. A CO_2-buffer ($KHCO_3$ 0.1 M: K_2CO_3 0.1 M) was used to keep constant the partial pressure of CO_2 in the gas phase. The intensity of the Warburg light source was approximately 5000 Lux.

For all statistical evaluation a 95% confidential limit was accepted. Mostly results of paired samples were analyzed using the group comparison technique (Woolf, 1968).

RESULTS

In Fig. 1 we have summarized the leaf growth of
1) etiolated seedlings

2) light-grown plants
3) 8-day-old etiolated plants fully exposed to continuous WL
4) 8-day-old etiolated plants with their leaves selectively exposed to continuous WL.

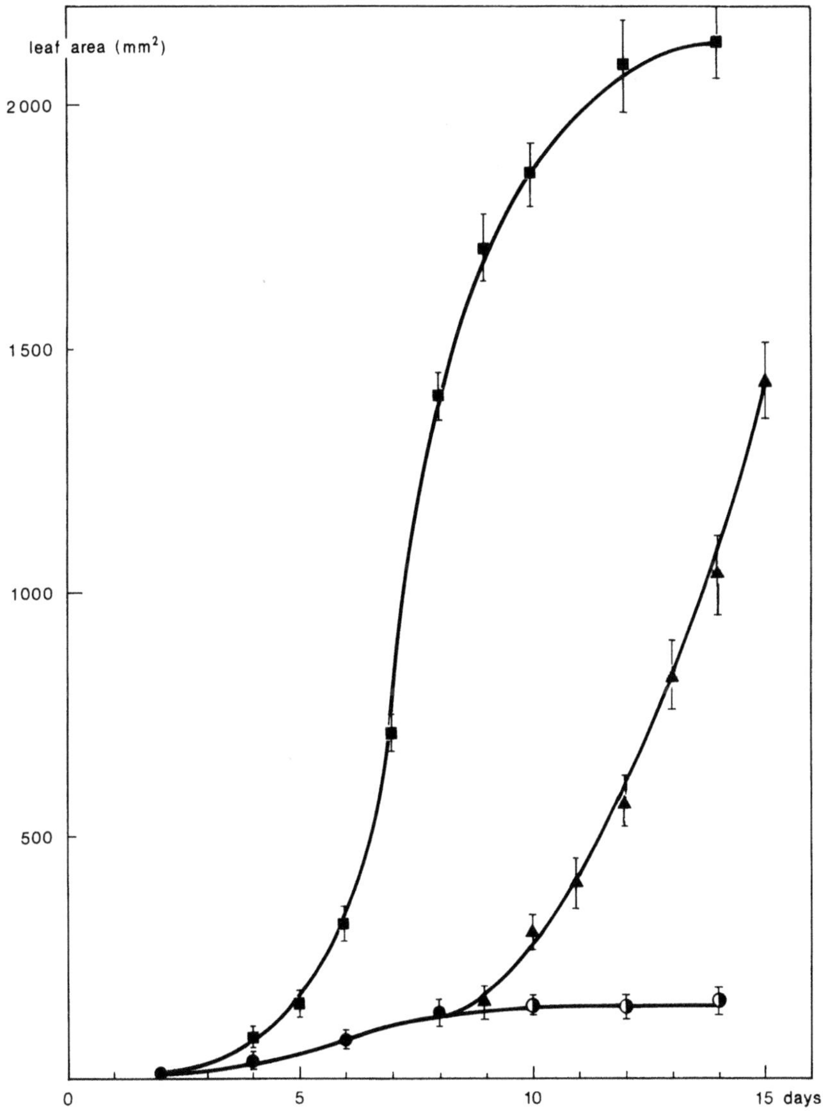

Fig. 1. Growth of primary leaves of bean seedlings (*Phaseolus vulgaris* L. cv. Limburg). Light-grown plants (■), etiolated seedlings (○ and ◑), 8-day-old etiolated seedlings fully exposed to continuous WL (▲) and 8-day-old etiolated seedlings with their leaves selectively exposed to continuous WL (◑).

For both etiolated and light-grown plants the increase of leaf area in function of seedling age shows a sigmoid curve, which is typical for developing stem leaves (Maksymowych, 1973). After 14 days of growth the light-grown plants developed leaves which were at least 10-fold larger than those of fully etiolated plants of the same age. The leaves of the latter did not differ in size from those of 8-day etiolated seedlings of which the leaves are selectively exposed to continuous WL for 6 days. When 8-day-old etiolated seedlings were fully illuminated for 6 days their leaves started to expand as those of WL-grown plants. Comparing the photosynthetic activities of the primary leaves (expressed as 0_2-evolution per unit of time and surface area) of 8-day-old etiolated seedlings, which were either fully exposed to WL or selectively illuminated on their leaves, we noticed a marked difference between days 1 to 6 (Fig. 2). After one day of light exposure the leaves of fully illuminated plants showed a steep

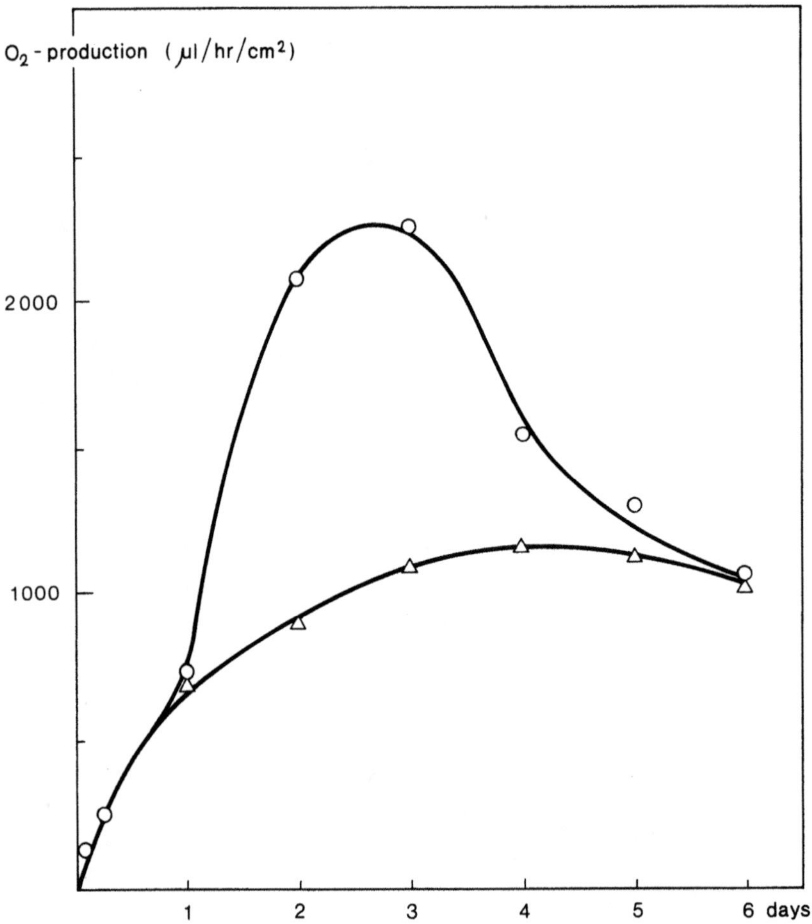

Fig. 2. Photosynthetic rate of primary bean leaves of 8-day-old etiolated seedlings either fully exposed to continuous WL (○) or selectively illuminated on their leaves with continuous WL (△).

upcline of 0_2-production reaching a maximum after 3 days and being twice as high as the maximal activity of leaves selectively exposed to the same light conditions for the same time. Thereafter the photosynthetic activity of the former declined and levelled off to the same activity of the latter.

In order to explain the temporal differences of photosynthetic activity between both experimental series we looked for changes occurring at the morphological level (Table 1). During their expansion the leaves of totally illuminated seedlings developed bigger palisade cells having a larger contact surface with the epidermis and being taller. After 3 days of illumination 10% of the palisade layer of the expanding leaves consisted of inter-cellular spaces, while the latter were completely absent in etiolated leaves and in leaves selectively illuminated with WL (Table 2). Within the same time course the plastid diameter of the expanding leaves doubled though the chlorophyll content per gram fresh weight stayed the same in both series (Table 1). The plastids of the plants with selectively illuminated leaves were much smaller, but structurally they were fully organized. The differences in diffusion resistance of the upper and lower epidermis (Table 2) were related to their stomatal frequencies (fewer stomata per cm^2 of photosynthetic surface with totally illuminated plants than with selectively illuminated leaves). The leaves showing the highest photosynthetic activity had also the highest diffusion resistance. This observation indicated that this physical factor was not limiting for the CO_2 availability in the intercellular spaces.

Since the differences in photosynthetic efficiency between both leaf types were transient and inherent in leaf expansion, we examined these parameters

Table 1. Influence of illumination mode on leaf, cell and plastid development.*

	D	$WL_{etiol.leaf}$	$WL_{etiol.plant}$	$WL_{cont.}$
Leaf area (cm^2)	1.56 ± 0.12	1.65 ± 0.15	10.20 ± 1.00	21.70 ± 2.00
Leaf weight (mg/cm^2)	20 ± 2	21 ± 1.5	21 ± 1.6	21 ± 1.4
Cell height (μm)	31.2 ± 2.6	32.0 ± 2.0	43.0 ± 2.6	47.0 ± 3.1
Cell area palisade (μm^2)	31 ± 2	29 ± 2	135 ± 12	221 ± 15
Plastid diameter (μm)	2.9 ± 0.1	3.9 ± 0.2	5.6 ± 0.2	6.0 ± 0.2
CHL a ($\mu g/g$ fresh weight)	–	2890 ± 110	2970 ± 100	1850 ± 25

* D : 8-day-old etiolated bean seedlings ; $WL_{etiol.leaf}$: the same seedlings with their leaves selectively exposed to WL for 6 days ; $WL_{etiol.plant}$: etiolated seedlings fully illuminated with WL for 6 days ; WL_{cont} : maximum values obtained under continuous WL.

Table 2. Leaf area, intercellular spacing and diffusion resistance of 8-day-old etiolated seedlings illuminated for 3 days with WL.*

	leaf area (cm^2)	intercellular space (%)	epidermal diffusion resistance (sec/cm) upper	lower
$WL_{etiol.leaf}$	1.55 ± 0.15	0	1	1
$WL_{etiol.plant}$	5.30 ± 0.50	10	5.6	2.5

* $WL_{etiol.leaf}$: the leaves of the seedlings were selectively exposed to WL ; $WL_{etiol.plant.}$: the seedlings were entirely irradiated with WL.

which could be responsible: plastid size, volume of the chlorenchyma cells and intercellular spacing.

As we mentioned in the introduction we were able to grow plants with the same leaf size but containing a different amount and volume of cells.

When 3- and 9-day-old etiolated bean seedlings were exposed for 9 days to continuous WL, they developed leaves of the same size and weight (Table 3). Their plastids were qualitatively and quantitatively equal. On a fresh weight basis there was no difference in chlorophyll content. The volume of air cavities between chlorenchyma cells amounted to 25% of the palisade layer for both leaf types, but their intercellular spaces had a different configuration. The leaves of the younger plants contained more palisade cells of smaller size than the older ones. Per unit leaf area there was thus the same volume of air space between cells but fewer palisade cells in the older leaves. The photosynthetic capacity of the latter however was ± 50% better than in the younger leaves in spite of their higher epidermal diffusion resistance (Table 3).

Table 3. Influence of seedling age on leaf morphology and photosynthesis after 9 days of continuous WL illumination.*

	3d D + 9d WL	9d D + 9d WL
Leaf area (cm^2)	20.50 ± 1.90	21.00 ± 1.90
Leaf weight (mg)	451 ± 17	443 ± 21
Cell area palisade (μm^2)	181 ± 26	378 ± 26
Cell height (μm)	45 ± 3.0	47 ± 4.0
Plastid diameter (μm)	6.15 ± 0.3	6.10 ± 0.3
CHL a (μg/g fresh weight)	1848 ± 15	1836 ± 15
Intercellular space (%)	25	25
Diffusion resistance (sec/cm)		
upper epidermis	9.9	22.5
lower epidermis	2.3	4.3
Photosynthesis (rel.units)	1.0 ± 0.1	1.43 ± 0.15

* 3d D + 9d WL : 3-day-old etiolated seedlings fully exposed to WL for 9 days.
9d D + 9d WL : 9-day-old etiolated seedlings fully exposed to WL for 9 days.

DISCUSSION

Under continuous white light the growth of the primary leaves of bean plants consists of three successive periods: a short period of cell expansion, a few days of intensive cell division and a period of cell expansion (Fig. 1). With etiolated seedlings the second period of leaf growth is longer since the division rate is much slower, but the ultimate period of cell expansion is absent. Therefore, at 8 to 9 days after sowing the etiolated leaf contains much less and smaller cells than does the light-grown one. When at that time the dark grown seedlings are transferred to continuous WL, the leaves turn green and grow by cell expansion only. However, when the etiolated leaves are exposed to WL, while all the other plant parts are kept in complete darkness, they also green normally, but they do not expand at all. Their etioplasts turn into structurally normal chloroplasts, but the plastid diameter increases only up to two-thirds of that of mature chloroplasts present in leaves of fully illuminated plants (Table 1).

Comparing the photosynthetic rate of both leaf types, we notice that the selectively illuminated leaves are less efficient than the leaves of totally irradiated plants though the higher photosynthetic efficiency of the latter is transient (Fig. 2). At the time of maximal activity (after 3 days of illumination) the expanding leaves of plants fully exposed to WL have larger palisade cells with bigger chloroplasts, but their chlorophyll content per unit fresh weight is the same as in the selectively illuminated leaves (Table 1). In spite of the presence of air cavities (10% in surface area) within the palisade layer the epidermal diffusion resistance of the expanding leaves is higher than in the leaves showing no expansion and without intercellular spaces (Table 2). The difference in epidermal structure can thus be ruled out as a causal factor since the diffusion resistance is much greater on both sides of the leaf having the highest photosynthetic efficiency. These results are in agreement with the data of Chartier (1970) for 12-days-old bean plants. This author found that the diffusion resistance for CO_2 transfer in the boundary layer and through stomata and cuticle is of little influence on the photosynthetic rate when compared with the diffusion resistance for CO_2 from intercellular spaces to carboxylation sites.

From all the parameters studied mainly the ratios plastid volume/size of palisade cell and size of palisade cell/total leaf volume seem to be considerably changed in expanding leaves.

The etiolated leaf has no intercellular spaces. The latter develop gradually in the palisade layer when the plants are fully exposed to WL, but they are absent when the leaves are selectively illuminated (Table 2). By the same token, it is well known that during cell expansion the ratio cell surface/cell volume is lowered, while there is an increase of the ratio cell surface/cytoplasmic volume owing to the vacuolation of the mesophyll cells. By these facts the intra- and intercellular environment of the photosynthetic tissue is markedly changed during leaf expansion.

A suitable approach to this problem is given by the technique which enabled us to grow plants with the same leaf area but different in leaf architecture (Table 3). When 3- and 9-day-old etiolated plants are cultivated for 9 days under continuous WL, the primary leaves have the same size and weight. Their plastid diameter and development is equal. They have the same chlorophyll content. The intercellular spaces of the palisade layer amounts to 25% for both series. Nevertheless the photosynthetic efficiency is about 50% higher for the older leaves, which have the most pronounced epidermal diffusion resistance (Table 3). The most conspicuous difference between the younger and older leaves resides in the volume of the palisade cells. They are bigger and fewer in number per unit area in the older leaves.

In considering leaf anatomy in relationship to photosynthesis it is obvious that photosynthesis increases as the mean cell size increases. This finding is at a first glance contradictory with the statement of Wilson & Cooper (1970) studying *Lolium* genotypes collected in various ecosystems. They concluded that mesophyll cell size is inversely proportional to net CO_2 assimilation and growth. In their research cell size is an inheritable trait that can be used as a criterion in plant breeding work. In our experiments cell size is the result of a modulation of plant photomorphogenesis determined by seedling age and illumination mode. Hence genetics, biochemistry and ultrastructure of the photosynthetic apparatus are invariables in our approach. By changing the cell morphology of the photosynthetic tissue it seems likely that the diffusion resistance for CO_2 through the mesophyll (from air spaces to carboxylation sites) is drastically altered. Bigger mesophyll cells have a larger contact surface with the intercellular spaces and a higher cell surface to cytoplasmic volume ratio. The large and small palisade cells contain both the same amount of cytoplasm. Due to the presence of a large central vacuole the plastids of the bigger cells are embedded in a narrower band of cytoplasm, adjacent to the cell surface, than those of the smaller cells. In view of these facts we may suppose that the leaf architecture of the older plants allows a better gas exchange between the intercellular cavities and the cytoplasm of the mesophyll cells, on the one hand, and within the cytoplasm by a more efficient viscous flow, on the other.

Further studies are necessary to prove that the diffusion resistance for CO_2 through the mesophyll is an important factor for photosynthetic efficiency in relation to internal leaf morphology.

If our findings can be confirmed with other plant species, studies on leaf architecture should be incorporated into plant improvement programs as a useful criterion in selection and breeding.

REFERENCES

Chartier, P. A model of CO_2 assimilation in the leaf. pp. 307–315, in: Proceedings of the IBP/PP Technical Meeting. Trebon, 14–21 Sept. 1969, Prediction and Measurement of Photosynthetic Productivity, Centre for Agricultural Publishing and Documentation, Wageningen, The Netherlands (1970).

De Greef, J.A. & R. Caubergs. Studies on greening of etiolated seedlings. I. Elimination of the lag phase of chlorophyll biosynthesis by a pre-illumination of the embryonic axis in intact plants. *Physiol. Plant.* 26: *157–165* (1972).

De Greef, J.A., R. Caubergs, J.P. Verbelen & E. Moereels. Phytochrome-mediated inter-organ dependence and rapid transmission of the light stimulus. pp. 295–316, in: Smith, H. (ed.), Light and Plant Development. Butterworths, London-Boston (1976).

Downs, R.J. Photoreversibility of leaf and hypocotyl elongation of dark grown red kidney bean seedlings. *Plant Physiol.* 30: *468–472* (1955).

Maksymowych, R. Analysis of leaf development. Abercrombie, M., D.R. Newth & J.G. Torrey (eds.), Developmental and Cell Biology Series. Cambridge University Press (1973).

Parker, M.W., S.B. Hendricks, H.A. Borthwick & F.W. Went. Spectral sensitivities for leaf and stem growth of etiolated pea seedlings and their similarity to action spectra for photoperiodism. *Am. J. Bot.* 36: *194–204* (1949).

Verbelen, J.P. Fotoregulatie van de blad-ontwikkeling bij *Phaseolus vulgaris* L. op orgaan-, weefsel en celniveau. Doctoral Thesis, University of Antwerpen (1976).

Wilson, D. & J.P. Cooper. Effect of selection for mesophyll cell size on growth and assimilation in *Lolium perenne* L. *New Phytol.* 69: *233–245* (1970).

Woolf, C.M. Principles of Biometry. D. Van Nostrand Inc., Princeton, New Jersey, Toronto, London, Melbourne (1968).

Authors' address:
Department of Biology
University of Antwerpen
R.U.C.A. & U.I.A.
Universiteitsplein 1
B - 2610 Wilrijk
Belgium.

SESSION II

PHOTOSYNTHESIS AND FLOWERING

PHOTOSYNTHETIC AND PHOTOMORPHOGENETIC EFFECTS OF HIGH PHOTON FLUX DENSITIES ON THE FLOWERING OF TWO LONG-DAY PLANTS, *ANAGALLIS ARVENSIS* AND *BRASSICA CAMPESTRIS**

D.J.C. FRIEND, J. DEPUTY & R. QUEDADO
(Honolulu, Hawaii, U.S.A.)

ABSTRACT

Flower induction in both species was strongly dependent on a high photon flux density. The number of days to floral initiation of *Brassica* was inversely related to the reciprocal of daylength when plants were grown continuously at daylengths ranging from 4 to 24 hr. For equal photon flux densities over the photosynthetically active spectrum (400 to 700 nm) the effect of daylength on flower initiation was the same for fluorescent as for incandescent radiation. Participation of the HIR (High Intensity Reaction) in flower promotion of *Brassica* was shown by the superior effectiveness of blue and far-red radiation over red radiation given during a single 16 hr dark period. The actions of red and far-red radiation on hypocotyl growth were antagonistic, indicating phytochrome activity, but the flower promoting effects of red and far-red radiation during the same period were additive.

In *Anagallis* the photosynthetic inhibitor DCMU reduced flowering when given at a concentration sufficiently high to inhibit CO_2 uptake during the 16 hr radiation treatment. High photon flux densities may also increase flowering through the activity of the HIR. Flowering was promoted by levels of broad-band far-red irradiation that were not sufficient to stimulate net CO_2 uptake. CO_2 fixation may however be required for the flower-promoting effect of the HIR as the absence of CO_2 reduced the effect of far-red radiation on the promotion of flowering. Far-red radiation reduced the level of CO_2 release as compared to darkness and there was more incorporation of $^{14}CO_2$. The increased flowering obtained by supplying glucose during far-red irradiation further indicates the need for photosynthates during the action of the HIR.

INTRODUCTION

Photosynthesis may be involved in the promotion of flowering of long-day plants through two main actions. Photosynthates are presumably needed to provide relatively non-specific substrates for mitochrondrial respiration, providing energy-rich compounds such as ATP and NADH for the metabolism associated with the perception, stabilization, transport and realization of the photoperiodic stimulus. In addition, the photosynthetic sequence could interact more directly with the perception or later stages of the photoperiodic response.

*Hawaii Agricultural Experiment Staion Journal Series No. *2282*. Part of this publication was taken from a thesis submitted by J. Deputy in partial fulfillment of the requirements for the M.S. degree, and part taken from a dissertation submitted by R. Quedado in partial fulfillment of the requirements for the Ph.D. degree, University of Hawaii.

We thank Dr. W. E. Evenson, Brigham Young University, Provo, Utah for most useful discussions on the relationship between daylength and flowering in *Brassica.*

Formation of ATP and NADPH during photosynthesis, as well as the activity of the electron transport chain and associated proton pumps may well interact with phytochrome activity, as both processes are associated with membranes (Hendricks & Taylorson, 1978). These interactions could affect P_{FR} or, in the case of the HIR (High Irradiance Reaction) some other active intermediate (Kendrick & Spruit, 1972; Fuad & Yu, 1977; Jose & Vince, 1978). Phytochrome is also known to affect the activities of several enzymes involved in the energetics of metabolism, both in respiration and in photosynthesis (Deitzer *et al*., 1977). Endogenous changes in the activities of these enzymes may further complicate interactions with phytochrome or the HIR (Vince, 1965; Wagner, 1977; Deitzer, Hayes & Jabben 1978).

Evidence for the influence of photosynthesis on flowering is of several types. The role of photosynthesis in the supply of carbohydrates is emphasized in experiments in which the uptake of CO_2 parallels increased flowering (Bodson *et al*., 1977). In *Anagallis* however, a further increase in flowering was obtained by increasing the photon flux density above that required to saturate CO_2 uptake (Quedado & Friend, 1978). Preventing CO_2 uptake by placing plants in a CO_2-free environment during irradiation lowers the effectiveness of floral induction in several long-day plants (Bavrina *et al*., 1969; Bodson *et al*. 1977). In some cases, sugar feeding can replace the need for externally applied CO_2, as in *Anagallis* (Quedado & Friend, 1978) as well as increase the level of flowering of plants irradiated at sub-optimal photon flux densities.

The role of photosynthesis in the supply of energy-rich intermediates has been investigated by feeding ATP or NADPH. The results show an increase in flowering under some conditions, and a reduction under others (Kandeler *et al*., 1975). While the results are difficult to interpret, they do at least show that the applied intermediates were indeed taken up into some compartment of the plant cell.

DCMU and other photosynthetic inhibitors reduced the level of flowering in some long-day plants (Bavrina *et al*., 1969; Evans, 1969), including *Anagallis* (Imhoff, 1973). It is not clear whether the effect of these inhibitors is directly related to interference with specific metabolic reactions, or to a general effect on lowered plant growth through a reduced supply of carbohydrates. DCMU for instance may inhibit photosystem 2 and prevent NADPH formation but will also lower total CO_2 uptake and reduce growth of the plant in dry weight. The methods of cell biology, where the action of an inhibitor may be overcome by simultaneous addition of the specific substance blocked, are often not useable when dealing with whole plants because of the difficulties of ensuring penetration of the added chemicals to the active site.

The experiments reported here deal with the quantitative relationship between daylength, photon flux density and flowering in the quantitative long-day plant *Brassica campestris* and the relationship between photosynthesis and the HIR in the qualitative long-day plant *Anagallis arvensis*.

Brassica campestris flowers more rapidly the longer the daylength, whether plants are grown continuously at different daylengths or are given single periods of supplementary irradiation (Friend, 1968a, 1969). The higher the photon flux density during a single irradiation the higher is the degree of flowering. The experiments reported here were designed to test the hypothesis that plants would initiate an inflorescence after a certain critical value of accumulated quanta had been received independently of daylength or quantum flux density. Daylengths were selected to include the lowest compatible with growth at any given intensity. The relative importance of photosynthetically active radiation (PAR as measured by the photon flux density between 400 to 700 nm) and the HIR, was studied by growing plants under equal PAR values of both fluorescent and incandescent radiation. Fluorescent light is low in far-red radiation and generally has a small effect in the HIR, while incandescent radiation is high in far-red radiation and generally has a large effect in the HIR. Evidence that typical reversible phytochrome reactions were present in plants at the same time as HIR reactions was sought from experiments on the effect of a single supplemental irradiation on both hypocotyl elongation and on flowering.

Anagallis was used to determine whether the flower-promoting effect of broad-band far-red radiation depended on the simultaneous presence of CO_2, and whether glucose provided during such irradiation could increase the flowering response. An attempt was made to distinguish between an effect of far-red radiation on lowering dark respiration and an effect on increasing the incorporation of CO_2.

MATERIALS AND METHODS

Seeds of *Brassica campestris* cv. Ceres were sown 25 in a 5 cm pot of vermiculite and germinated for 2 days under white fluorescent lamps with PAR equal to P_{FR} 8 μmol m^{-2} s^{-1} as measured by a Lambda L1–170 quantum meter. Plants were then moved to the stated intensity in growth rooms maintained at 25°C. Incandescent radiation was provided by tungsten filament spotlights and was filtered through 5 cm of water to remove longwave infra-red radiation. White fluorescent radiation was provided by Sylvania F96PG1 cool white tubes. Seedlings were grown without mineral nutrients in order to restrict plant size; this procedure does not affect the time of flower initiation (Friend, 1968). The number of days to 50% flower initiation was determined by periodic examination of the apical meristem of seedlings under the dissecting microscope.

The effect of spectral quality on flower intitiation was investigated with both broad-band and narrow-band spectra. Plants were grown under an 8 hr daylength with radiation of 270 μmol m^{-2} s^{-1} PAR from white fluorescent lamps. The temperature was maintained at 25°C. During the fifth dark period the plants were irradiated for 16 hr with broad-band spectra provided by the lamp and filter combinations given in Table 1. The radiant flux density was measured by a Yellow Springs Industry Model 6J radiometer. Narrow-band spectra were provided by inserting interference filters in the beam of a 500 watt incandescent projector fitted with an infra-red filter. Peak transmission of the filters was at 660 nm (red) 730 nm (far-red) and blue (426 nm) and the bandwidth was from 10 to 16 nm. Hypocotyl lengths were measured 12 days after sowing as differences between treated and control seedlings remained constant after the end of the 16 hr irradiation treatment. Plants were then dissected and examined for the presence of flower buds.

A clone of *Anagallis arvensis* L. subsp. *arvensis* was derived from cuttings taken from a single plant from seed lot 119-A provided by L. A. T. Ballard (1969). Plants were grown in pots of vermiculite watered with one-fifth strength Hoagland's nutrient solution (Hoagland & Arnon, 1939). Stock plants were cultivated in growth cabinets under an 8 hr daylength at 800 μmol m^{-2} s^{-1} PAR fluorescent light (G.E. cool white) and 80 μmol m^{-2} s^{-1} PAR incandescent radiation provided by standard incandescent bulbs. The temperature was maintained at 25°C. Experimental cuttings were taken with five pairs of fully expanded leaves and placed in water or other solutions during the specific irradiation treatment. The floral response of freshly made cuttings did not differ from that of previously rooted cuttings and facilitated the uptake of experimental solutions into the plants. After irradiation of cuttings, the basal ends were washed and replaced in containers of water under standard non-inductive 8 hr daylengths. Rooting occurred in about one week, and plants were then placed in one-fifth strength Hoagland's solution until the floral response was measured about three weeks later. Plants were dissected and examined under the microscope for the production of floral primordia. The flowering response was quantified by measuring both the percentage of plants that had floral initials and the mean number of flower buds per plant.

Table 1. Lamp and filter combinations used to provide broad-band spectra.

Color	Spectral peak (nm)	Lamp	Filters	Ratio P_{FR}/P_R
Blue	445–460	Cool white fluorescent	Cinemoid 20	0.44[1]
Red	660	Cool white fluorescent	Cinemoid 14	0.80[2]
Far-red	710–730	GE BJC incandescent	5 cm water	0.21[2]

[1] Calculated from data of Hartmann (1966).
[2] From Butler & Lane (1965).

Two main types of radiation were used to treat *Anagallis* cuttings for a single period of 16 hr after the 8 hr main light period. White incandescent radiation was provided by General Electric Company 150 watt cool beam flood lamps, and was filtered through 10 cm of water to remove longwave infra-red radiation. Broad band far-red radiation was provided by General Electric Company B.J.C. lamps and was similarly filtered through a water bath; the radiant flux density was measured by a Yellow Springs Industry Model 65 radiometer.

Rates of CO_2 exchange were measured using an infra-red gas analyzer as described in Friend (1975). Five cuttings were placed in the chamber and the rate of CO_2 production was measured in darkness and under broad-band far-red irradiation of 120 Merg m^{-2} s^{-1}. The incorporation of $^{14}CO_2$ was followed by injecting 29 μg $NaHCO_3$ containing 2.9 μCi $^{14}CO_2$ into a sealed 500 ml chamber of atmospheric air. The total CO_2 concentration was then 5.0 μM. Five cuttings were weighed and placed in glass vials of water. Oil was floated on the water to prevent direct evaporation and the rate of transpiration during irradiation was measured by weighing the vials before and after treatment. Calcium chloride was placed on the floor of the sealed chambers to maintain a low water potential. Three replicate chambers were placed in darkness and three under broad-band far-red radiation of 60 Merg m^{-2} s^{-1}. The duration of exposure was 2 hr, starting

8 hr after the beginning of the 16 hr dark period. Plants were boiled in methanol to dissolve sugar and other compounds including chlorophyll. Chloroform was used to remove chlorophyll and other quenching pigments; adding water separated the pigmented chloroform from the methanol-water mixture which was made up to 100 ml. Samples of 0.1 ml of the aqueous phase were then mixed with 2 ml of Aquasol and the radioactivity was measured in a scintillation counter.

RESULTS AND DISCUSSION

The quantitative nature of the photoperiodic response in *Brassica* is shown by the relationship between days to floral initiation and daylength at any one level of irradiation. The number of days to floral initiation (F) was linearly and inversely related to the reciprocal of the daylength (D). Regressions were fitted for the equation $F = A + B\ (D)^{-1}$ and at any one level of PAR a similar relationship was found whether plants were grown under incandescent or fluorescent irradiation (Fig. 1). The value of 0.0036 (A + B), representing the number of accumulated moles of PAR per m^2 required before floral initiation,

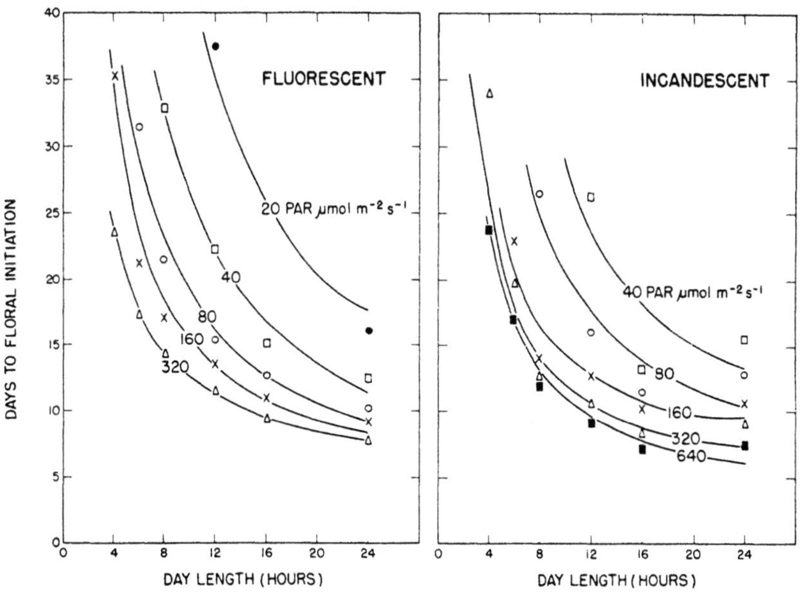

Fig. 1. Interaction of daylength and photon flux density (PAR) of incandescent and fluorescent radiation on the days to flower initiation of *Brassica campestris*. Plants grown continuously at stated daylength at 25°C. Lines are regressions, Days to floral initiation = A + B $(\text{Daylength})^{-1}$. Values of R for the fluorescent series were 1.00, 0.99, 0.99, 0.99, for PAR values of 320, 160, 80, and 40 μmol $m^{-2}\ s^{-1}$, and for the incandescent series 0.99, 0.97, 0.92, 0.93, and 0.76 for PAR values of 640, 320, 160, 80, and 40 μmol^{-2} s^{-1} respectively. Each point is a mean value derived from periodical dissection of about 60 plants per treatment.

decreased with increasing PAR, the response being similar for plants grown under either incandescent or fluorescent irradiation (Fig. 2).

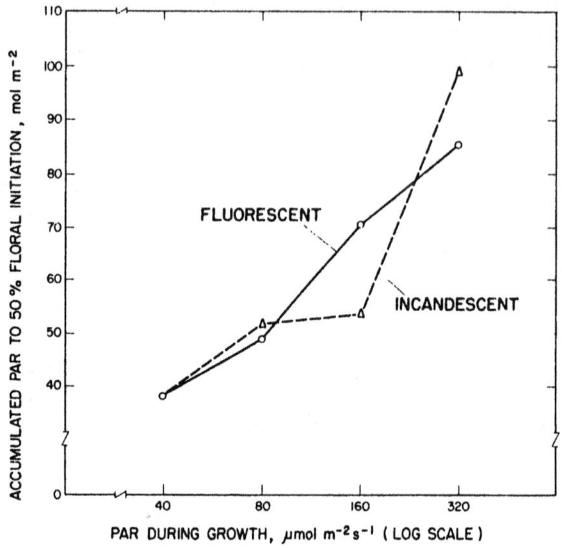

Fig. 2. Relationship between accumulated PAR received by *Brassica campestris* up to the time of 50% floral initiation (0.0036 (A + B)) PAR mol m^{-2} s^{-1} from the regressions given in Fig. 1), and the logarithm of the PAR during growth under either fluorescent or incandescent radiation. For plants under fluorescent radiation, R = 0.99, for plants under incandescent radiation R = 0.92.

When plants were irradiated for a single 16 hr period with narrow-band radiation, flowering was promoted to a greater extent by far-red and blue than by red irradiation (Fig. 3). This implicates the action of the HIR as already shown by the action spectrum for floral initiation in *Brassica* (Wassink *et al.*, 1950; Friend, 1968b). Absence of antagonism between the effects of simultaneously applied red and far-red radiation further suggests that flowering in *Brassica* is not controlled directly by the level of P_{FR} (Table 2). That the typical reversible phytochrome system was indeed operative in the plant during the same irradiation treatment was shown by the effect of radiation on hypocotyl extension. Whereas far-red radiation promoted elongation more than red radiation, simultaneous application of red radiation reduced the elongating effect of far-red (Table 2). This antagonism between the action of red and far-red irradiations was seen whether the main light period ended with white fluorescent light, when a high level of P_{FR} would be present, or whether the main light period ended with 15 min of wide-band far-red irradiation, when a low level of P_{FR} would be present.

The involvement of both the level of photosynthesis and the activity of the HIR is implicated from these results. Flower initiation took place once a certain number of PAR hours had been accumulated and at any one photon flux density this number was the same at any daylength over a very wide range of 4 to 24 hr.

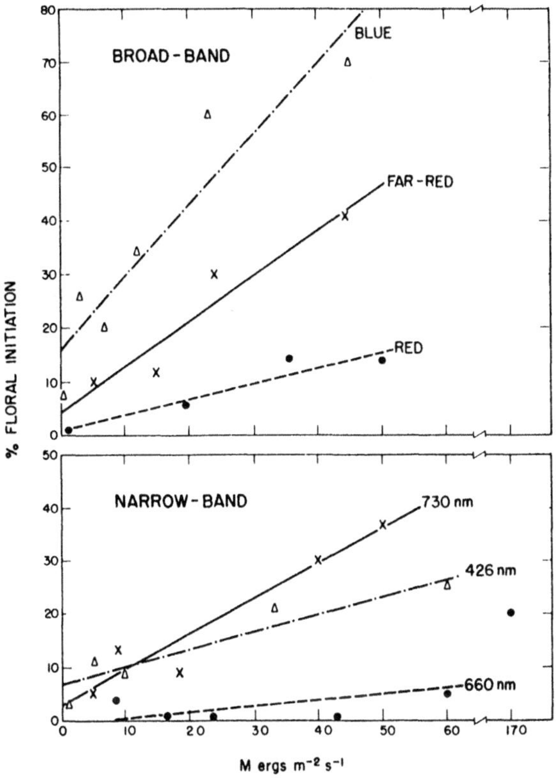

Fig. 3. Effect of broad and narrow band irradiations in blue, red and far-red spectral regions on floral initiation of *Brassica campestris*. Plants given 16 hr of irradiation 4 days from germination. Percentage floral initiation determined 12 days after sowing, mean values from about 60 plants per treatment.

Table 2. Effect of simultaneous red (R) and far-red (FR) irradiation on flower initiation and hypocotyl elongation of *Brassica*.

Radiation at the end of main light period	% flowering					Hypocotyl length, mm				
	R	F	R+F	D	S.E.	R	F	R+F	D	S.E.
					Broad-band irradiation					
White fluorescent	18	37	31	0	±3.7	11.3	15.5	13.4	10.0	±0.55
15 min. broad-band FR	27	44	56	0	±5.1	9.5	12.3	9.4	10.0	±0.53
	Narrow-band irradiation									
White fluorescent	0	25	50	0						

Incandescent radiation has a ratio of red to far-red radiation of about 1, compared to fluorescent radiation with a ratio of about 9. The similar flowering response of plants irradiated with either incandescent or fluorescent radiation of equal PAR values thus suggests that when plants were grown continuously under a range of different daylengths the activity of a HIR was not a major factor controlling floral initiation.

In experiments with *Anagallis*, several inhibitors of different partial processes of photosynthesis were used to further study the inter-relationships of photosynthesis and the flowering response during a 16 hr irradiation with 650 μmol m^{-2} s^{-1} PAR from incandescent lamps. A concentration of 100 μM DCMU totally inhibited CO_2 uptake for 24 hr but plants did not recover and eventually died (Table 3). A concentration of 4 μM DCMU inhibited CO_2 uptake about 50% during the 16 hr irradiation treatment but allowed recovery of the plants so that 4 wks later the fresh weight was not significantly different from that of untreated plants. Flowering was inhibited by DCMU treatment, the greater the inhibition of CO_2 uptake the greater was the inhibition of flowering (Table 3). This effect of DCMU agrees with that reported for another strain of

Table 3. A. Effect of DCMU on CO_2 uptake.

DCMU conc., μM	Rate of CO_2 uptake, % initial Time after DCMU treatment, hr 1–5	8–12	20–24	140–144
0	100	100	89	71
1	86	80	89	72
4	66	60	50	63
10	25	12	16	15
40	0	0	0	7
100	0	0	0	4

B. Effect of DCMU on ability of a single exposure to 16 hr supplementary incandescent irradiation (1900 μmol m^{-2} s^{-1} PAR) to promote flowering of *Anagallis*.

	Fresh wt. at 4 wks., % water control	% Survival	% Flowering	S.E.
0	100	100	55	±2.8
4	100	100	39	±1.9
10	56	50	18	±1.1
40	3	4	0	
100	0	0	0	

Anagallis (Imhoff, 1973). Neither sucrose nor glucose solutions relieved the inhibitory effect of DCMU on floral induction however (Table 4), although analyses of plants for total sugar content by the anthrone method showed that sugar feeding doubled the sugar content by the end of the 16 hr irradiation treatment as compared with water treated controls (Quedado & Friend, 1978). Solutions of 10 μM ATP, 10 μM NADPH or a mixture of the two did not reduce

Table 4. Inability of glucose and sucrose (10 kg m^{-3} in water) to overcome inhibitory effect of DCMU on flowering response of *Anagallis.**

Leaf spray	% Flowering					
	Stems placed in these solutions during 16 hr irradiation at 1900 μmol m^{-2} s^{-1} PAR				650 μmol m^{-2} s^{-1} PAR	
	Water	S.E.	Sucrose	S.E.	Water	Glucose
Water	55	±2.8	45	±5.0	34	19
4 μM DCMU	39	±1.9	27	±5.5	10	13
10 μM DCMU	18	±1.1	18	±1.5	–	–

*Short-day controls were all vegetative.

the inhibitory effect of DCMU on flowering. It is not known whether ATP and/or NADPH reached a carboxylation site however, as the rate of CO_2 uptake during the 16 hr irradiation treatment was not significantly affected. Three other inhibitors of photosynthesis also reduced CO_2 uptake during the 16 hr irradiation treatment and also reduced the level of flowering (Table 5). Fresh weights of plants measured 4 wk after treatment were only from 65 to 86% of the weight of water treated controls, so for these inhibitors it cannot be determined whether the reduced flowering was caused by inhibition of photosynthesis or a non-specific reduction in general growth.

Table 5. Effect of some inhibitors upon CO_2 uptake and % flowering of *Anagallis.**

Inhibitor	Mode of action	CO_2 uptake, % of control	% flowering of water controls	Fresh weight % of control
0.02 M salicylaldoxime	Inhibits both cyclic and non-cyclic photophosphorylation	58.5	46.0	65
0.4 M NH_4OH	Uncoupler of photophosphorylation	57.0	52.6	86
0.1 M $NaNO_2$	Inhibits NADP reduction	75.0	38.0	70

* Inhibitor given as a leaf spray before a single 16 hr incandescent irradiation at 1300 μmol m^{-2} s^{-1} PAR. Figures are means from about 20 plants.

The involvement of a HIR in the promotion of flowering of *Anagallis* was suggested by the continued increase in flowering obtained by raising the level of PAR above that required for the light saturation of photosynthesis (Quedado & Friend, 1978). This suggestion receives support from the promotion of flowering by two 16 hrs irradiations with broad-band far-red irradiation (30 Merg m^{-2} s^{-1}); a similar treatment with broad-band red irradiation (6 Merg m^{-2} s^{-1}) was ineffective (Table 6). Water-filtered incandescent radiation (40 Merg m^{-2} s^{-1}) gave a low level of flower promotion with one 16 hr period of treatment and a high

Table 6. Effect of incandescent, red and broad-band far-red radiation on flowering of *Anagallis*. Radiation given during one or two 16 hr dark periods of an 8 hr daylength.*

	% flowering			Mean flower buds per plant		
	Incandescent	Red	Far-red	Incandescent	Red	Far-red
1 irradiation	18	0	0	0.17	0	0
2 irradiations	89	0	72	2.4	0	1.13

* Levels of irradiation – water filtered incandescent 40 Merg m^{-2} s^{-1}
red 6 Merg m^{-2} s^{-1}
Far-red (broad band), water filtered,
30 Merg m^{-2} s^{-1}

level of flowering when two 16 hr periods were given. A single period of treatment with broad-band far-red radiation resulted in flowering in other experiments, and the optimal temperature was close to 25°C (Table 7). The participation of the HIR in Ballard's strain of *Anagallis* is thus similar to that reported for the Paris strain (Imhoff *et al.*, 1971).

Table 7. Promotion of flowering of *Anagallis* by 16 hr of broad-band far-red irradiation (30 Merg m^{-2} s^{-1}) given at different temperatures. Plants grown at 25°C.

	% flowering			Mean flower buds per plant		
	18	24	30°C	18	24	30°C
1 irradiation	0	25	16	0	0.31	0.16
2 irradiations	66	100	81	1.2	2.2	1.5

Is the flower-promoting effect of high photon flux densities of broad-band far-red radiation attributable to photosynthesis? Under treatment with 120 Merg m^{-2} s^{-1} of water filtered far-red irradiation the plants were below the compensation point, but the net CO_2 release was less (44 μg m^{-2} s^{-1}) than that of dark respiration (54 μg m^{-2} s^{-1}). The lowering of the rate of CO_2 production by far-red irradiation could be caused by a depression of mitochondrial respiration or by increased rates of fixation of recycled respiratory CO_2. Such fixation could occur either through the regular RBP oxygenase-carboxylase system or through the activity of PEP carboxylase. An attempt was made to distinguish between a reduction in mitochondrial respiration and a refixation of respired CO_2. Plants were given far-red irradiation for two hours at the mid-point of a 16 hr dark period in the presence of $^{14}CO_2$. Preliminary experiments established no difference in transpiration rates under far-red radiation or in darkness. Any observed differences in rates of $^{14}CO_2$ uptake could not then be caused by different stomatal resistances under the two conditions and related differences in the rates of exchange of labelled and unlabelled CO_2. The rate of $^{14}CO_2$ uptake under far-red irradiation was higher than in darkness, even though rates of transpiration were similar (Table 8). The higher rate of incorporation of $^{14}CO_2$ under far-red irradiation must therefore be attributed to an increased rate of CO_2 fixation rather than to a suppression of respiration. It is not at present known whether this incorporation should be regarded as accelerated dark fixation though a PEP carboxylase system or as photosynthesis though the activity of RBP oxygenase-carboxylase.

Table 8. $^{14}CO_2$ exchange and water vapor exchange during darkness or irradiation with 60 Merg m^{-2} s^{-1} broad-band far-red.

	Dark	Far-red
Relative $^{14}CO_2$ uptake	30	100
H_2O release % fresh wt. hr^{-1}	12.1	11.8

It seems likely that photosynthates participate early in the sequence of events leading from the action of far-red radiation on flower initiation. The absence of CO_2 during treatment with far-red radiation reduced the promotion of flowering (Table 9) as was found with incandescent radiation (Quedado & Friend, 1978). Feeding glucose during far-red irradiation also increased the flowering response (Table 10).

Table 9. Effect of removal of CO_2 during treatment with 60 Merg m^{-2} s^{-1} broad-band far-red irradiation on flowering of *Anagallis.*

	Atmospheric CO_2	O CO_2
Mean flower buds per plant	0.96	0.07
% flowering	43	7

Table 10. Effect of Glucose (10 kg m^{-3} in water) during 16 hr incandescent irradiation at 6 Mmole m^{-2} s^{-1} PAR.

	Water	Glucose
% flowering	47	66
Mean flower buds per plant	0.60	1.05

In summary, the participation of photosynthesis in the photoperiodic response of *Anagallis* was shown by the need for CO_2 for effective promotion of flowering by supplementary periods under white irradiation. The ability to increase flowering by feeding glucose during irradiation at sub-optimal levels of radiation suggests that the need for CO_2 is simply to provide photosynthates for general metabolism. The lower flowering level obtained when CO_2 was withheld during far-red irradiation implies however that an increased photosynthate supply is not the main effect of CO_2 giving during supplementary irradiation.

Plants were still below the compensation point under far-red irradiation. The increased incorporation of $^{14}CO_2$ compared to that of plants with equal stomatal resistances maintained in darkness shows that far-red irradiation increases CO_2 fixation rather than depresses mitochondrial respiration. Whether this fixation should be regarded as photosynthesis through the RBP oxygenase carboxylase system in this C_3 plant or as accelerated dark fixation presumably through PEP carboxylase (Levi *et al.*, 1978), is not at present known.

In either case, the fixation of CO_2 would reduce the energy charge of the cell by depleting NADPH and ATP. Far-red radiation would be expected to promote ATP synthesis through cyclic photophosphorylation, but the flower promoting effect of white incandescent radiation shows that the relative amounts of ATP and NADPH are probably not important in the interaction of photosynthesis with the HIR.

In *Brassica* the activity of the HIR was shown by the greater effectiveness of far-red than red radiation in promoting flower initiation when given during a single dark period and the additive effects of these spectral regions. The activity of typical phytochrome reactions in the same seedlings at the same time as the HIR was shown by the antagonism between the effect of red and far-red radiation on hypocotyl elongation. The HIR was not so evident when plants were grown continuously under different daylengths using either fluorescent or incandescent radiation. The time to floral initiation was inversely related to the reciprocal of the daylength. The total number of quanta received per unit area up to the time of floral initiation was the same at any one photon flux density whether incandescent or fluorescent radiation was used (Fig. 1) in spite of the greater effectiveness of incandescent radiation in the HIR. The non-reciprocal relationship between quantal flux density and the accumulated quanta received before floral initiation may reflect the decreasing effectiveness of CO_2 fixation as light saturation of photosynthesis is approached. The promotion of flowering by sucrose in the closely similar long-day species *Sinapis alba* has been shown in tissue culture even under short-day conditions (Deltour, 1967).

ABBREVIATIONS

DCMU = 3-(3,4-dichlorophenyl)-1, 1-dimethylurea
HIR = High irradiance reaction (Mohr)
PAR = Photosynthetically active radiation (400 to 700 nm)
PEP = Phosphoenolpyruvate
P_{FR} = The far-red absorbing form of phytochrome
P_R = The red absorbing form of phytochrome
RBP = Ribulosebisphosphate

REFERENCES

Ballard, L.A.T. *Anagallis arvensis* L. pp. 376–392, in: Evans, L. T. (ed), The induction of flowering, some case histories. Cornell University Press, Ithaca, New York (1969).

Bavrina, T.V., N.P. Aksenova & T.N. Konstantinova. The participation of photosynthesis in photoperiodism. *Soviet Plant Physiol.* 16: *315–322* (1969).

Bodson, M., R.W. King, L.T. Evans & G. Bernier. Photosynthesis and flowering in the long-day plant *Sinapis alba. Aust. J. Plant Physiol.* 4: *467–468* (1977).

Butler, W.L. & H.C. Lane. Dark transformations of phytochrome *in vivo.* II. *Plant Physiol.* 40: *13–17* (1965).

Deitzer, G.F., D.W. Hopkins, U. Haertle & E. Wagner. Effect of light on oscillations of enzyme activity during photomorphogenesis in *Chenopodium rubrum* L. *Photochem. Photobiol.* 27: *127–131* (1977).

Deitzer, G.F., R. Hayes & M. Jabben. Circadian rhythm in sensitivity of floral induction in winter barley to far-red light. (Abstract). *Plant Phys.* 61, Suppl: 13 (1978).

Deltour, R. Action du saccharose sur la croissance et la mise à fleurs de plantes issues d'apex de *Sinapis alba* L. cultivés *in vitro. C.R. Acad. Sc. Paris*, 264: *2765–2767* (1967).

Evans, L.T. *Lolium temulentum* L. pp. 328–349, in: Evans, L.T. (ed.), The induction of flowering, some case histories. Cornell University Press, Ithaca, New York (1969).

Friend, D.J.C. Photoperiodic responses of *Brassica campestris* L. cv. Ceres. *Physiol. Plant.* 21: *990–1002* (1968 a).

Friend, D.J.C. Spectral requirements for flower initiation in two long-day plants, rape (*Brassica campestris* L. cv. Ceres), and spring wheat (*Triticum x aestivum*). *Physiol. Plant.* 21: *1185–1195* (1968b).

Friend, D.J.C. *Brassica campestris* L. pp. 364–375, in: Evans, L.T. (ed), The induction of flowering, some case histories. Cornell University Press, Ithaca, New York (1969).

Friend, D.J.C. Adaptation and adjustment of photosynthetic characteristics of gametophytes and sporophytes of Hawaiian tree-fern (*Cibotium glaucum*) grown at different irradiances. *Photosynthetica* 9: *157–164* (1975).

Fuad, N. & R. Yu. An action spectrum of phytochrome binding for a subcellular fraction of maize coleoptiles. *Photochem. Photobiol.* 25: *491–496* (1977).

Harmann, K.M. A general hypothesis to interpret high energy phenomena of photomorphogenesis on the basis of phytochrome. *Photochem. Photobiol.* 5: *349–366* (1966).

Hendricks, S.B. & R.B. Taylorson. Dependence of phytochrome action in seeds on membrane organisation. *Plant Physiol.* 61: *17–19* (1978).

Hoagland, D.R. & D.I. Arnon. The water culture method for growing plants without soil. *Calif. Agr. Expt. Sta. Cir.* 347: *1–39* (1939).

Imhoff, C. Action d'un inhibiteur de la photosynthèse, le 3-(3,4-dichlorophenyl)-1, 1 – diméthylurée (DCMU), sur la floraison de l'*Anagallis arvensis. C.R. Acad. Sc. Paris,* 276 Série D: *3303–3306* (1973).

Imhoff, C., J. Brulfert & M.R. Jaques. Mise à fleurs, en éclairement monochromatique, de l'*Anagallis arvensis* L., plante de jours longs absolu: spectre d'action et rôle du phytochrome. *C.R. Acad. Sci. Paris*, 273, Série D: *737–740* (1971).

Kandeler, R., B. Hügel & T.H. Rottenburg. Relations between photosynthesis and flowering in *Lemnaceae*. pp. 161–169, in: Marcelle, R. (ed), Environmental and biological control of photosynthesis. Junk, The Hague (1975).

Kendrick, R.E. & C.J.P. Spruit. Phytochrome decay in seedlings under continuous incandescent light. *Planta* (Berl.) 107: *341–350* (1972).

Levi, C., J.T. Perchorowicz & M. Gibbs. Malate synthesis by dark carbon-dioxide fixation in leaves. *Plant Physiol.* 61: *477–480* (1978).

Purohit, A.N. & E.B. Tregunna. Carbon dioxide compensation and its association with the photoperiodic response of plants. *Can. J. Bot.*, 52: *1146–1148* (1974).

Quedado, R. & D.J. Friend. The participation of photosynthesis in floral induction of the long-day plant *Anagallis arvensis* L. *Plant Physiol.* in press (1978).

Vince, D. The promoting effect of far-red light on flowering in the long-day plant *Lolium temulentum. Physiol Plant.* 18: *474–482* (1965).

Wassink, E.C., C.M.J. Sluijsmans & J.A.J. Stolwijk. On some photoperiodic and formative effects of coloured light in *Brassica rapa, f. oleifera,* subf. *annua. Koninkl. Ned. Akad. Wetenschap. Proc. Sev.* C: *1466–1475* (1950).

Authors' address:
Department of Botany
Hawaii Agricultural Experiment Station
University of Hawaii
3190 Maile Way
Honolulu, Hawaii 96822
U.S.A.

FLOWERING OF *SINAPIS* AS INFLUENCED BY DIFFERENT TREATMENTS ACTING ON PHOTOSYNTHETIC ACTIVITY*

M. BODSON, G. BERNIER, J.M. KINET, A. JACQMARD & A. HAVELANGE

(Liège, Belgium)

ABSTRACT

Sinapis alba (a long day species) can be induced to flower by one long day or by one displaced short day, *i.e.*, by displacing by 10 hr the timing of one of the 8 hr photoperiod.

It has been shown previously that the floral response to one long day is influenced by treatments modifying photosynthetic activity. Results presented here show that photosynthetic input is also of importance for the induction of flowering by a displaced short day. The significance of these results is discussed in relation to the possible participation of photosynthesis in the initiation of flowering.

INTRODUCTION

Plants of *Sinapis alba*, a long day species, grown for 2 months in short days at irradiance of 22–25 W/m^2 can be induced to flower by one long-day cycle at the same irradiance (Bernier, 1963). The floral response to the long day can be modified by treatments which influence photosynthesis, such as CO_2 removal from the air (Kinet *et al.*, 1973) and variation in irradiance (Bodson *et al.*, 1977). Flower initiation has also been observed in plants kept in short days at high irradiance (Bernier, 1969; Bodson *et al.*, 1977) and in cultured shoot tips grown in short days on a medium containing a high ratio of sucrose to nitrogen (Deltour, 1967a, 1967b). It was therefore suggested that photosynthesis plays an important role in the initiation of flowering of this species (Bodson *et al.*, 1977).

It is known, however, that *Sinapis* plants exhibit a 24-hr cycle of sensitivity to light and that flowering will occur providing light is given during the photophil phase (Kinet *et al.*, 1973). So, floral initiation can be induced by displacing by 10 hr the timing of one of the 8-hr photoperiods in a 24-hr cycle (Kinet, 1972). By this photoperiodic treatment, which was called 'displaced short day', the plants are induced without increasing the duration of illumination as compared to vegetative plants. Therefore the generality of the conclusion that the photosynthetic input participates in flowering is questionable.

In this paper, our aim is to examine the effect of the level of irradiance, CO_2

*Financial support to G.B. from the F.N.R.S. (Fonds National de la Recherche Scientifique) and to M.B. and J.M.K. from the I.R.S.I.A. (Institut pour l'Encouragement de la Recherche Scientifique dans l'Industrie et l'Agriculture) is gratefully acknowledged.

supply and sucrose treatment on the floral response of plants induced to flower by a displaced short day.

MATERIAL AND METHODS

Growing conditions and criterion of flowering response.

Plants of *Sinapis alba* were grown from seeds at 20°C in short days of 8 hr given from 8.30 to 16.30. The term 'usual short day' refers to such a light period. The light irradiance was 22–25 W/m^2 at the top of the plants, provided by fluorescent white lamps (Phytor C.R.H. Lg). Plants grown for two months in these conditions remained vegetative. When 65 days old, they were subjected to a displaced short day and then returned to usual 8 hr days for two weeks before examination. The term 'day I' or 'day II' refers to the first or second usual short day following the displaced short day. The term 'subjective short day' refers to a period of 8 hr of darkness given to the plants at a time of the 24 hr cycle corresponding to the usual short day.

Fifteen plants were used in each experimental batch. Two weeks after the beginning of the inductive treatment, the apical buds were dissected and classified as vegetative or floral using the visibility of at least one flower primordium. Each experiment was repeated at least twice.

CO_2 removal from the air.

Plants were enclosed in a gas-tight glass chromatography tank for 8 hr. The compressed air before entering the tank was circulated through three successive bottles containing 10% KOH. A calcium hydroxide turbidity test showed that all the existing CO_2 in the air had been removed. The gas was admitted at the lower end of the tank and allowed to escape at the opposite end. At the beginning of the treatment the tank was immediately flushed with the gas at a flow rate of 200 liters/hr. Half an hour later the flow was reduced to about 100 liters/hr for the rest of the treatment period. Light intensity and quality and gas temperature were identical inside and outside the tank.

Sucrose treatment.

Sucrose solution at 2% was sprayed on the whole plant. The spray was repeated 4 hr later. A total of 12–20 ml was sprayed on each plant. A wetting agent (Tween-20) was added to the solution.

RESULTS

1. *Preliminary experiments to determine the duration of illumination required to induce a high floral response.*

Flowering as a function of the length of the displaced short day is given in Fig. 1. Floral response is almost nil when the light period of the displaced short day is eliminated. The percentage of flowering plants increases progressively with the duration of the light period with an almost maximal floral response observed with 8 hr of light.

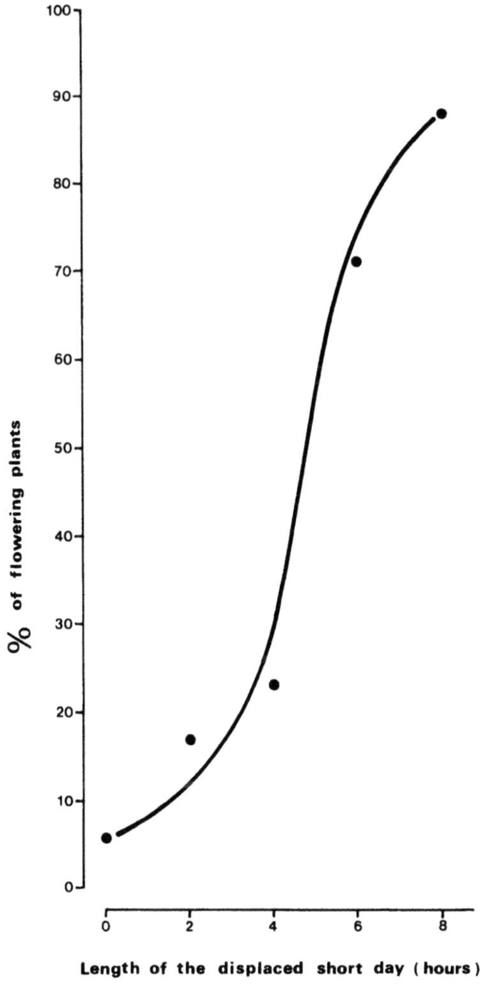

Fig. 1. Percentage of flowering plants as a function of the length of the displaced short day.

When the plants are exposed to a displaced short day of 8 hr but held in darkness on day I, the floral response is greatly reduced (Table 1, Fig. 2). Flowering increases with the duration of light period of day I, with 8 hr of light required to induce a high floral response (Fig. 2). Suppression of light during the day II does not inhibit flowering (Table 1).

2. *Effect of a low irradiance.*

A low irradiance (1.5 W/m^2) given during the displaced short day (Fig. 3 C)

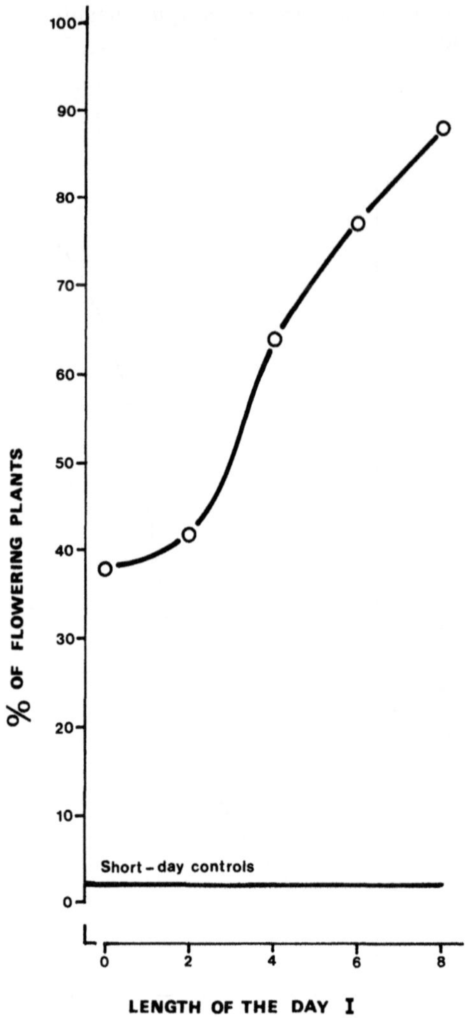

Fig. 2. Floral response of plants exposed to a displaced short day of 8 hr as a function of the length of the day I.

Table 1. Effect of suppression of day I or day II on the floral response of plants exposed to a displaced short day of 8hr.

Treatment	Percentage of flowering plants
Displaced short day controls	67
Suppression of day I	33
Suppression of day II	60

inhibits flowering more strongly than the same treatment given during day I (Fig. 3 D).

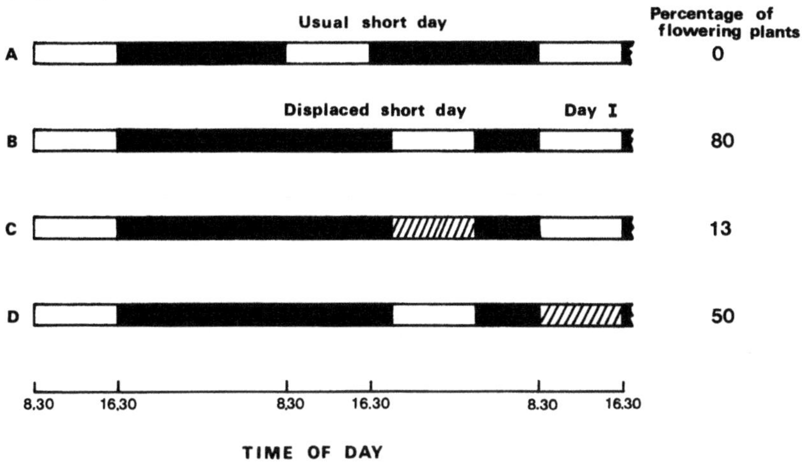

Fig. 3. Floral response of plants exposed to a low irradiance (1.5 W/m^2) during the displaced short day (C) or during the day I (D). Solid bars represent darkness. Hatched bars represent time of low irradiance.

3. *CO_2 removal from the air.*

Plants have been submitted to a displaced short day of 6 hr and at various times before and after the start of the displaced short day have been kept in a CO_2-free atmosphere for 8 hr (Fig. 4). CO_2 removal during the displaced short day or during day I greatly inhibits the floral response. Flowering is reduced to a lesser extent by CO_2 removal during the short night just following the displaced short day. Later on, the treatment does not affect significantly the floral response.

4. *Sucrose treatment.*

Sucrose treatments have been made in an attempt to obviate the light requirement of the displaced short day or day I.

a. *Elimination of light period of day I.* Batches of plants have received the displaced short day and have been held in darkness for day I (Fig. 5C, D, E). As reported above suppression of day I greatly reduces the floral response (Fig. 5C) relative to the controls (Fig. 5B). Sucrose treatment at the time of day I is without effect on the floral response (Fig. 5D). However the same treatment given at the time of the subjective short day, *i.e.,* just before the displaced short day, increases significantly the floral response (Fig. 5E).

b. *Elimination of light period of the displaced short day.* When the displaced short day is omitted, flowering is almost completely suppressed as shown above (Fig. 6C). This treatment is equivalent to the suppression of one usual short day cycle, *i.e.,* holding the plants in darkness for 40 hr. Sucrose treatments have been made at the time of the displaced short day or at the time of the subjective short day. Both treatments induce a significant percentage of plants to flower (Fig. 6E, D).

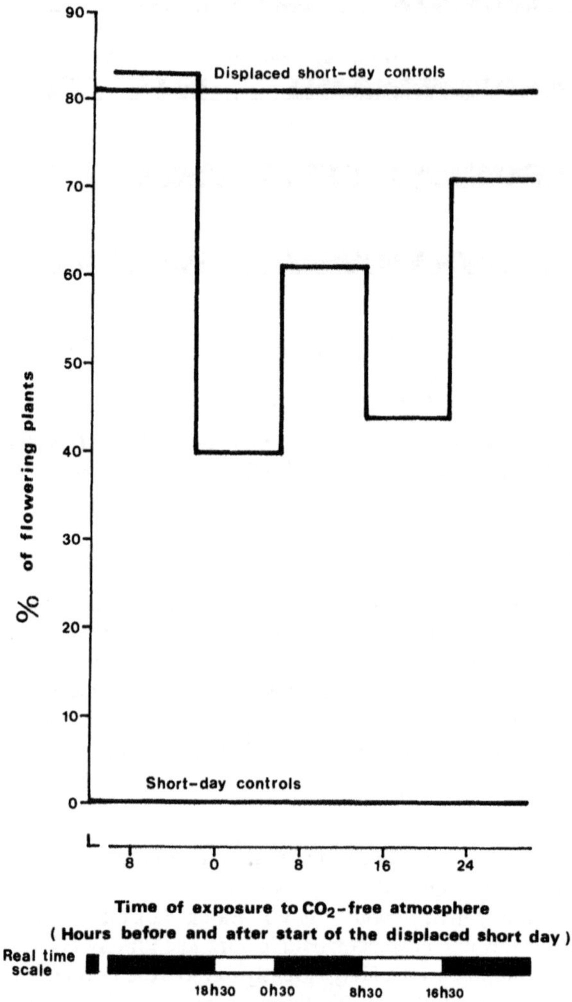

Fig. 4. Floral response of plants exposed to a displaced short day of 6 hr as a function of time of exposure to CO_2-free atmosphere for 8 hr.

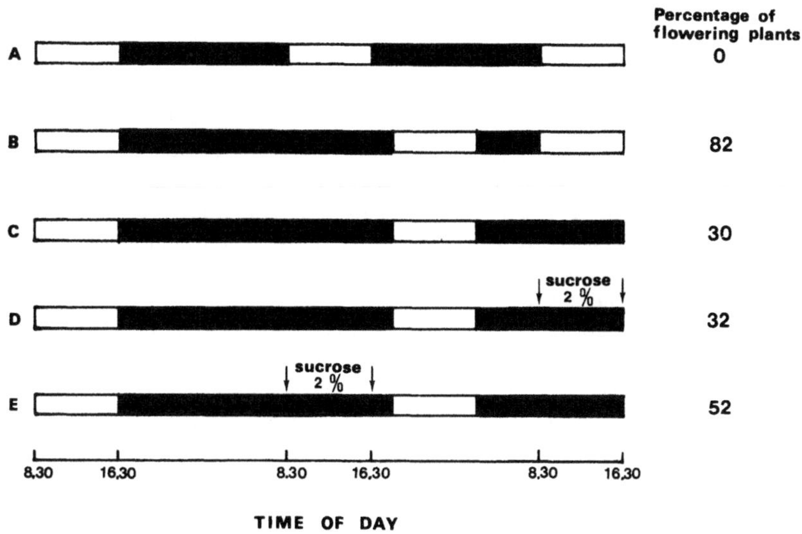

Fig. 5. Effect of sucrose treatment on the floral response of plants exposed to a displaced short day of 8 hr and kept in darkness during day I. A: short day controls; B: displaced short day controls; C : suppression of day I; D and E : supression of day I and sucrose treatments. Solid bars represent darkness.

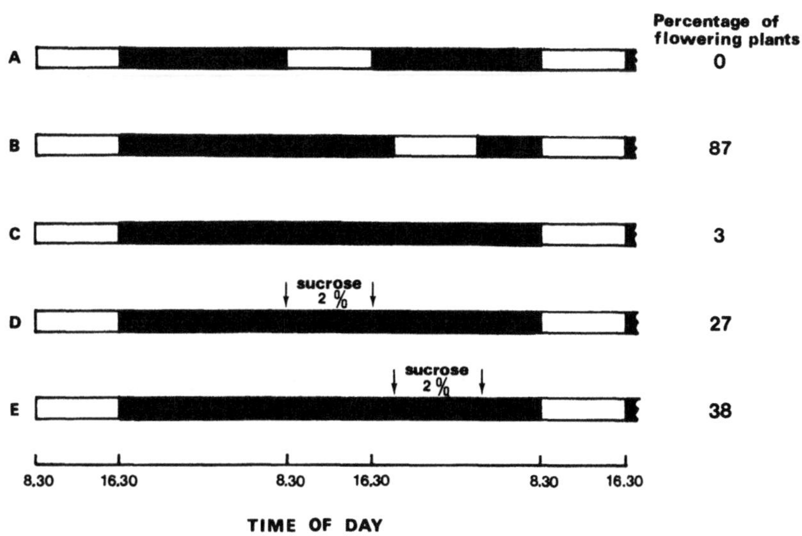

Fig. 6. Effect of sucrose treatment on the floral response of plants kept in darkness for 40 hr. A : short day controls; B : displaced short day controls; C : darkness for 40 hr; D and E : darkness for 40 hr and sucrose treatments.

DISCUSSION

From the experiments reported here, it is clear that flowering induced by a displaced short day is strongly dependent on photosynthetic activity. The displaced short day has to be followed by one usual short day at least in order to induce a high floral response. A similar conclusion was reached by Kinet *et al.*(1973). CO_2 assimilation is required during both light periods. As sucrose treatments are able to substitute partially for light, it is likely that carbohydrate availability is involved in the floral response induced by a displaced short day. Therefore the inhibition of flowering by light of low irradiance could result partially from low carbohydrate production. Further information is needed however to insure that the effect of sucrose application on flowering is a metabolic effect and not an osmotic one.

It must be pointed out that sucrose treatment in replacement of the day I enhances the effectiveness of the displaced short day only when it is given just prior to this displaced short day. This observation suggests again the notion that an optimal timing for photosynthetic input may be required for flowering (Bodson *et al.*, 1977). In addition, it suggests that the flower inducing effect of light in the photophil phase involves reactions influenced by the availability of carbohydrates. Cumming (1967) has already suggested that sucrose is a substrate for phytochrome P_{FR} action in the control of flowering of *Chenopodium rubrum.*

As light of low irradiance inhibits flowering more strongly than CO_2 removal from the air and also sucrose applications overcome only partially the suppression of light, an additional effect of light on some processes other than CO_2 assimilation must be suspected. It is worthwhile to question to what extent the additional effect of light could be mediated via the photosynthetic light reactions.

The importance of adenylate energy charge (Atkinson, 1969), *i.e.,* the amount of metabolically available energy stored in the adenylate pools, in the regulation of enzyme activity has already been shown (Atkinson *et al.*, 1975). From detailed studies of several endogenous rhythms in *Chenopodium rubrum* (Wagner & Cumming, 1970; Frosch *et al.*, 1973; Frosch & Wagner, 1973; Wagner & Frosch, 1974; Wagner *et al.*, 1974a, 1974b), it has been suggested that the relative concentration of adenine nucleotides and pyridine nucleotides, by regulating the metabolic sequences of energy metabolism, may be basic to the endogenous rhythmicity (Wagner, 1976).

It may therefore be possible that flowering depends on particular levels of ATP and NADPH and it would be worthwhile to compare adenylic nucleotide concentration and NADP/NADPH level in *Sinapis* plants exposed to an inductive photoperiodic treatment under conditions of low irradiance or CO_2 removal from the air to those observed in plants induced to flower in normal conditions.

The importance of an adequate level of ATP from photophosphorylation for the induction of flowering of *Lemna gibba* has already been suggested by Kandeler (1969). Also, an increase in ATP level in the leaf of *Spinacia oleracea* has been observed soon after the critical photoperiod is reached (Bonzon *et al.*, 1977).

Results presented here confirm the previous conclusion (Bodson *et al.*, 1977) that photosynthetic input influences flowering of *Sinapis alba*. It is also suggested that the effect of photosynthesis is partially mediated by sucrose availability. This is in agreement with previous results which have shown that an early increase in the sugar content occurs in the bud of plants induced to flower by a long day and by a displaced short day (Bodson, 1977.).

REFERENCES

Atkinson, D.E. Regulation of enzyme function. *Ann. Rev. Microbiol.* 23: *47–68* (1969).

Atkinson, D.E., P.J. Roach & J.S. Schwedes. Metabolite concentrations and concentration ratios in metabolic regulation. *Advances in Enzyme Regulation* 13: *393–411* (1975).

Bernier, G. *Sinapis alba* L., a new long-day plant requiring a single photoinductive cycle. *Naturwiss.* 50: *101* (1963).

Bernier, G. *Sinapis alba* L. pp. 305–327, in: Evans, L.T. (ed.), The induction of flowering. Macmillan, Publishers, Melbourne (1969).

Bodson, M. Changes in the carbohydrate content of the leaf and the apical bud of *Sinapis* during transition to flowering. *Planta* (Berl.) 135: *19–23* (1977).

Bodson, M., R.W. King, L.T. Evans & G. Bernier. The role of photosynthesis in flowering of the long-day plant *Sinapis alba. Aust. J. Plant Physiol.* 4: *467–478* (1977).

Bonzon, M., H. Greppin & J. De Greef. Contenu en ATP des feuilles de l'épinard sous différents régimes photopériodiques inducteurs ou non de la floraison. *Archives Intern. Physiol. Bioch.* 85: *953–954* (1977).

Cumming, B.G. Circadian rhythmic flowering responses in *Chenopodium rubrum*: effects of glucose and sucrose. *Can. J. Bot.* 45: *2173–2193* (1967).

Deltour, R. Action du saccharose sur la croissance et la mise à fleurs de plantes issues d'apex de *Sinapis alba* L. cultivés *in vitro. C.R. Acad. Sc. Paris.* 264: *2765–2767* (1967a).

Deltour, R. Action de l'azote minéral sur la croissance et la mise à fleurs de plantes issues d'apex de *Sinapis alba* L. cultivés *in vitro. C.R. Acad. Sc. Paris.* 265: *1932–1935* (1976b).

Frosch, S., E. Wagner & B.G. Cumming. Endogenous rhythmicity and energy transduction. I. Rhythmicity in adenylate kinase, NAD- and NAPD-linked glyceraldehyde-3-phosphate dehydrogenase in *Chenopodium rubrum. Can. J. Bot.* 51: *1355–1367* (1973).

Frosch, S. & E. Wagner. Endogenous rhythmicity and energy transduction. III. Time course of phytochrome action in adenylate kinase, NAD- and NADP-linked glyceraldehyde-3-phosphate dehydrogenase in *Chenopodium rubrum. Can. J. Bot.* 51 *1529-1535* (1973).

Kandeler, R. Förderung der Blütenbildung von *Lemna gibba* durch DCMU und ADP. Z. *Pflanzenphysiol.* 61: *20–28* (1969).

Kinet, J.M. *Sinapis alba*, a plant requiring a single long day or a single short day for flowering. *Nature* 236: *406–407* (1972).

Kinet, J.M., G. Bernier, M. Bodson & A. Jacqmard. Circadian rhythms and the induction of flowering in *Sinapis alba. Plant Physiol.* 51: *598–600* (1973).

Wagner, E. & B.G. Cumming. Betacyanin accumulation, chlorophyll content and flower initiation in *Chenopodium rubrum* as related to endogenous rhythmicity and phytochrome action. *Can. J. Bot.* 48: *1–18* (1970).

Wagner, E. & S. Frosch. Endogenous rhythmicity and energy transduction VI. Rhythmicity in reduced and oxidized pyridine nucleotide level in seedlings of *Chenopodium rubrum. J. interdiscipl. Cycle Res.* 5: *231–239* (1974).

Wagner, E., L. Stroebele & S. Frosch. Endogenous rhythmicity and energy transduction V. Rhythmicity in adenine nucleotides and energy charge in seedlings of *Chenopodium rubrum. J. interdiscipl. Cycle Res.* 5: *77–88* (1974).

Wagner, E., J. Tetzner, U. Haertle & G.F. Deitzer. Endogenous rhythmicity and energy transduction VIII. Kinetics in enzyme activity, redox state and energy charge as related to photomorphogenesis in seedlings of *Chenopodium rubrum* L. *Ber. Deutsch. Bot. Ges.* 87: *291–302* (1974).

Wagner, E. Kinetics in metabolic control of time measurement in photoperiodism. *J. interdiscipl. Cycle Res.* 4: *313–332* (1976).

Authors' address:
Centre de Physiologie Végétale
Appliquée et
Service de Physiologie Végétale
Départment de Botanique
Université de Liège, Sart Tilman
B - 4000 Liège
Belgium.

PHYTOCHROME INVOLVEMENT IN CO_2 EXCHANGE DURING GROWTH AND DEVELOPMENT OF A QUANTITATIVE SHORT DAY PLANT, *CHENOPODIUM POLYSPERMUM* IN DIFFERENT PHOTOPERIOD*

MARIANNE MOUSSEAU
(Gif-sur-Yvette, France)

ABSTRACT

A preferential short day plant, *Chenopodium polyspermum*, has been grown under short or long day conditions: in each case, the plants received daily the same total amount of energy (9 h). The night period was given either by darkness or by dim red or far-red radiation.

The different growth conditions induced large variations in the development of the plant, particularly in dry matter distribution.

No significant differences were found in photosynthetic activity in relation to the observed differences in development. However different time patterns of CO_2 output at night were found depending on the photoperiodic regime. Red light, acting through phytochrome, appeared to be responsible for the presence and timing of an increase of CO_2 output at the end of the night. This peak of CO_2 production also developed progressively under long night conditions conducive to flower formation.

Respiration rates were always larger in darkness than in red light; in particular a sharp decrease in the respiration rate was observed during night breaks of thirty minutes of red light, when applied at the end of the dark period.

INTRODUCTION

Most studies concerning plant response to photoperiod are generally related to flowering (Vince-Prue, 1975). However, variations in CO_2 exchange with photoperiod have been reported (Eaton, 1925; Tsylbul'ko, 1973; Mousseau, 1975; Burian & Winter, 1976) and consequently such changes are often assumed to be essential for flowering to occur (Imhoff, 1973; Bassi, Tregunna & Joliffe, 1976; Zeevaart, 1977; Bodson, King, Evans & Bernier, 1977; Quedado & Friend, 1978). It is therefore surprising that very little is known about CO_2 metabolism in relation to photoperiod. The purpose of the present investigations was to define possible changes in photosynthesis and respiration with photoperiod. In particular, previous work on *Sinapis alba* (Mousseau, 1976) emphasized the need for a precise study of the changes occurring in CO_2 output at night consequent upon daylength extension with low light, to show whether the duration or quantity of light controls the response.

Phytochrome is now generally recognized as the photoreceptor by which plants distinguish between night and day and detect their duration. The most

*I wish to thank Professor Paul Jarvis for valuable advice and for linguistic revision of the manuscript.

striking property of phytochrome is its photoreversibility, which confers upon it the capacity to detect changes in the quality of light in the 600–800 nm region through the establishment of different photoequilibria between P_r and P_{fr}. Thus, we may postulate the involvement of phytochrome in the changes in CO_2 metabolism promoted by changes in daylength. Unhappily, much of the work involving phytochrome in CO_2 exchange has been done either on etiolated tissues (Chia-Looi and Cümming, 1969) or with plants the organization of which is not typical of higher plants, *e.g., Lemna* sp. (Hillman, 1976a; Kandeler, 1976).

Chenopodium polyspermum was chosen for this work because the impact of phytochrome on its development is precisely known (Jacques, 1968) and changes in CO_2 metabolism with photoperiod have already been reported (Mousseau, 1977). It is a quantitative short day plant: a minimum of 7 short days is sufficient for total floral expression by a plant with 6 pairs of leaves (3 long nights are necessary for floral induction and 4 short days to express this induction into visible flowers) (Jacques, 1968).

The action of phytochrome on a process is most easily demonstrated by a night break of red light, which, by converting phytochrome to P_{fr}, interrupts its natural dark decay. Another way of showing the involvement of phytochrome in a phenomenon is to lower artifically the P_{fr} level at the end of the main light period, by applying then a short period of far red light (Downs, Hendricks and Borthwick, 1957). Phytochrome may be assumed to affect gas exchange through either a direct effect on metabolism or an indirect effect on metabolism as a result of morphogenetic changes during growth. In this study, phytochrome effects on morphogenesis have been experimentally produced by varying the quality of light during daylight extension. By using either red light or incandescent light, large differences in the energy flux densities at 660 and 730 nm can be obtained. This experimental design ensures a long photoperiod in which the plants remain vegetative but grow differently, as a result of the different wavelengths during daylight extension. Any differences in CO_2 exchange between these two types of plant will then be attributable to morphogenetic changes and not to the length of photoperiod.

A comparison of CO_2 exchange by plants with different developmental histories and the use of these well known red-far red reactions will be used in this paper to provide evidence for the action of phytochrome on CO_2 exchange. Moreover, as flower inducing conditions (long dark nights) have been shown to induce changes in CO_2 output at night by this species (Mousseau and Louason, 1976), the results may be interesting in relation to flowering processes.

MATERIAL AND METHODS

Culture conditions

Chenopodium polyspermum was grown on vermiculite in controlled environment rooms of the phytotron at Gif-sur-Yvette. All plants were grown from seed collected from mother plants grown under the same experimental conditions. Plants were watered automatically twice a day with nutrient solution or water. Growth conditions were as follows: air temperature 22°C day and night, relative humidity 70%. All plants received the same amount of daily energy (100 Wm^{-2} for 9 hr, from 08.45 to 17.45 hr) provided by a combination of fluorescent and incandescent lights (ζ = 4, 6*). This mainlight period was followed by 15 hr of continuous red or incandescent light (referred to as '9 + red' and '9 + inc.' respectively in text and figures) to give long day photoperiods. A short day photoperiod was achieved by giving 15 hr darkness after the mainlight period. Red light was provided by Philips TL 15 fluorescent lamps which gave a radiation flux of 1.2 Wm^{-2} between 600 and 700 nm (ξ being about 42*). The radiation flux from the incandescent lights was 15 Wm^{-2}, 9% of which was between 700 and 800nm. (ξ being about 0.55). Far red light, when required, was provided by a 40 W lamp enclosed in a box made of blue and red 'Plexiglas' (Altuglas S 200 and S 300).

Gas exchange measurements

Respiration and net photosynthesis of whole plants were measured as CO_2 evolved or absorbed with an open circuit infra-red gas analyzer in the cultivation conditions. The 'Plexiglas' assimilation chambers were varied in size to maintain a relatively constant ratio of chamber volume to plant size. The air flow was about 1000 cm^3 mn^{-1} and the air was mixed with a fan within the chamber to ensure a small and constant boundary layer resistance. A switching system allowed consecutive measurements to be made on each of 4 plants, 10 times an hour. Mean hourly rates of respiration or photosynthesis were then calculated.

Leaf areas were determined with an automatic area meter (Hayashi, Denko Co., Ltd) on detached leaves at the end of the experimental period.

Experimental procedure

The plants were grown from seed in long days conditions until they has six fully expanded leaves. Some were then transferred to short days or to the other long day treatment. In the long day conditions the plants remained vegetative. In short days (15 hr darkness following the main light period) the plants were rapidly induced to flower. From this time (day 0) samples of 10 plants were harvested every 2 days from each treatment for growth analysis. Gas exchange was measured on two plants from each treatment every day or every second day. Special treatments, such as night breaks, were given between days 7 and 14 after day 0.

RESULTS

Effects of photoperiod and quality of supplementary light on growth.

The effects of different light qualities during the daylength extension on the growth of *Chenopodium polyspermum* are shown in Figure 1a and 1b. There were no significant differences (P = 0.05) in total leaf area or total dry matter per plant before the sixth day of growth in all conditions. The supplementary red

*ζ = zeta = ratio of quantum fluxes at the specific wavelengths 660 nm and 730 nm (Smith & Holmes, 1977).

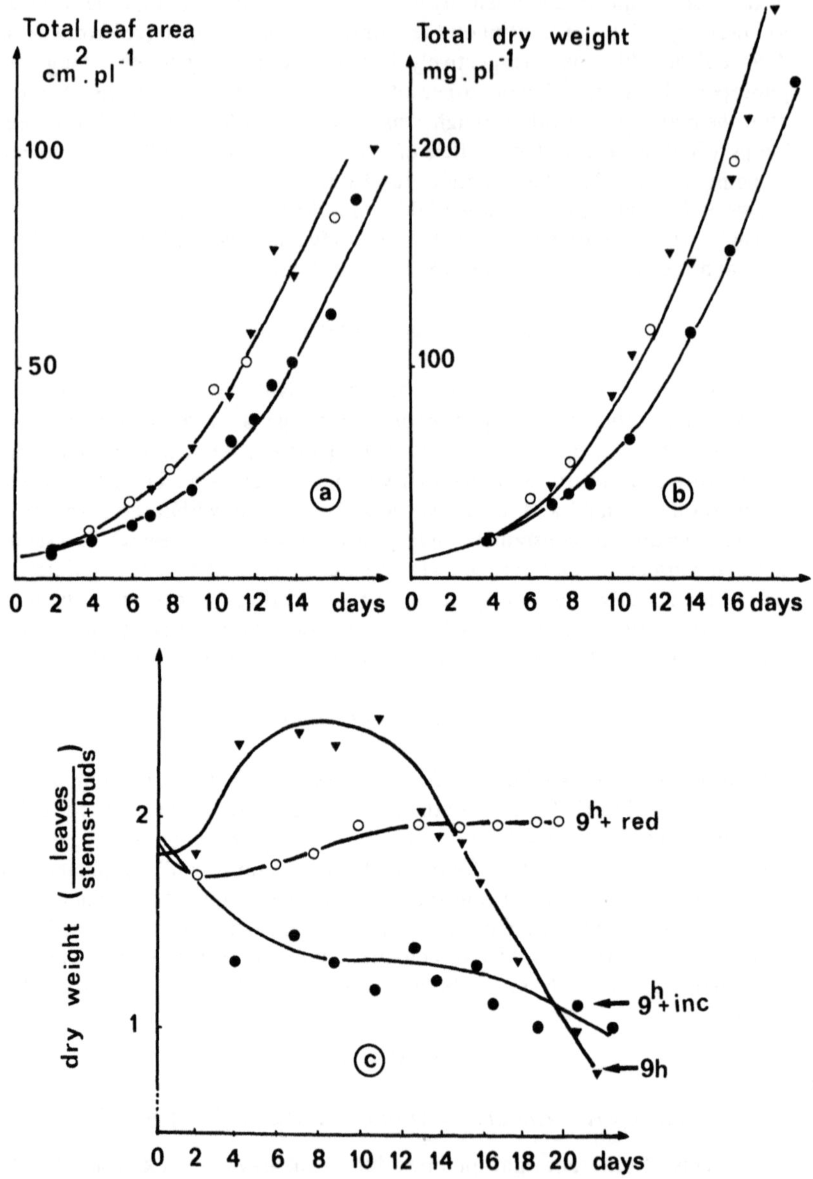

Fig. 1. The growth of *Chenopodium polyspermum* in three different photoperiod conditions (○—○ '9 + red' -▼—▼: short days. ●—● '9 + inc.').
Time 0 corresponds to a plant with 6 pairs of fully developed leaves grown in '9 + inc.' conditions.

light resulted in both larger leaf area and more dry matter production than incandescent light.

The distribution of dry matter was also altered considerably by the quality of supplementary light. Supplementary red light resulted in more dry matter accumulation in the leaves than in the stems (Fig. 1c), whereas incandescent light resulted in extensive growth of internodes and a relative increase in the stem dry weight. The leaves were also thinner and had a larger specific area in comparison with the '9 + red' leaves.

Figure 1 also shows that the distribution of dry matter is affected by whether the plant was going to flower. Initially, the plants in short inductive days channeled more dry matter into leaves than into stems. After flower induction, dry matter was required to build the floral axes and buds, with the result that the proportions of dry matter in the leaves decreased rapidly with time.

There were no differences in total leaf area or total dry matter between short day and long day plants when the long photoperiod was supplemented with red light (Fig. 1a and b). The situation was obviously different when short day grown plants were compared with vegetative plants receiving supplementary incandescent light. The overall growth of the latter was reduced, and the ratio of leaves to stem was always lower.

Effects of photoperiod on photosynthetic rate.

There were no differences in the mean hourly rate of CO_2 uptake per unit leaf area in plants of the same age grown in short days or in '9 + red' conditions. This rate was ca. 7 mg $dm^{-2}hr^{-1}$ between the fifth and the ninth day of growth. Similarly no differences were found in the general shape of the daily curve of photosynthesis. However, '9 + inc.' plants had a slightly lower rate of net photosynthesis per unit leaf area, probably as a consequence of the lower ratio of leaf to stem dry matter (Fig. 1).

Effects of photoperiod on CO_2 evolution.

Figure 2 shows the course of CO_2 output when red or incandescent light was added after the main 9 hr photoperiod. In all cases, CO_2 output was maximum at the end of the 9 hr photoperiod and then decreased for 3 hours. In the '9 + inc.' treatment, CO_2 evolution then remained constant until a second decrease occurred during the last four hours of the night period. In the '9 + red' treatment, there was a significant increase in respiration rate during the last four hours of the red period, reaching a maximum between 11 hr and 12 hr after the end of the main photoperiod.

Figure 3 shows that two red light breaks of 30 minutes instead of continuous red light were sufficient to induce the same maximum in CO_2 output between 11 hr and 12 hr after the onset of darkness.

Table 1 and figure 4 show that a brief exposure to far red light was sufficient to produce a shift in the timing of the maximum in the CO_2 output at the end of the night period towards earlier times, and that this effect was reversed by 15 minutes of red light following far red, the CO_2 output maximum being then reset to the previous time. The three kinds of observations reported above support the suggestion that a relationship exists between phytochrome and CO_2 evolution at night.

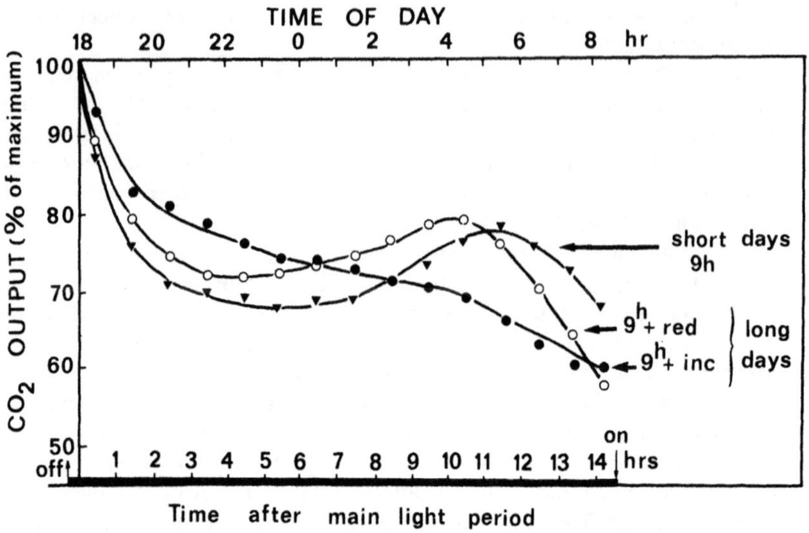

Fig. 2. The time course of CO_2 output of *Chenopodium polyspermum* grown in different photoperiods, during the period following the main 9 hr photoperiod. Mean maximum (100%) values of CO_2 output per plant were 2.4 mg h^{-1} for short day plants; 1.5 mg h^{-1} for '9 + red' plants; 1.2 mg h^{-1} for '9 + inc.' plants.

▼ 9hr light + 15 hr darkness

● 9hr light + 15 hr incandescent light

○ 9hr light + 15 hr red light.

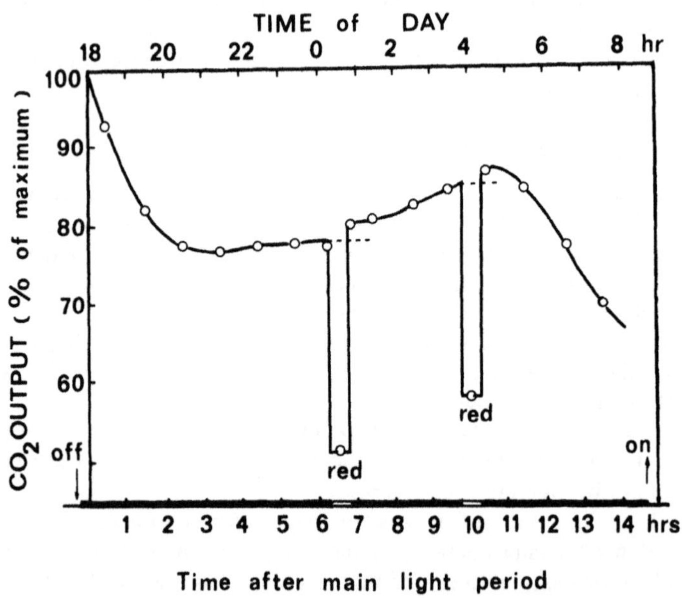

Fig. 3. Effects of two 30 minutes red light breaks in the dark period on CO_2 evolution of *Chenopodium polyspermum* (mean of 10 successive experiments on plants growing in these conditions). Timing of the red light breaks: 00.15 – 00.45 hr and 04.00 – 04. 30hr.

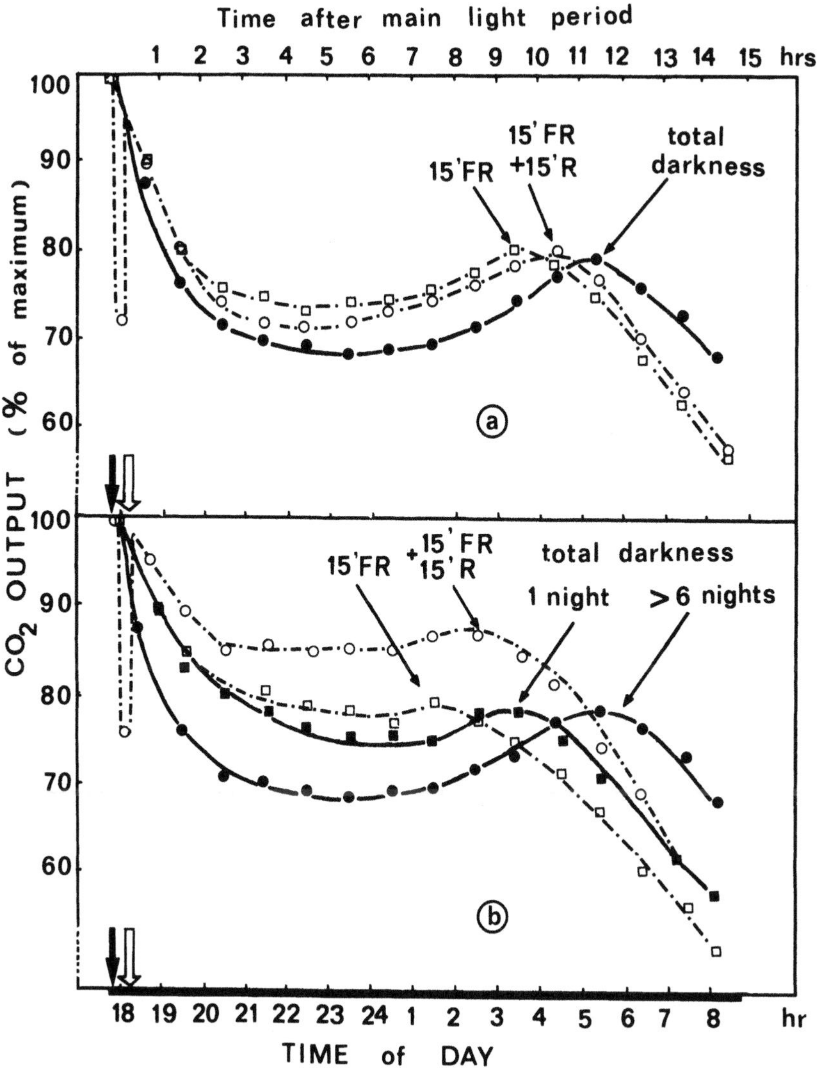

Fig. 4. Effect of red and far red light on the timing of the CO_2 output maximum at night. Black arrow: 15 min. F.R. (curve □—·—□) applied at 17.45 hr. White arrow: 15 min. red applied after F.R. (curve ○—·—○)

ⓐ plants grown in '9 + red' conditions.

ⓑ plants grown in '9 + inc.' conditions.

Darkened curves correspond to total darkness.

Table 1. Effect of a brief exposure to far red light on the timing of the subsequent CO_2 output maximum of long day grown plants and its reversal by red light.*

Growth conditions before experiment	Experimental conditions at the onset of darkness	Time after main 9hr photoperiod (hr)						
			7	8	9	10	11	12
	Red →**		0	0,26	0,21	*0,52*	0	
Long day "9 + red"	15 min.F.R		0	0,19	**0,5**	0,3	0,03	0
	15 min.F.R +15 min.Red		0	0,07	**0,25**	**0,66**	0	
	inc.	0	0	0	0	0	0	
Long day "9 + inc."	15 min.F.R	**0,8**	0,2	0	0	0		
	15 min.F.R +15 min.Red	0	**0,37**	**0,37**	0,25	0		

*–the numbers represent the fraction of the observations having the CO_2 output maximum in a particular time period (over 15 to 20 experiments)

**–arrow means that a given light treatment continues throughout the dark period.

CO_2 output under short days, conducive to flower formation.

When a well developed plant grown in '9 + inc.' conditions is transferred to short inductive days, a maximum in the CO_2 output is initiated towards the end of the night, generally between 9 and 10 hr after the onset of darkness (Mousseau, 1977).

Table 2 shows that the situation was gradually modified with number of short days given: the timing of the CO_2 output maximum was progressively shifted towards the end of the dark period. After 4 to 6 short days, a new curve resulted (Fig. 4b) with a maximum occurring around the twelfth hour after the onset of darkness. The same result was found with a '9 + red' plant transferred to short days, the shift of the maximum being rather smaller, since it was initially rather later in the night (Fig. 4a).

DISCUSSION

The reported effects of length of photoperiod on photosynthetic activity found in the literature (Denne and Smith, 1971; Burian and Winter, 1976) can always be attributed to indirect effects such as dry matter redistribution, since the response was measured a considerable time after the beginning of the treatment. A direct effect of phytochrome on CO_2 uptake itself has not been demonstrated, although the action of phytochrome on a large number of enzymes of the photosynthetic pathway, including RuBiP carboxylase, has been reported (Shopfer, 1977).

On the other hand, the fact that no changes in photosynthetic activity were found as a result of the photoperiod treatments does not necessarily mean that no differences occurred among its components (*e.g.*, true photosynthesis and photorespiration). For example, it has been shown that the ratio of CO_2 evolved to total CO_2 fixed in the light is lower and therefore more favorable in terms of the efficiency of net photosynthesis, in short days than in long days (Steward *et al*. 1971).

It has often been shown that CO_2 evolution in the dark may be a good indicator of photoperiodic timing (Chia Looi and Cümming, 1972; Hillman, 1976a). From the present data, it seems clear that the timing of CO_2 output at night is related to the level of active phytochrome at the onset of the night period (Table 2).

Table 2. Change in timing of the CO_2 output maximum with the number of short days given to a long day grown plant ("9 + inc.").*

Number of short days	Time after main 9hr photoperiod (hr) 7–8	8–9	9–10	10–11	11–12	12–13
1	0,11	0,16	**0,5**	0,22		
2		0,16	**0,55**	0,22	0,05	0
3	0,14	0,07	0,21	**0,57**	0	
4		0,12	**0,43**	**0,31**	0,12	
5		0,10	*0,47*	**0,37**	0,05	
6		0	0,28	**0,43**	0,14	0,14
7				0	**0,9**	0,09
8					**0,8**	0,2
9				0,2	**0,7**	0,09
more than 11					**0,45**	**0,48**

* The numbers represent the fraction of the observations having the CO_2 output maximum in a particular time period (over 10 to 20 experiments).

Some other results suggest that phytochrome may also have a direct effect on CO_2 output. Phytochrome has been reported to affect O_2 evolution of tissues of different species (Leopold & Guernsey, 1954) and the respiratory quotient of *Sinapis alba* cotyledons (Hock & Mohr, 1965). These authors found that red light appeared to reduce the respiration rate of short day plants while it enhanced the respiration rate in long day plants. Similarly, our results show an instantaneous drastic reduction in dark CO_2 output when red light was applied (Fig. 3 and 4). In addition there was a significant increase in CO_2 evolution on the return to dark conditions after the night break. No change in CO_2 output was found when far red light was given for a short time (Fig. 4). Whether these results are specific to red light action on respiration or are the result of some photosynthesis inside the leaf, reassimilating the CO_2 produced, is not known and has to be tested. It must be remembered that CO_2 output does not necessarily mean 'respiratory' CO_2, in the oxidative phosphorylation sense. Furthermore, the respiration rate measured may be rate of *net* respiration. For example, the effect of red light on CO_2 evolution might be the stimulation of some carboxylase at low energies and may not reflect a change in rate of respiration at all.

As *Chenopodium polyspermum* is known to be induced by one long night if its vegetative development is sufficient, it is tempting to correlate the shift of the maximum CO_2 output (Table 2) with the occurrence of floral induction. In *Lemna* (Hillman, 1976b) a rhythm in CO_2 evolution was also found and compared with the flowering behaviour. Many of the characteristics of the CO_2 output rhythm have been found to be similar to those of flowering. In this study, the timing of the maximal output after 12 hours darkness seems to be a characteristic of the floral state. The fact that we found a natural shift in the maximum with ageing towards 6.00 hr in '9 + red' plants is consistent with this assumption, because *Chenopodium polyspermum* is known to flower in this long day condition after a long time, being a preferential and not a strict short day plant. Another fact supporting this assumption is that two red light breaks in the second part of a long night are known to inhibit flowering in this species (Jacques, 1968). This treatment always resets the maximum CO_2 output earlier in the night (Fig. 3).

Cümming *et al.* (1965) pointed out that the phase of the flowering rhythm in another species of *Chenopodium* seemed to be influenced by phytochrome. They found that a short exposure to far red light at the end of the day advanced the phase of the first circadian oscillation and resulted in floral initiation with a shorter dark period. The results presented in this paper are obviously related to such a rhythmic pattern, as already stated (Mousseau, 1977). The characteristic of the CO_2 output pattern reflect the several components of the effect of photoperiod such as phytochrome and circadian rhythmicity.

It must not be forgotten also that respiration furnishes the energy required for floral stimulus or for the synthesis of required hormones (flowering is known to be depressed by application to the leaves of respiratory inhibitor or uncouplers of phosphorylation during the second half of the inductive period) (Salisbury, 1963). The increase of respiration at the end of the inductive night may be a marker of such events.

Lastly, we may raise the question as to whether the patterns of CO_2 output found in the present work are related to stomatal aperture in the dark at the end of the night. Stomatal aperture was found to depend upon the previous photoperiodic regime (Schwabe, 1952) and has been correlated with an underlying dark rhythm (Martin & Meidner, 1971). But little is known about stomatal control of dark CO_2 evolution and phytochrome action on stomatal aperture.

REFERENCES

Bassi, P.K., E.B. Tregunna & P.A. Joliffe. Carbon dioxide exchange and phytochrome control of flowering in *Xanthium pennsylvanicum. Can. J. Bot.* 54: *2881–2887* (1976).

Bodson, M., R.W. King, L.T. Evans & G. Bernier. The role of photosynthesis in flowering of the long-day plant *Sinapis alba. Aust. J. Plant Phys.* 4: *467–478* (1977).

Burian, K. & C. Winter. Die Wirkung verschieden langer Licht perioden auf die Photosynthese einiger Gräser. *Photosynthetica* 1: *25–32* (1976).

Chia-Looi A.S. & B.G. Cümming. Circadian rhythms of dark respiration, flowering, net

photosynthesis, chlorophyll content and dry weight changes in *Chenopodium rubrum. Can. J. Bot.* 50: *2219–2222* (1972).
Cümming, B.J., S.B. Hendricks & H.A. Borthwick. Rhythmic flowering responses and phytochrome changes in a selection of *Chenopodium rubrum. Can. J. Bot.* 43: *825–853* (1965).
Denne, M.P. & C.J. Smith. Daylength effects on growth, tracheid development, and photosynthesis in seedlings of *Picea sitchensis* and *Pinus sylvestris. J. exp. Bot.* 22: *347–361* (1971).
Downs, R.J., S.B. Hendricks & H.A. Borthwick. Photoreversible control of elongation of Pinto beans and other plants under normal conditions of growth. *Bot. Gaz.* 118: *199–208* (1957).
Eaton, F.M. Assimilation respiration balance as related to length of day reactions of soybeans. *Bot. Gaz.* 77: *311–321* (1925).
Hillman, W.S. A metabolic indicator of photoperiodic timing. *Proc. Nat. Acad. Sci. USA* 73: *501–504* (1976a).
Hillman, W.S. Light timer interactions in photoperiodism and carbon dioxide output patterns: towards a real-time analysis of photoperiodism. pp. 383–397 in: H. Smith (ed.), Light and plant development, Butterworths, London (1976b).
Hillman, W.S. Control of plant respiration through non photosynthetic light action. *Nature* 266: *833–835* (1977).
Hock, B. & H. Mohr. Die Regulation der O_2-aufnahme von Senfkeimlingen (*Sinapis alba* L.) durch Licht. *Planta* (Berl.), 61: *209–228* (1964).
Imhoff, C. Action d'un inhibiteur de la photosynthèse le (3,4-dichlorophenyl)-1, 1 diméthylurée (DCMU) sur la floraison de l'*Anagallis arvensis. C.R. Acad. Sc. Paris,* 276: *3303–3306* (1973).
Jacques, R. Action de la lumière par l'intermédiaire du phytochrome sur la germination, la croissance et le développement de *Chenopodium polyspermum. Physiol. Vég.* 6: *137–164* (1968).
Kandeler, R., B. Hügel & T.H. Rottenburg. Relation between photosynthesis and flowering in Lemnaceae pp. 161–169 in: R. Marcelle (ed.), Environmental and biological control of photosynthesis. Junk, The Hague (1975).
Leopold, A.C. & F.S. Guernsey. Respiratory responses to red and infra red light. *Phys. Plantarum* 7: *30–40* (1954).
Martin, E.S. & H. Meidner. Endogenous stomatal movements in *Tradescantia virginiana. New Phytol.* 70: *923–929* (1971).
Mousseau, M. The effect of daylength on daily CO_2 balances of *Sinapis alba*. pp. 135–147. in : R. Marcelle (ed.), Environmental and biological control of photosynthesis, Junk, The Hague (1975).
Mousseau, M. & G. Louason. Action de la photopériode sur la respiration nocturne. Cas d'une plante de jours courts, le *Chenopodium polyspermum. C.R. Acad. Sc. Paris* 282: *719–722* (1976).
Mousseau, M. Night respiration in relation to growth, photosynthesis and development of *Chenopodium polyspermum* in long and short days. *Pl. Sc. Letters* 9: *339–346* (1977).
Quedado, R.M. & D.J.C. Friend. Photosynthesis in flowering of *Anagallis,* in press.
Salisbury, F.B. The flowering process. Pergamon Press (1963).
Schwabe, W.W. Effects of photoperiodic treatment on stomatal movement. *Nature* 169: *1053–1055* (1952).
Shopfer, P. Phytochrome control of enzymes. *Ann. Rev. Pl. Phys.* 28: *223–330* (1977).
Smith, H. & M.G. Holmes. The function of phytochrome in the natural environment. III. Measurement and calculation of phytochrome equilibrium *Photochem. Photobiol.* 25: *539–547* (1977).
Steward, F.C., G.H. Craven, P.R. Weerasinghe & R.G.S. Bidwell. Effects of prior environmental conditions on the subsequent uptake and release of carbon dioxide in the light. *Can. J. Bot.* 49: *1999–2007* (1971).
Tsybul'ko, V.S. Respiration of plants in connection with the nature of their photoperiodic response. *Sov. Pl. Phys.* 19: *981–984* (1973).
Vince-Prue, D. Photoperiodism in plants. McGraw-Hill Book Company UK (1975).

Zeevaart, J.A.D. Physiology of flower formation. *Ann. Rev. Pl. Phys.* 27: *321–348* (1976).

Author's address:
Laboratoire du Phytotron
91190 Gif-sur-Yvette
France.

PHOTOSYNTHETIC ASSIMILATION AND NUTRIENT DIVERSION AS CONTROLLING FACTORS IN FLOWER INITIATION IN *BOUGAINVILLEA* (SAN DIEGO RED) AND *NICOTIANA TABACUM* CV. WIS. 38.

R.M. SACHS, W.P. HACKETT, A. RAMINA & C. MALOOF
(Davis California, U.S.A. and Padova, Italy)

ABSTRACT

Studies with a day neutral tobacco cultivar and a short day *Bougainvillea* cultivar reveal that reproductive development, whether expressed as earliness to flower or numbers of inflorescences developing is promoted in direct relationship to light flux and leaf area. Removal of young leaves and cytokinin applications to apical buds also promote early flower initiation and increased rates of development. In *Bougainvillea* there is an inverse relationship between photosynthetic area and the first node to develop inflorescences. These results support the hypotheses that 1) photosynthetic assimilate supply to the apical meristematic plays a critical role in the transition from the vegetative to reproductive mode of development and 2) the effects of some daylength, manipulative and hormonal treatments on this transition can be explained by their control of the direction and flux of assimilate transport. The studies showing cytokinin induced promotion of flowering in tobacco are of particular interest with respect to Wardell's (1976, 1977) findings that a DNA fraction isolated from flowering tobacco plants causes flowering in vegetative plants held under non-inductive conditions. Since we employed Wardell's tobacco strain and similar assay procedures, it appears reasonable to suggest that 1) the DNA fraction used by Wardell may have been a source of cytokinins and/or 2) cytokinins may promote the synthesis of a 'floral' DNA.

INTRODUCTION

In *Bougainvillea* (T'se *et al*, 1974; T'se, 1975) there is clear evidence that assimilate supply to the inflorescence axes increases before morphological change is observed. For many plants low photon flux density, although considerably in excess of the compensation point for photosynthesis, delays the onset of reproductive development. Conversely, high photon flux density often promotes floral initiation and eliminates the requirement for special photoperiodic requirements (Bodson, 1977; Quedado & Friend, 1978). It was proposed that assimilate supply may limit floral initiation and development and that some environmental and chemical influences on reproductive development are the result of their influences on assimilate supply to the apical meristem (Halevy, 1975, 1977; Sachs & Hackett, 1977). Specifically promotion of reproductive development may result from treatments that increase assimilate supply to the apical meristem. Modified assimilate supply is the essence of 'nutrient diversion' hypotheses, such as those proposed to account for apical dominance (Phillips, 1975), fruit set and abscission (Crane, 1969) root versus shoot growth, *etc*. A nutrient diversion hypothesis is difficult to distinguish experimentally from one that proposes that specific morphogenetic agents control developmental processes. For example, in the case of flower initiation, a flower inducing substance (florigen, anthesin) has long been considered the limiting factor for floral initiation. This substance is phloem mobile; hence it moves from leaves to apical meristems along with photosynthetic assimilates. As a consequence in all experiments in which assimilate flow is diverted from the receptor apical meristem inhibition of flower initiation results (Zeevaart *et al.*, 1977). We have begun testing the concept of nutritional limitations to flower initiation and development by employing techniques that modify quantitatively assimilate flow to receptor buds. That is, borrowing from the lexicon of source/sink studies, we have modified 1) source activity by changing leaf area and photon flux density, 2) sink activity by changing the number of developing receptor buds, and 3) distribution of assimilates by applications of cytokinins. Work with cytokinins has, in addition, led us to examine Wardell's (1976, 1977) recent work on promotion of flower initiation in a day neutral tobacco using DNA extracted from flowering plants. A brief summary and discussion of these studies follows.

MATERIALS AND METHODS

Bougainvillea

Rooted cuttings of *Bougainvillea* 'San Diego Red' were grown in a 26°/21° (approximate average day and night temperatures) greenhouse until they had at least 10 fully expanded leaves, or approximately 35 nodes. Fertilization with half strength Hoagland's solution occurred with every irrigation; supplemental N in the form of 'Osmocote' (26–0–0) was added at the rate of 4g per 10 cm pot.

Long day conditions, 4 hours of incandescent light (*ca* 100 lux at plant tops) from 2200 to 0200 daily, were maintained unless otherwise noted. Short day conditions were achieved by shading plants with black cloth from 1630 to 0830 daily.

Leaf area was varied by cutting 6 expanded leaves to conform to rectangular templates of known areas since many of the leaves on partially defoliated plants continued growing during treatments; a final leaf area was determined at the end of each experiment. Fresh weight, dry weight (70° C, 24 hr) and per cent dry weight of the plants and the component parts (including leaves; stems above and below the uppermost leaf, inflorescence and the root system) were determined initially (for a sample group of 5 plants) and at the end of the treatment period. Soluble solids was obtained by weighing the residue in the 80% methanol extract of the dried stem tissues. Three methanolic extractions were made, the extracts combined and dried before weighing the residue.

Young leaf removal for the 'defoliated' plants was as described previously (T'se *et al.*, 1974). In decapitated plants young leaves were removed from developing branches as they unfolded from the terminal bud. Approximately 11 tightly clustered, unfolded leaves remained as part of the terminal bud even in 'defoliated' plants.

Inflorescence development was evaluated by counting the number and nodal position of inflorescences greater than 2 mm diameter. Earlier studies have shown that these inflorescences will proceed to anthesis.

Cytokinin treatment of *Bougainvillea* was with a 100 ppm solution of N-(phenyl methyl)-9-tetrahydro-wH-pyran-2yl)-9H-purin-6-amine (PBA, SD 8399 obtained from Shell Development Co., Modesto, CA). Ten microliters of the solution was applied at 3-day intervals to the apical bud. Benzyl adenine solutions were used to treat tobacco plants. Details of the treatments for tobacco are found in the results section.

Gibberellin treatments were with 100 ppm solutions of gibberellic acid (GA_3); 10 microliters were applied to the apical bud at weekly intervals.

Tobacco

Seed of *Nicotiana tabacum* 'Wisconsin 38' obtained from Dr. Harold Smith (Brookhaven National Lab., Upton, N.Y.), were germinated in a greenhouse under natural light conditions and transferred at the 4 to 5 leaf stage to shaded greenhouse conditions in which the maximum photon flux density was 1500 lux. Some experiments were performed in constant temperature (23° C), fluorescent lighted chambers where controlled photon flux density was varied. Wardell's bioassay was repeated as follows: 26 cm tall plants were decapitated to a 15 cm stem piece with 4 fully expanded leaves remaining. A 2 cm portion of the stem above the uppermost axillary bud was hollowed out and cytokinin or other solutions were added to the cavity. All other axillary buds were removed, the leaf subtending the one bud remaining was excised, and all leaves were removed from the axillary shoot as it was elongating. Observations were made daily, the date of anthesis and node number to the first inflorescence were recorded.

RESULTS AND DISCUSSION

Bougainvillea

Inflorescence development in *Bougainvillea* as a function of leaf area, cytokinin and gibberellin treatment is shown in Tables 1 and 2. The first order linear regression lines were computed and provided convincing evidence of correlation between earliness to flower (or number of inflorescences developing) and leaf area. As expected dry weight gain during the experimental period, 21 days, was a function of leaf area. Direct correlations were found between inflorescence number and of dry weight percentage and soluble solids in the developing stem (above the leaves). At all leaf areas, cytokinin treatments of the shoot apex reduced the node to first flower, increased the number of inflorescences, increased % dry weight; and increased soluble solids. Note, that the plants with the smallest leaf area (10.8 cm^2) developed inflorescences to anthesis when treated with PBA whereas the control plants did not.

GA_3 treatments completely inhibited inflorescence development regardless of leaf area, but did not reduce % soluble solids or % dry weight in the developing shoot. In fact there was a small significant increase in some treatments.

Thus, although the data for cytokinin treatment and leaf area support the hypothesis that nutrient supply to meristematic regions of *Bougainvillea* is an important factor in controlling reproductive development. The GA_3 data indicate that merely measuring soluble solids in developing shoots may not

Table 1. Inflorescence development in *Bougainvillea* as a function of leaf area, cytokinin (PBA), and gibberellin[1].

Leaf area (cm²)	Node to first inflorescence	No. inflorescences	% Dry wt[2]	% Soluble solids[3]
Control				
9.5	vegetative	vegetative	12.5 ± 0.8	0.71 ± 0.11
22.6	19.3 ± 0.9	1.8 ± 1.4	14.1 ± 0.9	0.97 ± 0.09
31.7	17.0 ± 1.4	8.3 ± 1.1	17.1 ± 0.8	1.23 ± 0.13
45.9	14.5 ± 0.7	12.0 ± 2.0	18.8 ± 1.0	1.31 ± 0.08
PBA				
10.8	17.6 ± 1.3	8.0 ± 2.0	15.5 ± 0.5	0.91 ± 0.13
22.9	14.0 ± 1.8	13.0 ± 1.5	17.9 ± 1.1	1.33 ± 0.12
33.9	12.2 ± 1.5	15.8 ± 2.0	20.9 ± 0.7	1.62 ± 0.14
44.4	9.3 ± 1.8	19.3 ± 1.8	23.3 ± 1.3	1.81 ± 0.11
GA				
9.1	vegetative	vegetative	13.2 ± 0.7	0.62 ± 0.14
23.1	vegetative	vegetative	14.7 ± 0.9	1.12 ± 0.09
31.1	vegetative	vegetative	18.5 ± 0.8	1.40 ± 0.13
45.7	vegetative	vegetative	19.8 ± 1.1	1.51 ± 0.11

[1] Single branched plants.
[2] (Dry wt stem above uppermost leaf/Fresh wt) X 100 = % Dry wt.
[3] (Wt of residue of methanolic extracts of stem/Fresh wt) X 100 = % Soluble solids.

Table 2. Inflorescence development in *Bougainvillea* as a function of number of branches, cytokinin (PBA), and gibberellin[1].

No. branches	Node to first inflorescence	No. inflorescences	% Dry Wt[2]	% Soluble solids[3]
Control				
1	6.0 ± 1.5	12.2 ± 1.7	17.4 ± 0.5	1.27 ± 0.01
2	7.8 ± 1.3	10.0 ± 1.5	18.2 ± 0.7	0.90 ± 0.02
3	9.3 ± 1.8	7.0 ± 2.0	12.5 ± 0.8	0.85 ± 0.05
4	11.8 ± 1.0	3.7 ± 1.7	11.3 ± 0.6	0.62 ± 0.01
5	13.8 ± 1.3	2.2 ± 1.2	8.3 ± 0.3	0.47 ± 0.01
PBA				
1	2.2 ± 1.3	19.5 ± 1.4	17.8 ± 0.4	1.72 ± 0.01
2	3.8 ± 0.9	17.8 ± 1.3	17.0 ± 0.5	1.43 ± 0.02
3	5.8 ± 1.1	13.8 ± 1.0	14.3 ± 0.6	1.25 ± 0.02
4	8.2 ± 1.2	11.2 ± 1.5	12.4 ± 0.4	1.03 ± 0.02
5	10.7 ± 1.0	8.8 ± 1.0	9.9 ± 0.6	0.73 ± 0.01
GA				
1	vegetative	vegetative	17.3 ± 0.4	1.48 ± 0.02
2	vegetative	vegetative	15.8 ± 0.6	1.41 ± 0.02
3	vegetative	vegetative	13.4 ± 0.7	0.73 ± 0.01
4	vegetative	vegetative	12.7 ± 0.4	0.71 ± 0.01
5	vegetative	vegetative	8.5 ± 0.5	0.51 ± 0.02

[1] Leaf area approximately 40 cm² for all plants.
2, 3 as in Table 1.

provide an adequate measure of the nutritional status of the developing reproductive axes and/or indicate a strong specific morphogenetic role for GA_3 in the reproductive axes prior to floret initiation. More sensitive techniques for estimating nutritional levels in the presumptive reproductive meristems must be developed.

Experiments studying inflorescence development as a function of the number of axillary branches developing (Table 2) supported the notion of a nutritional limitation to development. The inverse relationship between branch number and inflorescence initiation indicated that increased partitioning of assimilates coming from the leaves and root system was an effective way of delaying reproductive development. Note, too, the reduced % dry wt and soluble solids with increased branch number. Cytokinin treatment strongly promoted and GA_3 treatment completely inhibited reproductive development regardless of branch number. Again, the GA_3 induced inhibition occurred without a parallel decrease in % dry wt or soluble solids in the branches, indicating once again the limited value of these factors in evaluating the action of agents with strong action on meristematic regions.

Tobacco

Young leaf removal and/or apical applications of 250 ppm BA (10 microliters at 3-day intervals) to tobacco caused a measurable advance in the date of anthesis as well as a decrease in the node to flowering (Table 3) at photon flux densities of 1500 lux. At 4500 lux the major (perhaps only) influence of young leaf removal and/or BA was to advance the rate of development of flowers and/or increase the velocity of node formation. Notice that regardless of treatment the mean node to first flower was approximately the same. Thus, as Wardell reported, photon flux densities in excess of 1500 lux will permit flowering in Wisconsin 38. However, there is reason to believe from our data that at 1500 or 4500 lux the intact plants do not initiate leaves as rapidly as those from which the young leaves are removed. At 1500 lux in the intact plant it appeared as if leaf initiation was very nearly terminated by the 75th day of treatment. Some of the defoliated plants, and all of those treated with cytokinin, advanced to a

Table 3. Flower initiation in *N. tabacum* 'W38' as a function of photon flux density (lux), defoliation and cytokinin.

Treatment	Node to flower				No. days to flower bud			
Lux	700	1500	4500	15000	700	1500	4500	15000
Intact								
Exp 1	–	V[1]	35	–	–	–	–	–
Defoliated								
Exp 1	–	64	35	–	–	–	–	–
Exp 2	V	V	34	22.5	V	V	24	13
Defoliated + cytokinin								
Exp 1	–	38	34	–	–	–	–	–
Exp 2	V	39.5	33	23.5	V	24	19.5	13

[1] V = vegetative.

reproductive stage 40 to 60 days after treatment began. We believe, then, that further experiments on the role of young leaf removal and cytokinin in directing greater assimilate supply to the apical meristem are warranted. These results tend to support the hypothesis developed as a result of the work with *Bougainvillea.*

Wardell developed a flowering bioassay to test the flower promoting ability of DNA fractions extracted from tobacco plants. He found that a heated-cooled DNA fraction from flowering 'Wisconsin 38' plants, but not that from vegetative plants, promoted flowering in the bioassay. Using the same bioassay system we applied BA solutions (0.5 ml of 500 ppm solution at weekly intervals) and observed the development of reproductive structures to anthesis on the developing axillary shoot (Table 4). Note that cytokinin treatment advanced the average date to anthesis by roughly 10 days as well as reduced the average node to flowering by 10 nodes. All of the cytokinin treated plants flowered whereas 67% of the controls flowered – the experiment was terminated 65 days subsequent to the initial treatment owing to insect control difficulties in the greenhouse. Also, variable light in excess of 1500 lux and light quality undoubtedly affected the performance of these plants and makes interpretation difficult. A repeat of this experiment in fluorescent lighted, temperature controlled chambers is essential. Nevertheless, a tentative conclusion is that cytokinin applied to 'Wisconsin 38' in Wardell's bioassay advances flowering in a manner similar to that caused by the DNA fractions isolated by Wardell. It is possible that BA promotes formation of a specific flower promoting DNA or that the DNA promotes cytokinin biosynthesis. Considering Wardell's data for DNA-ase inactivation of the extract, it is not likely that in the bioassay his DNA fraction is completely degraded to cytokinin-like compounds before exerting its action on the flower initiating processes.

Table 4. Flower initiation in 'Wardell's' *N. tabacum* 'W38' bioassay as a function of cytokinin treatment.

Treatment	Node to flower	No. days to flower bud
Cytokinin		
250 ppm	48	31
500 ppm	48	33
Control	54	48

REFERENCES

Bodson, M. *et al*. The role of photosynthesis in flowering of the long-day plant *Sinapis alba. Aust. J. Plant Physiol.* 4: *467–478* (1977).

Halevy, A. Light intensity as a factor in flower formation and development. Proc. 3rd MPP Meeting. Ed. Y. Vardar, K. H. Sheikh, & M.A. Ozturk. Ege, U., Bornova, Izmar, Turkey (1976).

Quedado, R.M. & D.J.C. Friend. Photosynthesis in flowering of *Anagallis. Plant Physiol.* (In press.)

Sachs, R.M. Nutrient diversion: an hypothesis to explain the chemical control of flowering. *Hort Science* 12: *220–222* (1977).

Sachs, R.M. & W.P. Hackett. Chemical control of flowering. *Acta Horticulturae* 68: *29–49* (1977).
T'se, A. Inflorescence development in *Bougainvillea* 'San Diego Red' in relation to assimilate accumulation and distribution. Ph.D. Dissertation. Univ. California, Davis (1975).
T'se, A., A. Ramina, W.P. Hackett, & R.M. Sachs. Enhanced inflorescence development in *Bougainvillea* 'San Diego Red' by removal of young leaves and cytokinin treatments. *Plant Physiol.* 54: *404–407* (1974).
Wardell, W.A. Floral activity in solutions of deoxyribonucleic acid extracted from tobacco stems. *Plant Physiol.* 57: *855–861* (1976).
Wardell, W.A. Floral induction of vegetative plants supplied with a purified fraction of desoxyribonucleic acid from stems of flowering plants. *Plant Physiol.* 60: *85* (1977).

Authors' addresses:

Department of Environmental Horticulture
University of California
Davis, CA 95616
U.S.A.

A. Ramina
Istituto di Coltivazioni Arboree
Università di Padova
Padova
Italy

SIGNIFICANCE OF PHOTOSYNTHESIS, N-DEFICIENCY, ABA AND pH FOR SYNTHESIS OF MALATE IN *LEMNA**

R. KANDELER & R. HELDWEIN
(Vienna, Austria)

ABSTRACT

During the light phase, malate accumulates in *Lemna paucicostata* 6746. The light effect is blocked by 10^{-6}M DCMU, which suggests that it is mediated by photosynthesis. The finding, that the light effect cannot be substituted for by addition of 1% sucrose to the medium, excludes a mechanism, in which photosynthesis-produced carbohydrates may act as a source for malate formation.

The malate accumulation in *L. paucicostata* depends on the composition of the nutrient medium. Transition from pH 4.8 to 6.0 and withdrawal of nitrate promote malate synthesis. (The two factors may act each separately or only in combination, depending upon the state of the plant material). The effect of N-deficiency can be replaced largely by addition of 10^{-6}M abscisic acid (ABA).

After transfer of the plants to a NO_3^--free medium at pH 6 the light induced malate accumulation proceeds in two phases: During the first 8–10 hours the malate content of the plants increases, then falls slightly during 2–7 hours, and finally increases again.

In *Lemna gibba* G1, NO_3^--withdrawal has no promoting effect on malate accumulation within 24 hours, but addition of 10^{-6}M ABA at pH 6 causes a drastic increase of malate content.

The discussion of these results leads to the assumption that in *Lemnaceae* (C_3-plants) malate is synthesized as a product of photosynthesis in addition to carbohydrates. Two factors may lead to an enhancement of this malate production. Firstly malate synthesis acts as an intracellular pH-stat in the case protons are consumed by NO_3^--reduction and proton influx from the medium is kept low by high pH. Secondly N-deficiency and possibly other stress conditions act via an increase of the ABA level upon malate production.

INTRODUCTION

The role of malate as an intermediate in photosynthesis of C_4-plants and CAM-plants is well known. In contrast to this, relatively few investigations are dealing with the fact that malate appears as an end product of photosynthesis in C_3-plants. Calvin (1949) has shown that besides PGA and PEP, malic acid is one

*We thank Dr. A. Zohner, now at Chemie Linz A.G., Linz, for having adapted the enzymatic malate determination for *Lemnaceae*, as well as Dr. H. Löppert and Dr. W. Kronberger for great support by repeated discussions and advices. Moreover we are indepted to Dr. Cornelia Ullrich-Eberius, Botanical Institute, School of Technology, Darmstadt, FRG, having communicated some unpublished experimental results on malate determinations in *Lemna paucicostata* 6746. Financial support from the 'Fonds zur Förderung der wissenschaftlichen Forschung' of the Republic of Austria is greatly acknowledged.

of the first CO_2-fixation products in *Scenedesmus*. Graham & Whittingham (1968) have reported that in *Chlorella pyrenoidosa* CO_2-fixation proceeds normally within the Calvin cycle. However, after pre-culture of the algae with 5% CO_2 and reduction of CO_2 concentration to $0.33 \cdot 10^{-6}$ moles/1 during the experiment, a β-carboxylation predominates, which presumably is mediated by PEP-carboxylase. Ting and Osmond (1973a,b) have investigated the characteristics of PEP-carboxylases in higher plants. In leaves of C_3-plants they have characterized a special form of the enzyme, which is not identical with PEP-carboxylases in roots or in the leaves of C_4-plants and CAM-plants. The authors conclude that malate is synthesized as a product of photosynthesis in leaves of C_3-plants as appearing from unpublished findings made by Osmond. Finally, Böcher & Kluge (1977) have studied the CO_2-fixation in leaf slices of *Spinacea oleracea*. In experiments at pH 3.5, the pattern of ^{14}C-labelling corresponds to the C_3-type. At pH 7.5, however, an increased portion of label appears in malate and other compounds of the C_4-pathway.

The investigations on malate metabolism in *Lemnaceae* reported at present, were accomplished in order to elucidate the connection between acid metabolism and the process of flower induction. Soon it became apparent that malate synthesis in *Lemna* is dependent on photosynthesis and that N-metabolism has a substantial influence on the extent of malate accumulation. Possibly our data indicate, that the photosynthetical production of malate plays an important role in C_3-plants under certain ecological conditions.

MATERIAL AND METHODS

Lemna paucicostata 6746 and *Lemna gibba* G1 were cultivated under axenic conditions in the medium of Pirson & Seidel (1950) which contained iron as EDTA-complex instead of $FeSO_4$. The plants were illuminated during 16 hours per day ('long-day condition') with an irradiance of $20 Wm^{-2}$; an Osram HQIL (400W) has been used as light source. The temperature was 26° C during light phase and 22° C during night.

The experiments were carried out under continuous illumination or darkness at a temperature of 26° C. The bathing solution was the same as the culture medium, from which KNO_3 has been omitted in some cases. The pH was adjusted to 6.0 or 4.8 by addition of NaOH or HCl. Additions of sucrose, ABA and DCMU were made in some experiments as will be indicated.

Malate was determined enzymatically according to the method of Hohorst (1970) which has been modified in some respects. The plants were removed from the bathing medium and after determination of the fresh weight (at least 50 mg) they were immediately frozen. Malate was extracted with 1.5 ml $HClO_4$ (6% w/v) at a temperature of 5° C during 30 minutes. (Preliminary experiments have shown that homogenization of the tissue is unnecessary). The samples were then neutralized by addition of 200 μl 5 M KOH. To remove compounds, absorbing in the range of 340 nm, the samples were treated with about 15 mg charcoal each during 2 hours at 5°C. The plant tissue, $KClO_4$ and the charcoal were sedimented by centrifugation. For the consecutive enzymatic determination, 1 ml aliquots from the supernatant were mixed with 1 ml 0.4 M hydrazine-glycine buffer (pH 9.5), containing 2 mg NAD; absorbency at 340 nm was measured immediately. 30 minutes after addition of 10 μl of a suspension of MDH (approximately 5 mg protein per ml) the measurement of the absorbancy was repeated. A known addition of malate and a further measurement of the absorbancy was used to calculate the malate content of the samples. The malate content was always determined from 3 parallel samples and the received values are recorded in the figures. Each experiment has been accomplished at least twice, yielding qualitatively the same results.

All reagents were of analytical grade; MDH (E.C.No.1.1.1.37), NAD, ABA (mixed isomers) and DCMU were purchased from Sigma Chemical Co. (St. Louis, Mo.).

RESULTS AND DISCUSSION

In preliminary experiments with *Lemna paucicostata* 6746 it was found, that malate content increases during light phase and decreases to a low level within 1.5 to 2.5 hours after transition to darkness. The difference in malate content between light and dark conditions appeared more pronounced when NO^-_3 was omitted from the culture medium and the pH was set to 6 (Fig. 1). For this experiment the plants were kept in darkness for 24 hours (in complete culture medium at a temperature of 22° C). At the end of the dark phase the malate content was extremely low (0.11 μmoles per g FW); now the plants were

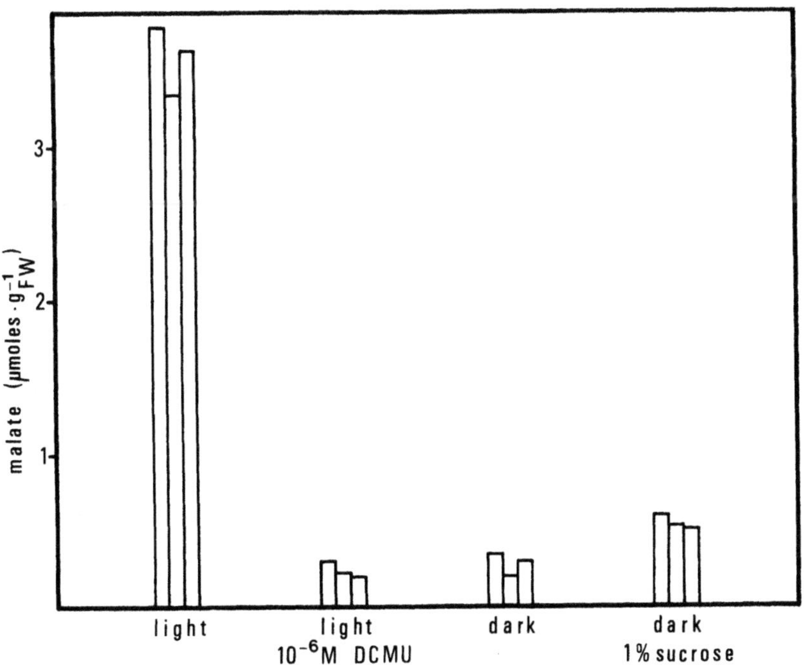

Fig. 1. The effects of continuous light, DCMU and sucrose on malate accumulation in *Lemna paucicostata* 6746 in Pirson-Seidel-medium without nitrate, at pH 6. Malate content of plants at the beginning of the 24-hours-lasting experiment: 0.11 μmoles/g FW.

transferred to fresh nutrient medium composed as indicated and illuminated continuously for 24 hours. The samples for the malate determination were withdrawn after this period. 10^{-6}M DCMU decreased the malate content to the dark level. Therefore it can be assumed that noncyclic electron transport is necessary for malate formation. However, the latter is not dependent upon photosythetically produced carbohydrates as sucrose could not substitute for light.

To elucidate the effect of NO_3^--withdrawal and increased pH these two parameters were altered systematically (experiment, shown in Fig. 2). Increased pH or NO_3^--deficiency applied separately caused already enhancement of the malate level, but not to the same extent as the combined application of the two parameters did; compared to this the malate level was substantially lower at full Pirson-Seidel-medium at pH 4.8. It should be mentioned in this context that an identical experiment carried out in winter instead of spring or summer yielded somewhat different results. Neither pH nor NO^-_3-withdrawal, given separately, had a promoting effect. A large increase of malate content was reached however at combined application of the two factors. As an example the following data may be cited: 5.04 (pH 4.5, + NO_3^-); 6.55 (pH 6.3, + NO_3^-); 5.75 (pH 4.5, − NO_3^-); 15.04 μmoles per g FW (pH 6.3, − NO_3^-).

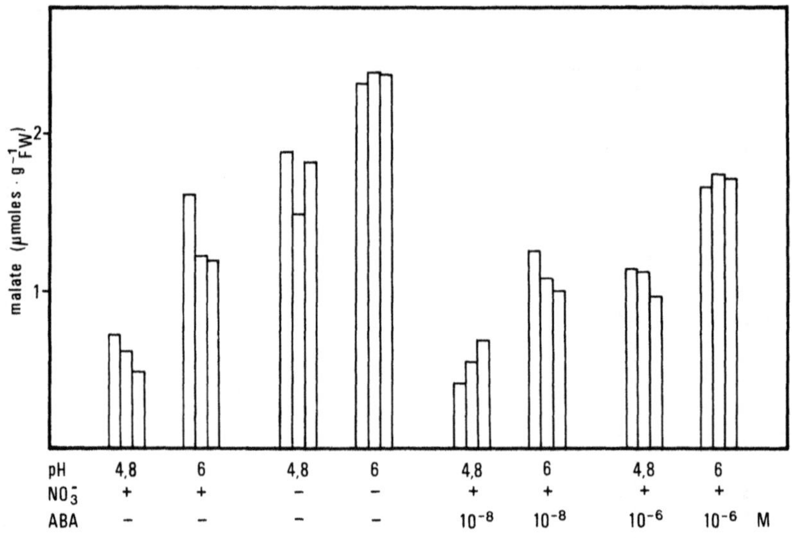

Fig. 2. The effects of pH, nitrate withdrawal and ABA on malate accumulation in *Lemna paucicostata* 6746. Duration of the experiment: 24 hours.

The pH effect on malate synthesis can be interpreted in the following way. The light-dependent process of nitrate reduction, proceeding in the mesophyll, causes an alkalinization of the cytoplasma by consumption of protons. A subsequent delivery of protons from the medium is possible – especially in water plants –, if the H^+-concentration of the medium is sufficiently high. In this case the uptake of protons functions as a cotransport with nitrate ions. For *L.*

paucicostata we have shown that the rates of the net ion movements exert a stoichiometry of 1:1 (Löppert *et al.*, 1977). On the other hand, if the H^+-concentration in the external solution is low, the constancy of cytoplasmic pH is maintained by a regulatory mechanism, which induces malic acid synthesis for delivery of protons (Davies, 1973; Raven & Smith, 1974). Therefore, under conditions, which allow nitrate reduction, transition of pH 4.8 to 6 in the medium causes an increased activity of the so-called 'biochemical pH-stat' and with that an enhanced malate accumulation. Böcher & Kluge (1977) have assumed that alkalinization of the medium effects a shift in the CO_2/HCO_3^--equilibrium in favour of the HCO_3^--concentration, which means that the activity of the PEP-carboxylase (substrate: HCO_3^-) is enhanced in comparison to the activity of the RuDP-carboxylase (substrate: CO_2). This assumption is not applicable to our problem, for the CO_2 portion of the equilibrium preponderates at pH 4.8 (97%) as well as at pH 6 (70%). Furthermore it must be considered that the cytoplasmic pH of plants is scarcely affected by the pH of the nutrient medium.

The effect of nitrate withdrawal on malate synthesis can be made understandable, if it is assumed that plants respond to this stress condition with an alteration of hormone metabolism. Goldbach *et al.* (1975) have shown that N-deficiency enhances the ABA-content in sunflower plants. The content of cytokinins in roots and leaves of sunflower and also in leaves of birch seedlings is diminished by N-withdrawal (Wagner & Michael, 1971; Wareing 1976). In *Spirodela polyrrhiza* the production of turions, induced by N-deficiency, can also be obtained by application of ABA (see Scheiner *et al.*, 1978, for references). Hence we have tested, whether malate synthesis in *L. paucicostata* can be enhanced by ABA in NO_3^--containing medium. In the right hand part of Fig. 2 the results of such an experiment are shown. The concentration of 10^{-8}M ABA has no effect, but 10^{-6}M ABA induces a significant increase in malate accumulation both at pH 4.8 and at pH 6. Thus the effect of NO_3^--withdrawal can be substituted by ABA application. In addition we have ascertained that the ABA-induced malate synthesis is also inhibited by DCMU (data not shown). Therefore ABA influences especially the photosynthetic malate production. In wheat seedlings ABA causes a decrease of RuDP-carboxylase activity and an increase of PEP-carboxylase activity (Sankhla & Huber, 1975). Presumably the enhancement of malate accumulation by ABA in *Lemnaceae* may be explained in this way.

The foregoing considerations implicate that the high malate accumulation, caused by NO_3^--withdrawal and increased pH, may be attributed to two different regulatory processes. The time course of malate accumulation under these conditions (Fig. 3), confirms this statement. Two experiments indicate clearly that malate level does not increase continuously within the 24 hours of light exposure, but rather in two separate phases. In the experiment of Fig. 3 the first phase covers about 8 hours, the second phase begins after 10 hours. In the other experiment the first phase of malate increase lasts 10 hours, the second phase starts after 17 hours. In both experiments malate content decreases slightly between the two phases of synthesis. Presumably, the activity of the 'biochemical pH-stat' mediates the first phase of malate synthesis. During this time, nitrate reduction and, therefore, H^+-consumption continues with nitrate stored in the vacuole. Subsequently certain nitrogen compounds, yet unknown, are exhausted. So the synthesis of ABA is deblocked and ABA induces the second phase of malate synthesis.

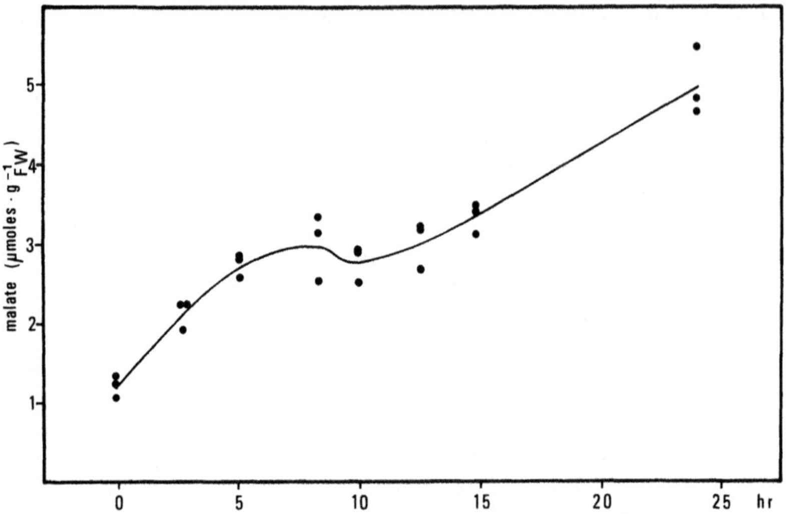

Fig. 3. Kinetics of malate accumulation in *Lemna paucicostata* 6746 in Pirson-Seidel-medium without nitrate, pH 6, under continuous light.

All investigations, so far communicated, were carried out with the short-day plant *Lemna paucicostata* 6746. For comparison some experiments were made with the long-day plant *Lemna gibba* G1 under quite the same conditions as in the experiments shown in Fig. 2. The results of one of these experiments are recorded in Fig. 4. Compared with *L. paucicostata, L. gibba* showed different behaviour in two special points. Firstly NO_3^--withdrawal had no effect on malate production within 24 hours. Possibly N-deficiency in this plant did not become effective so rapidly as in *L. paucicostata*. Secondly, addition of 10^{-6}M ABA is effective only at pH 6. This pH effect could be interpreted by the assumption that malate production, induced by ABA, occurs only in case protons can be extruded into the external medium by activity of a proton pump. It might be, that *L. gibba* – unlike *L. paucicostata* – cannot excrete protons into the external medium at the lower pH of 4.8.

In summary, the presented results may indicate that the significance of photosynthetically produced malate in C_3-plants increases under certain stress conditions. Another example for the shift of RuDP-carboxylase/PEP-carboxylase activities in favour of PEP-carboxylase under stress conditions, that is the infection of tobacco plant with tobacco mosaic virus, was given recently by Makovcová and Šindelář (1978). Under conditions of N-deficiency the plants obviously take advantage of producing assimilates, which firstly can be transported rapidly in all parts of the plant already during the light phase, and secondly can be exchanged easily against osmotically effective, N-containing compounds from the vacuole.

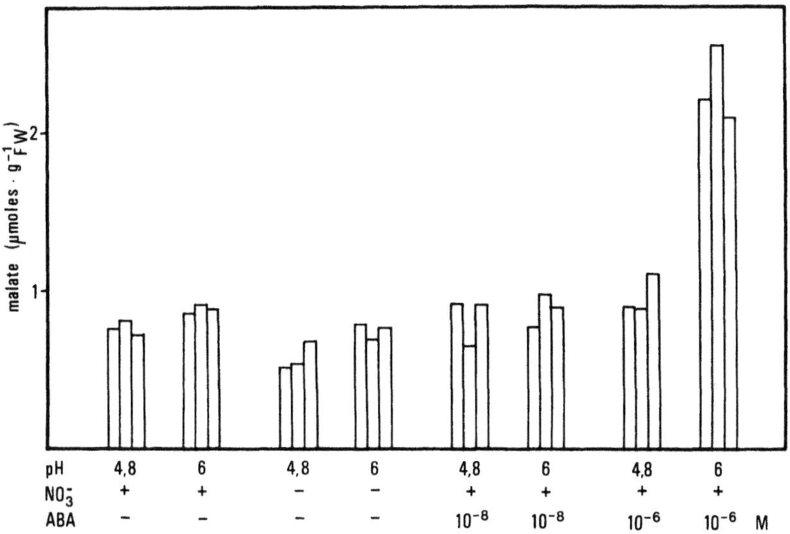

Fig. 4. The effects of pH, nitrate withdrawal and ABA on malate accumulation in *Lemna gibba* G1. Duration of the experiment: 24 hours.

ABBREVIATIONS

ABA = abscisic acid
CAM = Crassulacean acid metabolism
DCMU = 3-(3,4-dichlorophenyl)-1,1-dimethyl urea
FW = fresh weight
MDH = malate dehydrogenase
PEP = phosphoenolpyruvate
PGA = 3-phosphoglyceric acid
RuDP = ribulosediphosphate.

REFERENCES

Böcher, M. & M. Kluge. Der C_4-Weg der C-Fixierung bei *Spinacea oleracea* I. ^{14}C-Markierungsmuster suspendierter Blattstreifen unter dem Einfluß des Suspensionsmediums. *Z. Pflanzenphysiol.* 83: *347–361* (1977).

Calvin, M. The path of carbon in photosynthesis, VI. *J. Chem. Ed.* 26: *639–657* (1949).

Davies, D.D. Control of and by pH. *Symp. Soc. Exp. Biol.* 27: *513–529* (1973).

Goldbach, E., H. Goldbach, H. Wagner & G. Michael. Influence of N-deficiency on the abscisic acid content of sunflower plants. *Physiol. Plant.* 34: *138–140* (1975).

Graham, D. & C.P. Whittingham. The path of carbon during photosynthesis in *Chlorella pyrenoidosa* at high and low carbon dioxide concentrations. *Z. Pflanzenphysiol.* 58: *418–427* (1968).

Hohorst, H.-J. L-(-)-Malat. Bestimmung mit Malat-Dehydrogenase und NAD. pp. 1544–1549, in: Bergmeyer, H.U. (ed.), Methoden der enzymatischen Analyse, vol. III. Verlag Chemie, Weinheim/Bergstr. (1970).

Löppert, H.G., W. Kronberger & R. Kandeler. Correlation between nitrate uptake and alkalinisation by *Lemna paucicostata* 6746. pp. 283–288, in: Thellier, M. *et al.* (eds.),

Transmembrane Ionic Exchanges in Plants, Editions du C.N.R.S. et de l'Université de Rouen, Paris et Mont Saint Aignan (1977).
Makovcová, O. & L. Šindelář. Changes in phosphoenolpyruvate carboxylase and ribulosebiphosphate carboxylase activities in tobacco plants infected with tobacco mosaic virus. *Biol. Plantarum* (Praha) 20: *135–137* (1978).
Pirson, A., & F. Seidel. Zell- und stoffwechselphysiologische Untersuchungen an der Wurzel von *Lemna minor* L. unter besonderer Berücksichtigung von Kalium- und Kalziummangel. *Planta* (Berl.), 38: *431–473* (1950).
Raven, J.A. & F.A. Smith. Significance of hydrogen ion transport in plant cells. *Canad. J. Bot.* 52: *1035–1048* (1974).
Sankhla, N. & W. Huber. Regulation of balance between C_3 and C_4 pathway: role of abscisic acid. *Z. Pflanzenphysiol.* 74: *267–271* (1975).
Scheiner, O., F. Pittner, O. Bollmann & R. Kandeler. Die Wirkung von Stickstoffmangel und anderen Faktoren auf die Phytinsäurespeicherung in *Lemna gibba* G1. *Z. Pflanzenphysiol.* 88: *295–303* (1978).
Ting, I.P. & C.B. Osmond. Photosynthetic phosphoenolpyruvate carboxylases. Characteristics of alloenzymes from leaves of C_3 and C_4 plants. *Plant Physiol.* 51: *439–447* (1973).
Ting, I.P. & C.B. Osmond. Multiple forms of plant phosphoenolpyruvate carboxylase associated with different metabolic pathways. *Plant Physiol.* 51: *448–453* (1973).
Wagner, H. & G. Michael. Der Einfluß unterschiedlicher Stickstoffversorgung auf die Cytokininbildung in Wurzeln von Sonnenblumenpflanzen. *Biochem. Physiol. Pflanzen* 162: *147–158* (1971).
Wareing, P.F. Endogenous cytokinins as growth regulators. pp. 103–109, in: Sunderland, N. (ed.), Perspectives in Experimental Biology. 2. Botany, Pergamon Press, Oxford (1976).

Authors' address:
Botanical Institute
Agricultural University
Gregor Mendel Strasse 33
A-1180 Vienna
Austria.

PHOTOSYNTHESIS AND CROP GROWTH

D.A. CHARLES-EDWARDS
(Littlehampton, U.K.)

ABSTRACT

The absolute rate of leaf or crop photosynthesis can be written as the product of an intrinsic photosynthetic activity and leaf or crop size. An examination of a variety of data indicates that the rate of photosynthesis per unit leaf volume is a better indicator of intrinsic activity than the rate of photosynthesis per unit leaf area. In many instances specific leaf area provides an adequate measure of leaf thickness, and the specific photosynthetic rate (*i.e.* rate of photosynthesis per unit leaf dry weight) a useful indication of intrinsic photosynthetic activity.

The effects of aerial environment on plant growth can be separated into effects first, on intrinsic photosynthetic activities, and second, on dry-matter partitioning both within leaves, and between leaves and other plant parts. Such an analysis of the role of photosynthesis as a determinant of the growth rate of a dicotyledonous crop is presented, the analysis explicitly accounting for the variation in photosynthetic characteristics of leaves from different horizons within the crop canopy. Examination of published data leads to the conclusion that aerial environment affects growth and development primarily through effects on dry-matter partitioning, and not through effects on intrinsic photosynthetic activity.

INTRODUCTION

The absolute rate of leaf or crop photosynthesis can be written as the product of an intrinsic photosynthetic activity and the size of the leaf or crop. Whereas the rate of photosynthesis per unit leaf area is commonly used as a measure of intrinsic activity, it does not take into account the considerable variation in the size of the assimilatory apparatus beneath unit leaf area which arises with changes in the growth environment. The rate of photosynthesis per unit leaf volume is a better indicator of intrinsic photosynthetic activity. It follows that the role of photosynthesis as a determinant of plant growth cannot be studied in isolation from studies of the partition of dry matter.

A proper mechanistic approach to the problems of dry-matter partitioning within the plant involves physiological concepts such as priorities, sources and sinks, and active and passive modes of transport. Thornley (1977) has outlined how these concepts can be defined mathematically and then incorporated into models of dry-matter partitioning. Such models, however, require detailed information on matters such as the levels of carbon and nitrogen substrates in both shoot and root tissues, and the resistances to transfer of materials between shoots and roots; information that is not readily available, nor amenable to collection. Alternatively, there are certain features of plant growth that must obey general rules which are independent of detailed mechanisms, and an analysis at the organ level can provide useful interpretative relationships between plant growth rate, leaf and root masses and their respective activities. For example, if we consider a vegetative crop which consists of leaf, root and stem tissues, the total crop dry weight per unit ground area, M_T, is

$$M_T = M_L + M_R + M_S \tag{1}$$

where M_L, M_R and M_S denote leaf, root and stem dry weights per unit ground area. The crop growth rate, $\Delta M_T/\Delta t$, can be written as

$$\frac{\Delta M_T}{\Delta t} = \frac{M_L \sigma_C}{f_C} = \frac{M_R \sigma_N}{f_N} = \frac{(M_L + M_R)\sigma_C \sigma_N}{f_C \sigma_N + f_N \sigma_C} \tag{2}$$

where σ_C is the leaf specific activity with respect to carbon (C) and σ_N the root specific activity with respect to nitrogen (N) such that

$$\sigma_C = \frac{1}{M_L}\frac{\Delta C}{\Delta t} \text{ and } \sigma_N = \frac{1}{M_R}\frac{\Delta N}{\Delta t},$$

f_C and f_N are the incremental compositions of the plant in terms of carbon and nitrogen, so that

$$f_C = \Delta C/\Delta M_T \text{ and } f_N = \Delta N/\Delta M_T$$

and t is the time variable (Charles-Edwards, 1976; Thornley, 1977). Eqn (2) provides a simple description of crop growth rate in terms of leaf and root specific activities and dry matter partitioning.

At the whole plant level, specific leaf activity, σ_C, is a weighted average of the individual activities of all leaves on the plant integrated over the 24 hour day/night period. It can be written as

$$\sigma_C = \frac{0.273}{M_L}\int_0^L \int_0^{86400} (F + R')\, d\underline{t}\, dL \tag{3}$$

where 0.273 converts $g(CO_2)$ to g (carbon), L is the leaf area index of the canopy, F the net CO_2 exchange rate per unit leaf area, R′ a correction to allow for dark respiration by stem and root tissue and $\underline{t}$ the time variable in seconds. Eqns (2) and (3) relate crop growth rate to photosynthesis at the leaf level.

In the following sections the effects of growth environment on photosynthesis and dry-matter partitioning are examined. Firstly, photosynthesis is examined at the leaf level, and the effects of environment are analyzed into effects on intrinsic photosynthetic activity and dry-matter partitioning within the leaf. Secondly, the analysis is extended to describe canopy photosynthesis in terms of processes occurring at the leaf level. Thirdly, the effects of environment on dry-matter partitioning are examined. In the fourth section the effects of environment on crop growth rate are analyzed.

For studies at the crop level, the rate of photosynthesis per unit leaf area, F, is usefully described by

$$F = \alpha_m I\, (\tau C - \beta O) \,/\, (\alpha_m I + \tau C) - R_D \tag{4}$$

where I is the light flux density incident on the leaf surface, C and O are the ambient CO_2 and O_2 concentrations, α_m is the leaf light utilization efficiency at high CO_2 or low O_2 concentrations, τ is an overall carboxylation conductance, β is an overall oxygenation (photorespiratory) conductance and R_D the rate of dark respiration per unit leaf area (Charles-Edwards, 1978a). The parameters τ and β can be replaced by the products $h\tau'$ and $h\beta'$, where h denotes leaf thickness (Charles-Edwards, *loc. cit.*). The rate of light-saturated photosynthesis per unit leaf area, F_m, is then given by

$$F_m = h\,(\tau' C - \beta' O) - R_D \simeq h\,(\tau' C - \beta' O), \tag{5}$$

and the rate of light saturated photosynthesis per unit leaf volume, P_m, by

$$P_m = F_m / h \simeq (\tau' C - \beta' O). \tag{6}$$

If τ' and β' are proportional to the carboxylation and oxygenation constants for RuBP carboxylase it might be expected that they are tightly coupled, and measurements of the ratio β'/τ' for a wide range of C_3 plant species indicate a remarkable constancy in the ratio measured at a particular temperature (Charles-Edwards, 1978b).

Changes in F_m, h and P_m with growth light environment for three species are shown in Tables 1a and 1b. In each case, both F_m and h increase with the growth light level, whereas P_m, in contrast, remains fairly constant. These data suggest that the constants τ' and β' are independent of the growth light environment, and that the adaptive response of F_m to growth light level is attributable directly to changes in the size of the assimilatory apparatus beneath unit leaf area. However, in *Plectranthus parviflorus* (Table 1a) both F_m and P_m respond to the growth temperature level over the range 6.5 – 28.5° C. Similar effects of growth temperature on P_m have been observed in *Lolium* over the range 11 – 24° C, although above 24° C P_m declined with increasing growth temperature. Further, whereas P_m in *Lolium* increased over the growth temperature range 11 –17° C, F_m declined, the decline being attributable to a marked decrease in leaf thickness over this temperature range (Charles-Edwards, Charles-Edwards and Sant, 1974). Since P_m is affected by the growth temperature, some variation in P_m with growth light level might be expected in plants grown at the same nominal bulk air temperature, since the radiation load on the leaves may affect their temperature.

As a first approximation, leaf area per unit leaf fresh weight, s*, will be inversely proportional to leaf thickness. The data in Table 2a, for tomato leaves from plants grown at different light levels, show that s* declines with the growth light level, whilst F_m increases. The rate of light-saturated photosynthesis per unit leaf fresh weight, ν^*_m, remains almost constant. These data are analogous to those in Table 1, and can be compared with those in Table 2b, which were obtained for leaves from different positions within a tomato canopy growing

Table 1a. Response of leaf photosynthesis in *Plectranthus parviflorus* to plant growth environment (Nobel, 1977).

Daily light integral E m^{-2}	Growth temperature °C	h mm	F_m mg m^{-2} s^{-1}	P_m g m^{-3} s^{-1}
1.3	17.5	0.28	0.30	1.07
6.5	17.5	0.51	0.56	1.10
22.9	17.5	0.83	0.78	0.94
6.5	6.5	0.48	0.34	0.71
6.5	17.5	0.51	0.56	1.10
6.5	28.5	0.54	0.72	1.33

Table 1b. Response of leaf photosynthesis in *Abutilon theophrasti* and *Gossypium hirsutum* to plant growth light environment (growth temperature 31 °C day, 25 °C night) (Patterson, Duke & Hoagland, 1978).

Species	Daily light integral E m^{-2}	h* mm	F_m mg m^{-2} s^{-1}	P^*_m g m^{-3} s^{-1}
A. theophasti	4.9	0.07	0.36	5.14
	17.3	0.09	0.55	6.11
	40.5	0.12	0.66	5.50
G. hirsutum	4.9	0.11	0.24	2.18
	17.3	0.19	0.43	2.26
	40.5	0.28	0.79	2.82

h – leaf thickness; h* – mesophyll tissue thickness; F_m – light saturated photosynthesis per unit leaf area; P_m – light saturated photosynthesis per unit leaf volume; P^*_m – light saturated photosynthesis per unit mesophyll tissue volume.

under natural daylight in a glasshouse. Leaves of similar age, but from shaded and unshaded positions within the canopy (compare the two leaves 20 and leaves 8 and 9) have different values of F_m and v^*_m, but similar values for s*. The differences in F_m and v^*_m cannot be attributed to leaf age effects, since an old leaf from a partially shaded position at the lower edge of the canopy has a higher rate of photosynthesis on both a leaf area and fresh weight basis than a much younger leaf from a shaded position near to the top of the canopy (compare leaf 9 and the shaded leaf 20). As all leaves initially developed under similar light levels, it seems reasonable to suppose that the difference in F_m and v^*_m reflect an adaptation of τ' and β' to the immediate light environment of the particular leaf. The data in Table 2 therefore suggest two levels of adaptation by the leaf to the growth light level: firstly, during expansion of the leaf the growth environment affects the size of the assimilatory apparatus formed beneath unit leaf area, but not the intrinsic activity of that apparatus; secondly, once the leaf is fully expanded, shading leads to a progressive decrease in the intrinsic activity of the assimilatory apparatus.

Changes in F_m are also observed during leaf ontogenesis, although it is generally not clear whether the leaf has been subjected to a constant light environment throughout development, or has been subjected to shading by later leaves. F_m increases during leaf expansion, reaching a maximum value at about 70 per cent of full expansion, it then declines as the leaf ages (*cf.* Catsky, Ticha & Solarova, 1976; Mokronosova and Nekrasova, 1976). Changes in both F_m and P_m during ontogenesis of bean leaves (Phaseolus) are shown in Figure 1 (data from Catsky *et al., loc. cit.*). In contrast to the changes F_m, P_m declines almost linearly throughout the life of the leaf. From the data presented for these experiments, it is apparent that throughout the period of the experiments the stomatal conductance was almost an order of magnitude greater than the combined mesophyll and carboxylation conductances of the leaf. It seems reasonable to suppose that the decline in P_m during leaf ontogenesis reflects a

Table 2a. Photosynthesis of uppermost leaves from tomato plants grown in controlled environment cabinets at 20°C and three light levels (after Charles-Edwards & Ludwig, 1974).

Daily light integral MJ m^{-2}	s^* m^2 kg^{-1}	F_m mg m^{-2} s^{-1}	υ_m^* μg g^{-1} s^{-1}
1.2	6.6	0.48	3.13
2.9	5.1	0.79	3.98
4.6	3.6	0.89	3.20

Table 2b. Photosynthesis by leaves from different positions within a tomato canopy grown under natural daylight (~ 4 MJ d^{-1} PAR) (after Acock *et al.* 1978).

Leaf number	Leaf environment	s^* m^2 kg^{-1}	F_m mg m^{-2} s^{-1}	υ_m^* μg g^{-1} s^{-1}
20	unshaded, top of canopy	3.6	1.49	5.4
20	shaded, top of canopy	4.7	0.43	2.0
8	shaded, base of canopy	5.2	0.14	0.7
9	partially shaded, base of canopy, at edge	5.0	0.79	4.0

Table 2c. Photosynthesis by leaves from different horizons within a tomato canopy (after Acock *et al.* 1978).

Leaf area index above leaf, L	s^* m^2 kg^{-1}	F_m μg m^{-2} s^{-1}	υ_m^* μg g^{-1} s^{-1}	% incident light transmitted to leaf
1.0	3.4	1.33	4.5	39.5
2.6	2.6	0.34	0.9	15.2
4.3	2.1	0.11	0.2	3.1

s^* – leaf area per unit fresh weight; F_m – light saturated photosynthesis per unit leaf area; υ_m^* – light saturated photosynthesis per unit leaf fresh weight.

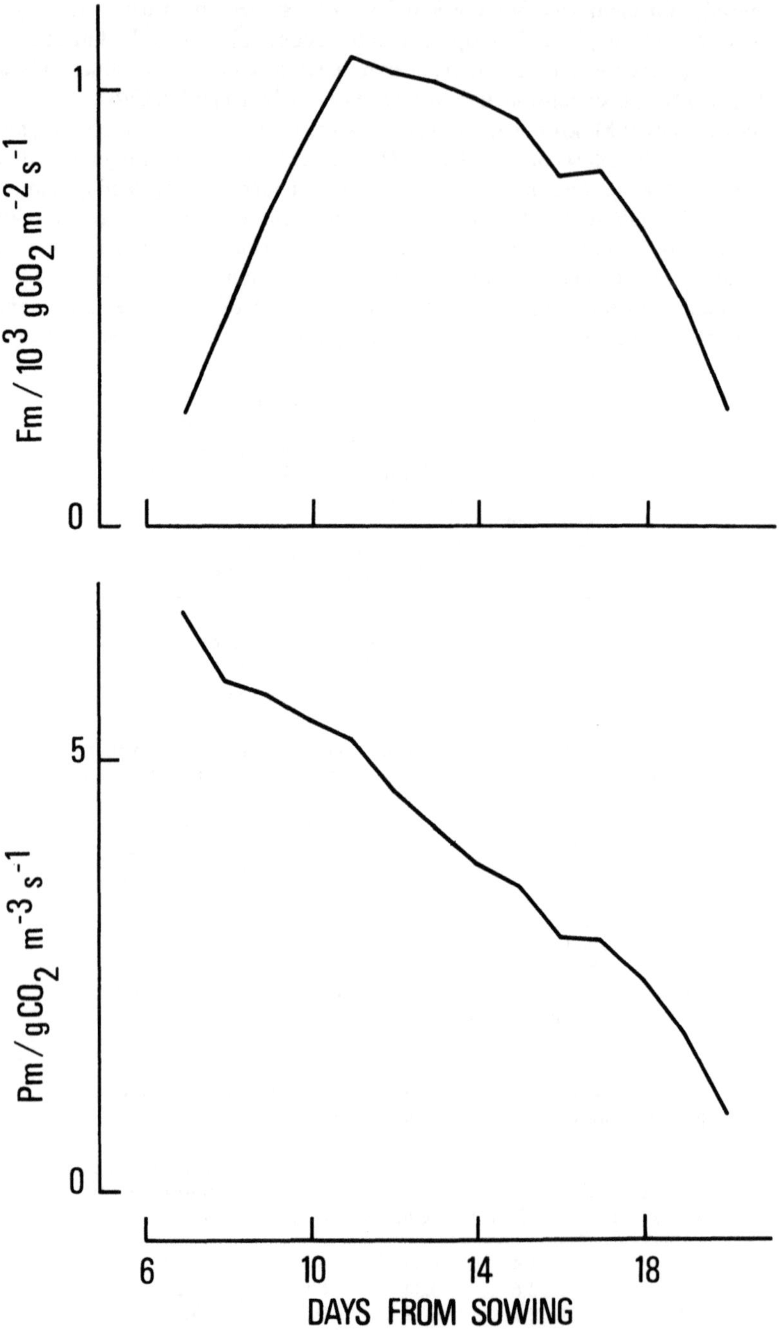

Fig. 1. Changes in bean-leaf photosynthesis during ontogenesis (after Catsky *et al.* 1976). F_m – light-saturated photosynthesis per unit leaf area; P_m – light-saturated photosynthesis per unit leaf volume.

decline in the *concentrations* of carboxylating enzymes in the leaf tissues rather than an ageing effect on the physical properties of the leaf. The increase in F_m during leaf expansion clearly reflects the increase in the size of the assimilatory apparatus beneath unit leaf area during the period of leaf thickening.

Provided that the dry weight to volume ratio of leaf tissue remains reasonably constant across growth environments, the specific leaf area, s, will be inversely proportional to leaf thickness, so that

$$s = 1/\rho h \tag{7}$$

where ρ is the dry-matter density of leaf tissue. It then follows that the specific photosynthetic rate at light saturation, ν_m, can be written as

$$\nu_m = sF_m \simeq (\tau'C\text{-}\beta'O)/\rho. \tag{8}$$

Caution needs to be exercised in the use of specific leaf area as an indicator of leaf thickness. It has been shown that whilst s decreases during the expansion of young alfalfa leaves (analogous to leaf thickness increasing) it increases sharply if the leaves are subsequently shaded (Wolf & Blaser, 1972). However, in both *Phaseolus vulgaris* and *Zea mays* it has been observed that the rate of light saturated photosynthesis per unit leaf area, F_m, of leaves grown at different light levels is inversely related to the specific leaf area (Louwerse & Zweerde, 1977). It is also frequently observed that genotypic differences in F_m are negatively correlated with s (*cf.* Pearce, Carlson, Bar Hart & Hanson, 1969; Dornhoff & Shibles, 1970; Criswell & Shibles, 1971). This may be used as evidence that much of the apparent genetic variability in F_m is due to simple differences in the sizes of the assimilatory apparatus beneath unit leaf area rather than differences in the intrinsic photosynthetic activities of contrasting genotypes (Charles-Edwards, 1978a).

CANOPY PHOTOSYNTHESIS

The evidence presented in the previous section suggests that whilst τ' and β' are independent of the growth light environment of the leaf during its expansion, they do respond to changes in the light environment on cessation of expansion. Using simpler leaf photosynthesis models it has been shown that $h\tau'$ can be an order of magnitude less for leaves from the lowest horizon in the canopy when compared with leaves from the uppermost horizon (Hozumi, Kirita & Nishioka, 1972; Acock, Charles-Edwards, Fitter, Hand, Ludwig, Warren Wilson & Withers, 1978). Indeed, photosynthesis per unit ground area by a closed stand of tomato plants has been successfully predicted from single leaf photosynthesis measurements by assuming that the decline in leaf conductance, $h\tau'$, with depth in the canopy is related to the decline in proportion of incident light transmitted through the canopy (Acock *et al.*, *loc. cit.*). Data from both these sources suggest that there is little, if any, change in leaf light utilization efficiency, α_m, with leaf horizon within the canopy.

The light flux density on a leaf surface at a leaf area index of L below the canopy surface, denoted I, can be written as

$$I = \frac{I_o K \exp(-KL)}{(1-m)} \tag{9}$$

where I_o is the light flux density incident at the top of the canopy, K is a canopy light extinction coefficient and m a leaf light transmission coefficient (Saeki, 1960). Values for F_m and ν_m^* for leaves from different horizons within a tomato canopy are shown in Table 2c. As a first approximation both F_m and ν_m^* are proportional to the percentage of the light incident at the top of the canopy that is transmitted through the canopy to the horizon containing the particular leaf. Denoting the overall carboxylation and oxygenation constants of unshaded, fully expanded leaves at the top of the canopy by τ'_o and β'_o, we can then write as a first approximation that

$$\tau' = \tau'_o \frac{I}{I_o} = \frac{\tau'_o K \exp(-KL)}{(1-m)} \tag{10}$$

and

$$\beta' = \frac{\beta'_o K \exp(-KL)}{(1-m)} \tag{11}$$

where τ' and β' are the overall constants for leaves at a leaf area index of L below the canopy surface. The rate of net CO_2 exchange per unit leaf area by a leaf in this horizon can then be written as

$$F = \frac{\alpha_m K I_o h (\tau'_o C - \beta'_o O) \exp(-KL)}{(1-m)(\alpha_m I_o + h\tau'_o C)} - R_D. \tag{12}$$

Integrating eqn (12) over the entire leaf area of the canopy gives the rate of net CO_2 exchange per unit ground area, F_c, as

$$F_c = \frac{\alpha_m I_o h(\tau'_o C - \beta'_o O)(1-\exp(-KL))}{(1-m)(\alpha_m I_o + h\tau'_o O)} - R_c \tag{13}$$

where R_c is the rate of dark respiration per unit ground area by the whole crop. Eqn. (13) describes canopy photosynthesis in terms of five leaf characteristics, three of which are independent of the light environment under which the canopy has developed (α_m, τ'_o and β'_o), the other two (h and m) depending on dry-matter partition within the leaf, which *is* dependent on the light environment under which the canopy has developed (m has been shown to be correlated with specific leaf area in forage grasses (Sheehy, 1975) and changes in m with season for a wide range of evergreen and deciduous trees and shrubs are associated with changes in specific leaf area (Kasanaga & Monsi, 1954)). Eqn. (13) also describes a hyperbolic dependence of F_c on the light flux density incident at the top of the canopy, and it has been shown that such a simple model can provide as good a description of photosynthesis by stands of green peppers as more complex canopy photosynthesis models (Acock, Hand, Thornley & Warren Wilson, 1976).

DRY-MATTER PARTITIONING

In the present analysis three aspects of dry-matter partitioning are of particular interest: partition of dry-matter within the leaf, that is leaf thickness and specific leaf area, partition of dry-matter between leaves and other plant parts (M_L/M_T), and the carbon content of dry-matter (f_c).

Some effects of growth light and temperature environments on the specific leaf areas of both chrysanthemum and tomato leaves are shown in Figure 2. These data can be described by the empirical equation

$$(1/s) = a_o + a_1 H + a_2(1/\bar{T}) \tag{14}$$

where a_o, a_1 and a_2 are constants, H is the average daily light integral under which the plants developed MJ m^{-2}) and $\bar{T}$ the average growth temperature (°C). Values of a_o, a_1 and a_2 obtained by regression of (1/s) on H and $(1/\bar{T})$, for both crops seperately and then combined, are shown in Table 3. Whilst it is no more than coincidental that the estimated values of the three constants are

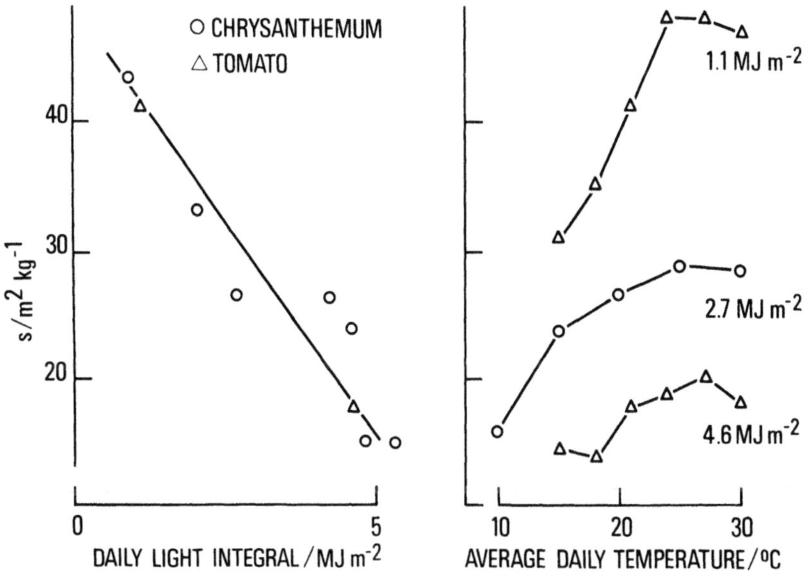

Fig. 2. Effects of growth light and temperature on specific leaf area, s, of tomato (Hurd – personal communication) and chrysanthemum (Charles-Edwards & Acock, unpublished data).

Table 3. Coefficients, with their standard errors, for the regressions of 1/s on growth light level (H) and the reciprocal of growth temperature $(1/\bar{T})$ for the data shown in Fig. 2, using the regression equation $1/s = a_0 + a_1 H + a_2 (1/\bar{T})$

(a) tomato	(b) chrysanthemum	(c) combined data
$a_0 = -0.0097 \pm 0.0056$	$a_0 = -0.0085 \pm 0.0109$	$a_0 = -0.0069 \pm 0.0057$
$a_1 = 0.0100 \pm 0.0007$	$a_1 = 0.0084 \pm 0.0019$	$a_1 = 0.0092 \pm 0.0009$
$a_2 = 0.4973 \pm 0.1079$	$a_2 = 0.4557 \pm 0.1483$	$a_2 = 0.0429 \pm 0.0959$
% variance accounted for = 95.2	% variance accounted for = 70.7	% variance accounted for = 84.8

(data for tomato from R. Hurd (personal communication) and for chrysanthemum from Charles-Edwards & Acock (unpublished)) nearly identical for the two crops, the combined regression illustrates more clearly the light × temperature interaction on specific leaf area. Increasing growth light levels lead to a decrease in s, analogous to the increase in leaf thickness shown in Table 1. Increasing growth temperatures lead to an increase in s, which is analogous to a decrease in leaf thickness. In *Lolium*, for example, increasing growth temperatures in the range 11 – 28° C lead to such a decrease in leaf thickness (Charles-Edwards, Charles-Edwards & Sant, 1974).

Leaf to total dry-matter ratio, M_L/M_T

Eqns. (2) are readily re-arranged to give

$$\frac{M_L}{M_T} = \frac{\sigma_N f_C}{\sigma_N f_C + \sigma_C f_N} (1 - \frac{M_S)}{M_T} \qquad (15)$$

which relates the leaf to total dry-matter ratio to the specific activities of the leaves and roots, the chemical composition of the plant and the stem to total dry-matter ratio. Eqn (15) is a functional rather than a mechanistic description of the processes of dry-matter partitioning. It predicts, for example, that if all other factors remain constant an increase in specific root activity will lead to an increase in the leaf to total dry-matter ratio, whereas an increase in specific leaf activity will lead to a decrease in the ratio.

Increasing the nitrogen and phosphorus available to the grass *Phalaris tuberosa*, and hence increasing the specific root activity, has been shown to increase the leaf to total dry-matter ratio by some 30 per cent (Williams, 1946). However, in a chrysanthemum crop grown under a range of natural light levels and at different temperatures the ratio remained remarkably constant (see Table 4). It is also interesting to note that in the chrysanthemum the ratio of the flower to total dry-weight of the plant is closely reflected in the developmental stage of the flower (Cockshull & Hughes, 1967), and that over a wide range of light, CO_2 and temperature levels the flower opens fully only when the flower to total dry-matter ratio reaches 0.5.

These data serve to make the point that the partition of dry-matter between plant parts depends on a number of factors, and the detailed mechanisms of partition are not understood. Eqn. (15), however, may allow the pattern of dry-matter partition to be predicted from certain basic measurements.

Carbon Content of dry-matter, f_C

Ho (1976) has demonstrated that the carbon content of tomato leaves declines with leaf age, so that leaves from different positions within an established canopy have different carbon contents. The younger leaves at the top of the canopy have higher values for f_C than older leaves near to the base of the canopy. In a growing canopy, where there is constant introduction of new leaves at the top and loss of old leaves from the bottom, it seems reasonable to suppose that the carbon content averaged over all leaves will remain fairly constant with time. This is found to be the case with a chrysanthemum crop growing from a leaf area index of 2 to one of about 8, although it may not be true when growth is determined and the crop flowers, since the age distribution of leaves within the canopy will change with time. In the chrysanthemum, the carbon content of dry-matter changes with the growth light level and temperature environment of the crop. It can be seen in Table 4 that as the light level increases there is an increase in f_C, but that f_C declines with increasing growth temperatures. These data suggest that changes in either the products of carbon metabolism or in the amounts of minerals acquired by the crop occur with changes in the crop environment. Similar changes in the carbon contents of tomato leaves are observed in response to changes in the growth light level (Ho, personal communication).

Table 4. Effect of growth temperature and light levels on dry-matter partitioning in chrysanthemum (Charles-Edwards & Acock, unpublished data).

Daily light integral MJ_m^{-2}	Growth temperature °C	$\triangle M_L/\triangle M_T$ × 100	f_C
1.0	20	45.8 ± 3.1	0.35
2.1	20	47.4 ± 2.1	0.37
2.7	20	46.3 ± 1.2	0.37
4.3	20	46.2 ± 1.0	0.38
4.6	20	47.0 ± 0.5	–
4.9	20	49.3 ± 2.3	0.42
5.3	20	46.0 ± 0.7	0.41
2.7	10	45.3 ± 1.5	0.39
2.7	15	45.2 ± 1.3	0.38
2.7	20	46.3 ± 1.3	0.37
2.7	25	46.4 ± 0.9	0.36
2.7	30	49.3 ± 0.9	0.36

CROP GROWTH RATE

In order to simplify the analysis we will consider the effects of environment on the growth of a closed crop under a constant light flux density, I_o, during a light period of t_D hours. Further, we will assume that the respiratory loss of CO_2 by the whole crop is directly proportional to the gross amount of CO_2 assimilated by the crop during the light period. From eqns. (2), (3) and (4) it then follows that

$$\frac{\Delta M_T}{\Delta t} = \frac{0.273\ Y t_D \alpha_m I_o h\ (\tau_o' C - \beta_o' O)\ (1 - \exp(-KL))}{f_C\ (1-m)\ (\alpha_m I_o + \tau_o' C)} \tag{16}$$

where Y is a yield factor allowing for respiratory loss of CO_2 by the crop. Replacing h by $1/\rho s$, and L by sM_L gives

$$\frac{\Delta M_T}{\Delta t} = \frac{0.273\ Y t_D \alpha_m I_o\ (\tau_o' C - \beta_o' O)\ (1 - \exp(-KsM_L))}{f_C\ (1-m)\ (s\rho \alpha_m I_o + \tau_o' C)} \tag{17}$$

which describes the crop growth rate in terms of intrinsic leaf photosynthetic activity and dry-matter partitioning.

From the data presented in the previous sections it is apparent that the growth light environment of the crop will affect its growth rate primarily through effects on s and f_C, that is through effects on dry-matter partitioning. The growth temperature will affect growth rate through effects on s, f_C, τ'_o and β'_o although the data suggest that at low growth light levels the effects of temperature will be primarily on s, the specific leaf area.

CONCLUSIONS

Crop growth rate can be analyzed in terms of the intrinsic photosynthetic activity of leaves and dry-matter partitioning both within leaves and between leaves and other plant parts. Eqns. (16) and (17) explicitly allow for the intra-canopy variation in leaf photosynthetic characteristics, and, combined with eqn. (14) and some similar description of the changes in τ'_o and β'_o with the growth light and temperature level experienced by the crop, they can be used to describe the seasonal adaptation of crop growth rate.

The brief survey of published data presented in the earlier sections indicate that the aerial environment of the crop affects its growth rate in several ways, but that the primary effects are through dry-matter partitioning. Indeed, the data suggest that leaf intrinsic photosynthetic activities, approximated by the rate of light saturated photosynthesis per unit leaf volume, are not affected by the light environment under which the leaf develops, and not particularly markedly affected by the growth temperature over the range 10 – 30° C. Specific leaf area or leaf thickness, both indicators of the size of the assimilatory system beneath unit leaf area, are, however, markedly affected by the growth aerial environment.

MAIN SYMBOLS

M	dry weight	g
(suffixes L, R, S and T denote leaf, root, stem and total)		
t	time	
σ_C	specific leaf activity	$g(C)g^{-1}\ d^{-1}$
σ_N	specific root activity	$g(N)g^{-1}\ d^{-1}$
f_C	carbon/dry-matter ratio	$g(C)g^{-1}$
f_N	nitrogen/dry-matter ratio	$g(N)g^{-1}\ d^{-1}$
L	leaf area index	
F	photosynthesis rate per unit leaf area	$g(CO_2)m^{-2}\ s^{-1}$
P	photosynthesis rate per unit leaf volume	$g(CO_2)m^{-3}\ s^{-1}$
υ	photosynthesis rate per unit leaf dry weight	$g(CO_2)g^{-1}\ s^{-1}$
υ^*	photosynthesis rate per unit leaf fresh weight	$g(CO_2)g^{-1}\ s^{-1}$
(suffix m denotes measurement at light-saturation)		
a_m	leaf light utilization efficiency	$g\ J^{-1}$
τ	leaf carboxylation conductance	$m\ s^{-1}$
β	leaf oxygenation conductance	$m\ s^{-1}\ g(CO_2)g(O_2)^{-1}$
h	leaf thickness	
τ'_o	leaf carboxylation constant at full expansion	s^{-1}
β'_o	leaf oxygenation constant at full expansion	$s^{-1}\ g(CO_2)g(O_2)^{-1}$
τ'	leaf carboxylation constant after full expansion	s^{-1}
β'	leaf oxygenation constant after full expansion	$s^{-1}\ g(CO_2)g(O_2)^{-1}$
ρ	dry matter density of leaf tissue	$g\ m^{-3}$
I	light flux density on leaf surface	$W\ m^{-2}$

I_0	light flux density incident at top of canopy	W m^{-2}
H	average daily light integral	MJ m^{-2}
C	ambient CO_2 concentration	$g(CO_2)\ m^{-3}$
O	ambient O_2 concentration	$g(O_2)\ m^{-3}$
$\bar{T}$	average growth temperature	°C
s	specific leaf area	$m^2\ g^{-1}$
K	canopy light extinction coefficient	
m	leaf light transmission coefficient	

REFERENCES

Acock, B., D.W. Hand, J.H.M. Thornley & J. Warren Wilson. Photosynthesis in stands of green peppers. An application of empirical and mechanistic models to controlled environment data. *Ann. Bot.* 40: *1293–1307* (1976).

Acock, B., D.A. Charles-Edwards, D.J. Fitter, D.W. Hand, L.J. Ludwig, J. Warren Wilson & A.C. Withers. The contribution of leaves from different levels within a tomato crop to canopy net photosynthesis: An experimental examination of two canopy models. *J. exp. Bot.,* in press.

Catsky, J., I. Ticha & J. Solarova. Ontogenetic changes in the internal limitations to bean-leaf photosynthesis. 1. Carbon dioxide exchange and conductances for carbon dioxide transfer. *Photosynthetica*, 10: *394–402* (1976).

Charles-Edwards, D.A., J. Charles-Edwards & F.I. Sant. Leaf photosynthetic activity in six temperate grass varieties grown in contrasting light and temperature environments. *J. exp. Bot.*, 25: *715–724* (1974).

Charles-Edwards, D.A. & L.J. Ludwig. The basis of expression of leaf photosynthetic activities. pp. 37–44, in: Marcelle, R. (ed.), Environmental and Biological Control of Photosynthesis. Junk, The Hague (1975).

Charles-Edwards, D.A. Shoot and root activities during steady-state plant growth. *Ann. Bot.* 40: *767–772* (1976).

Charles-Edwards, D.A. 1978a. An analysis of the photosynthesis and productivity of vegetative crops in the United Kingdom. *Ann. Bot.* 42: *717–731* (1978a).

Charles-Edwards, D.A. Leaf carbon dioxide compensation points at high light flux densities. *Ann. Bot.* 42: *733–739* (1978b).

Cockshull, K.E. & A.P. Hughes. Distribution of dry-matter to flowers in *Chrysanthemum morifolium. Nature* 215: *780–781* (1967).

Criswell, J.G. & R.M. Shibles. Physiological basis for genotype variation in net photosynthesis of oat leaves. *Crop Sci.* 11: *550–553* (1971).

Dornhoff, G.M. & R.M. Shibles. Varietal differences in net photosynthesis of soyabean leaves. *Crop Sci.* 10: *42–45* (1970).

Ho, L.C. Variation in the carbon/dry-matter ratio in plant material. *Ann. Bot.* 40: *163–165* (1976).

Hozumi K., H. Kirita & M. Nishioka. Estimation of canopy photosynthesis and its seasonal change in a warm-temperate evergreen oak forest at Minamata (Japan). *Photosynthetica* 6: *158–168* (1972).

Kasanaga, H. & M. Monsi. On the light-transmission of leaves and its meaning for the production of matter in plant communities. *Jap. J. Bot.* 14; *304–324* (1954).

Louwerse, W. & W.v.d. Zweerde. Photosynthesis, transpiration and leaf morphology of *Phaseolus vulgaris* and *Zea mays* grown at different irradiances in artificial and sunlight. *Photosynthetica* 11: *11–21* (1977).

Mokronosova, A.T. & G.F. Nekrasova. Ontogenetic aspects of photosynthesis (on the example of the potato leaf). *Fiz. Rastenii* 24: *458–465* (1977).

Nobel, P.S. Internal leaf area and cellular CO_2 resistance: photosynthetic implications of variations with growth conditions and plant species. *Physiol. plant.* 40: *137–144* (1977).

Patterson, D.T. S.O. Duke, & R.E. Hoagland. Effects of irradiance during growth on adaptive photosynthetic characteristics of velvetleaf and cotton. *Plant physiol.* 61: *402–405* (1978).

Pearce, R.B., G.E. Carlson, D.K. Barnes, R.H. Hart & C.H. Hanson. Specific leaf weight and photosynthesis in alfalfa. *Crop Sci.* 9: *16–23* (1969).
Saeki. T. Interrelationship between leaf amount, light distribution and total photosynthesis in a plant community. *Bot. Mag. Tokyo* 73: *55–63* (1960).
Sheehy, J.E. Some optical properties of leaves of eight temperate forage grasses. *Ann. Bot.* 39: *377–386*, (1975).
Thornley, J.H.M. Root: shoot interactions. In Jennings, D.H. (ed.), Integration of Activity in the Higher Plant. The University Press, Cambridge (1977).
Williams, R.F. The physiology of plant growth with special reference to the concept of net assimilation rate. *Ann. Bot.* 10: *41–72* (1946).
Wolf, D.D. & R.E. Blaser. Growth rate and physiology of alfalfa as influenced by canopy and light. *Crop Sci.* 12: *23–26* (1972).

Author's address
Glasshouse Crops Research Institute
Littlehampton
W. Sussex. BN16 3PU
United Kingdom.

DAILY CYCLES OF PHOTOSYNTHESIS, RESPIRATION AND TRANSLOCATION

CRAIG J. PEARSON
(Sydney, Australia)

ABSTRACT

Carbon acquisition and distribution are dynamic processes which change throughout a day in plants growing within chambers having square-wave irradiance patterns. Maximum rates of photosynthesis and photosynthate translocation are proscribed during leaf development and are usually achieved in the first and second half of the light period respectively. Periods of lower activity coincide with low levels of photosynthetic intermediates, as during photosynthetic induction and slow translocation early in the light period, or they are unexplained but perhaps related to ion availability, as towards the end of the light period. Trends in stomatal conductance, which may vary between leaf surfaces, also contribute to the dynamics of photosynthesis. Daily cycles of root respiration fall into two categories: those having small amplitude, and those having a complex pattern with two peaks as in festucoid grasses. Dry weight data suggest that photosynthate entry into the root, the process which is assumed to control root respiration, is greatest in the middle of the light period. Daily cycles of photosynthesis, respiration and translocation probably underlie the patterns imposed by current environment on crops which are grown in glasshouse and field.

INTRODUCTION

Carbon fixed during photosynthesis is sequestered in the chloroplast, transported to cytoplasm or vacuole or translocated out of the cell. A proportion, 20 to 50% of photosynthesis for a crop canopy, is respired (McCree & Troughton, 1966; King & Evans, 1967; Heichel, 1971; Pearson & Hunt, 1972; Robson, 1973; Minchin & Pate, 1974; Penning de Vries, 1975). In contrast to the biochemistry which underlies these processes, the daily dynamics of photosynthesis, respiration and translocation receive little attention. Yet, if we can describe and understand such daily dynamics we may mechanistically predict whole-plant growth and development and understand the sensitivity of development to non-catastrophic environmental fluctuations.

I will deal firstly with the daily cycling of rates of photosynthesis, respiration and photosynthate translocation in controlled environments, that is in chambers having square-wave irradiance patterns. Here daily carbon dioxide budgets predict dry-weight increments to within 10% error (Heichel, 1972; Pearson & Hunt, 1972). Here also the elaboration of dry-weight is not uniform throughout the day: it has a peak in the middle of the light period (Fig. 1). Later I will touch on cycling in field environments, where there is circumstantial evidence which suggests residual cycles of photosynthetic capability but a dearth of data which are not site-specific. Throughout, I deal only with C3 and C4 angiosperms, believing that we know little enough about the development of these agricultural species without adding the complexities of flexible crassulacean acid metabolism.

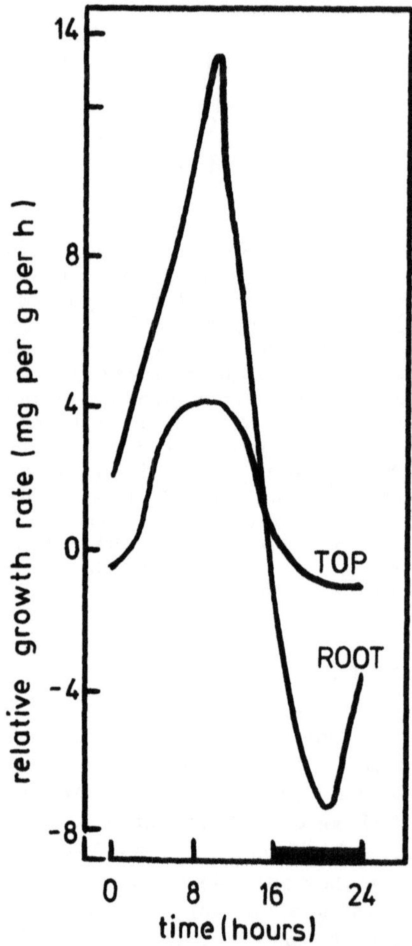

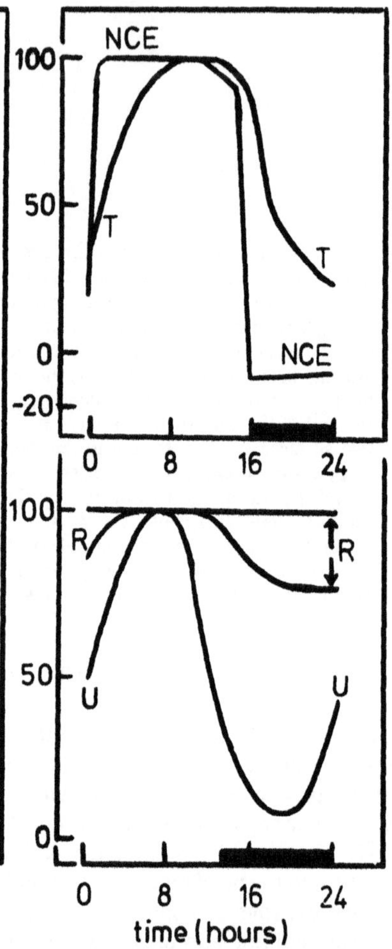

Fig 1. (left): Daily cycles of dry-weight increment in tops and roots of *Zea mays* XL45. Seedlings were grown at 22° C in a 16hr light period with a constant irradiance of 360 μE $m^{-2}s^{-1}$. Abscissa shows hours from start of light period, and duration of dark as a thick line. Adapted from data of Pearson, Volk & Jackson (unpublished), North Carolina 1978.

Fig 2. (right): Stylized daily cycles of net carbon dioxide exchange (NCE) and photosynthate translocation (T) in leaves, and of respiration (R) and nitrate uptake (U) in roots, of dicotyledons.
All values are expressed as percentage of maximum rates. Adapted from Pearson & Hunt (1972), Pearson (1974), Pearson & Steer (1977).

The cycling of leaf net photosynthetic rates is shown in Fig. 2. First, there is an induction period during which rates increase for 0 to 2 hours and up to 4 hours after the start of the light period, then a relatively constant plateau when rates are within 10% of the maximum. In this plateau period there may be oscillations having a periodicity of anything from 3 minutes to 3 hours (*e.g.*, Troughton, 1969; Hopmans, 1971; Cowan, 1972; Pearson & Hunt, 1972). The oscillations are common above 25° C when transpiration is high and transient water deficits are likely to occur. There is later a decline in net photosynthetic rates which precedes the end of the light period (Hopkinson, 1964; Pharis *et al.*, 1967; Barta & Hodges, 1970; Flech-Gerndt, 1971; Osman & Milthorpe, 1971; Pearson & Hunt, 1972; Upmeyer *et al.*, 1973; Pallas *et al.*, 1974; Pearson, 1974; Mahon *et al.*, 1976).

The photosynthetic induction period coincides with a time of low but increasing gas-phase and liquid-phase conductances. The liquid-phase or mesophyll conductance dominates the rate of increase in net photosynthesis (Osman & Milthorpe, 1972; Pearson, 1974). This is consistent with the idea that light-induction of photosynthesis in chloroplasts results from a need to accumulate intermediates of the Calvin cycle to steady state working concentrations (Walker, 1976). However, we should not be too hasty to generalize about the relative unimportance of stomata during induction. There are species in which photosynthetic induction has not been measured but in which stomatal conductance is a controlling component of whole-leaf conductance (Downes, 1970; Ludlow & Wilson, 1971; McPherson & Slatyer, 1973). In one such species, *Pennisetum americanum*, the conductance of the lower surface (which contains most stomata) may take several hours in relatively low light to reach its peak value whereas the upper surface stomata respond very rapidly to lights-on (Table 1). Moreover, it now appears that stomata are linked metabolically to the underlying mesophyll (Dittrich & Ruschke, 1977).

Table 1. Daily changes in gas-phase conductance to water vapor transfer from leaf surfaces of pearl millet (*Pennisetum americanum*) Tift 23B.* Values x 10^{-3} are cm s^{-1} from 5 plants.

	Hours after start of light period					
	0.5	3.5	7.0	10.0	12.0	LSD
Upper surface	44.3	33.3	31.8	35.6	39.9	6.28
Lower surface	20.1	38.7	58.6	64.4	54.0	14.1
Upper ÷ Lower	2.2	0.93	0.54	0.55	0.75	0.316

*Plants were grown at 22°C in a 14h light period with a constant irradiance of 360 $\mu E\ m^{-2}\ s^{-1}$.

The plateau photosynthetic rate is a product of both stomatal aperture and internal factors such as leaf architecture and ion nutrition. These are proscribed by genotype and by environment during leaf development (see other chapters of this symposium; also Table 2). The plateau rate is, however, generally quite sensitive to short-term preconditioning (*e.g.*, Kishitani & Tsunoda, 1974;

Pearson, 1974) and to manipulation of other plant parts (*e.g.*, King *et al.*, 1967; Rawson *et al.*, 1976). Exceptions exist, such as *Manihot esculenta* (cassava) where the plateau rate at saturating irradiance is low (70 ng CO_2 cm^{-2} s^{-1}; Mahon *et al.*, 1977) and stomatal conductance is insensitive to gross manipulation of photosynthate source/sink balance: conductance did not change significantly when 95% of other leaves were rolled-up on plants in the field (Pearson & Kriedemann, unpublished data, Indonesia 1977). Rhythmic oscillations in photosynthetic rate during the plateau period are related to stomatal attempts to maintain leaf turgor (Cowan, 1972; Shirazi *et al.*, 1976 and references therein) although cycles in non-stomatal conductance are reported in *Citrus sinensis* (Bravdo, 1977).

The decline in net photosynthetic rate before the end of the light period, which may cause an appreciable loss in potential productivity (Mahon *et al.*, 1976), remains unexplained. The decline is more pronounced at high temperatures (Pharis *et al.*, 1967; Pearson & Hunt, 1972), which suggests stomatal fatigue. However, it is also related in part to non-stomatal conductance (Upmeyer *et al.*, 1973). One explanation for the decline in net photosynthetic rate is that the affinity for oxygen of RuDP carboxylase/oxygenase might decrease due to the sharp decline in uptake of inorganic ions towards the end of the light period (Fig. 2). This is consistent with potassium deficiency depressing photosynthesis and stimulating oxygen uptake in *Phaseolus vulgaris* (Ozbun *et al.*, 1965). Another possibility is that the decline is correlated with an increase in respiration. Either hypothesis could explain, or be explained by, the daily cycles in energy charge and NADPH/NADP (Wagner *et al.*, 1974; Wagner & Frosch, 1974; *cf.* Bomsel & Pradet, 1968) and in epidermal ATPase (Raghavendra *et al.*, 1976).

RESPIRATION

Daily cycles of leaf or canopy respiration are of course masked by photosynthesis. There are therefore few, mostly indirect, measurements of respiration during the light period. Total respiration, estimated from the carbon dioxide compensation concentration 7, follows a rhythmic cycle with a trough in the middle of each light period in *Arachis hypogaea* (peanut) (Pallas *et al.*, 1974). The light-independent component of total respiration has a rate which is much the same as that of respiration in the dark according to isotope efflux (Webb & Zaerr, 1976) and enzyme (Chapman & Graham, 1974) analyses. It may remain constant (Hardwick *et al.*, 1968) or decrease (Pallas *et al.*, 1974) throughout the light period. Light-dependent respiration may therefore increase towards the end of the light period (Pallas *et al.*, 1974) and it contributes to a post-illumination 'outburst' and ceases shortly after the start of the dark period (Jackson & Volk, 1970). Respiration of leaves and canopies declines throughout the dark period (*e.g.*, Hopkinson, 1964; Iyama *et al.*, 1964; Pearson, 1974; Rawson & Hackett, 1974) although it may, less commonly, increase by up to 60% during 9 hour dark periods (Pearson & Hunt, 1972; Pallas *et al.*, 1974; Mahon *et al.*, 1976). Such divergent trends with time and variable rates are partially explained by differences in prior photosynthetic rates and in photosynthate levels in the respiring tissues (Fig. 3).

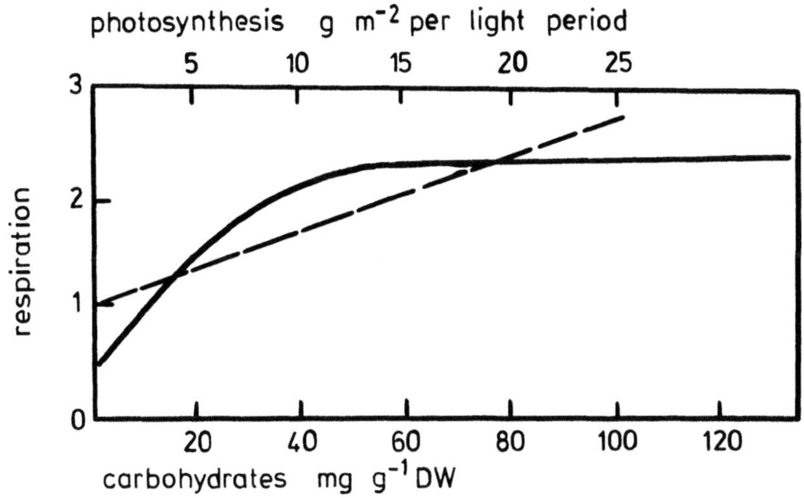

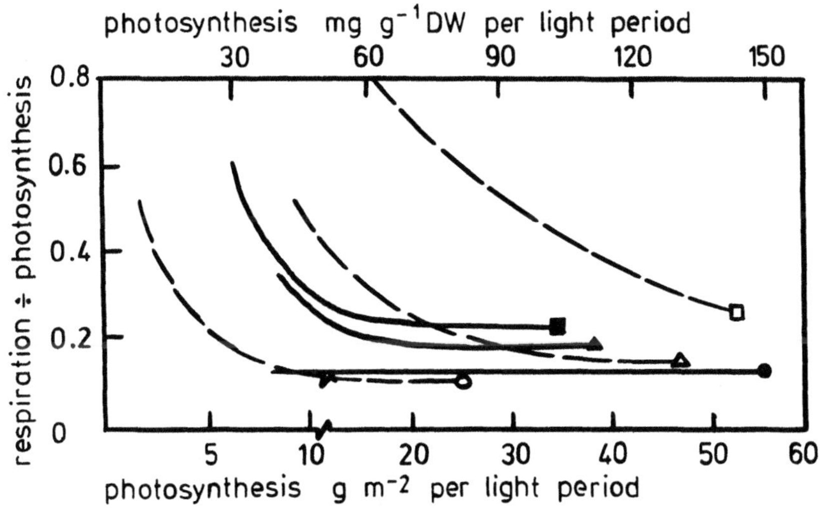

Fig 3. (upper): Dark respiration increases with increasing prior net photosynthesis and increasing soluble carbohydrate levels.
Dashed line shows a linear relationship between leaf respiration (g m^{-2} d^{-1}) and prior net photosynthesis (same units, upper abscissa). Solid line shows a curvilinear correlation between shoot respiration (mg g^{-1} h^{-1}) and total soluble carbohydrates (lower abscissa). Redrawn from Ludwig *et al.* (1974), Cunningham & Syvertsen (1977).

Fig 4. (lower): Dark respiration as a proportion of prior net photosynthesis decreases with increases in net photosynthesis brought about by changing irradiance (dashed lines; net photosynthesis on lower abscissa) or by changing irradiance and soluble carbohydrate levels (solid lines; upper abscissa). Irradiance effects are shown for leaves (○), shoots (■) and canopies in controlled environments (□) and for canopies in the field (△). Irradiance effects are the same but the respiration/net photosynthesis ratio increases as soluble carbohydrates are elevated. Calculated from data of: ○ Ludwig *et al.* (1974), ■ Cunningham & Syvertsen (1977), □ King & Evans (1967), △Koh & Kumura (1975).

Fruit respiration has a daily rhythm (*e.g.* Walker & Ho, 1977). This presumably reflects daily cycles of translocation to the fruit and rhythmic photosynthetic capacity. Fruit photosynthesis may contribute at least 10% of the carbon gained by the developing fruit (Steer & Pearson, 1976).

Daily cycles in root respiration are of two types. A small daily trend with a long plateau during the light period is found in dicotyledons (Fig. 2): in non-nodulated and nodulated legumes (Pearson & Hunt, 1972; Minchin *et al.*, 1977), cassava (Mahon *et al.*, 1976) and by inference in tobacco (where mean rates were the same during light and dark periods: Rawson & Hackett, 1974). A rather complex pattern having two peaks at 3–6 and 12–16 hours after the start of the light period is reported for festucoid grasses (Neales & Davies, 1966; Hansen, 1977) and *Zea mays* (Brouwer, 1977). In the simpler pattern much but usually not all of the daily trend is removed when account is taken of changes in root temperature. The lack of uniformity in daily trends reflects technical difficulties as well as generic differences. Respiration rates per unit root weight remain constant during seedling growth (Brouwer, 1977) but they decrease as plants age and non-functional biomass increases (Schiebe & Meyer zu Drewer, 1969; Pearson & Hunt, 1972). Apparent rates are very affected by solution pH, microbial contamination and respiration of detached roots; early long-term studies which did not check these variables (Huck *et al.*, 1962) may be misleading. Furthermore, there may be appreciable loss of organic compounds from roots when they are growing in soil (Martin, 1977) which we should attempt to measure along with respiration in future carbon budgets.

Root respiration depends on the downflow of photosynthate into the root, which itself may be far from constant throughout a day (Fig. 1). In support of this widely-accepted hypothesis, root oxygen consumption is linearly related to root dry weight increments in *Helianthus annuus* (Hatrick & Bowling, 1973); root respiration adapts to changes in photosynthetic rate within 10 hours (*e.g.*, Hansen, 1977); and phloem excision eliminates the metabolic component of root cell transmembrane potential within 5 hours (Graham & Bowling, 1977). However, it remains to be shown how tight the coupling is between photosynthate downflow and root respiration. Daily cycles of photosynthate entry into the root have not been directly measured and Fig. 1 represents the only indirect attempt to my knowledge. Here, the peak in root growth in the middle of the light period occurs during the probable peak in photosynthate export from leaves (Fig. 1 *cf.* Fig. 2). This requires further study; our knowledge of the daily entry of photosynthate and its regulatory effects on root metabolism is skimpy compared to information about metabolic pathways within roots (Fowler, 1975).

TRANSLOCATION

The translocation out of the leaf of recently-fixed carbon, distinguished by isotope labelling from that carbon fixed earlier or subsequently, exhibits an exponential decline with time. The decline is usually subdivided into two linear phases the rates of which are faster in light than in darkness (Hartt & Kortschak, 1967; Habeshaw, 1969; Plaut & Reinhold, 1969; Pearson, 1974; Moorby *et al.*, 1974; Troughton *et al.*, 1977; *cf.* Wardlaw & Moncur, 1976). The daily cycle of translocation in Fig. 2 was derived from estimates of the rate of ^{14}C export after exposure at four times of day coupled with continuous measurement of the net photosynthetic rate (Pearson, 1974). In the species studied, *Vicia faba*, there was a linear increase in translocation for about 4 hours after the start of the light period, when much of the photosynthate being exported was that which had been fixed into low-turnover substances in the preceding day. This was followed by rapid translocation during most of the day and a very steep decline at the onset of the dark period. Translocation in the dark depends on mobilization of substances whose turnover times are low (Moorby & Jarman, 1975; see also Ho, 1976, for an elegant demonstration that speed of export can be maintained by starch degradation when photosynthesis is very low). There is a need to run daily analyses of translocation in a range of species: ^{14}C export in *Capsicum annuum* at different times during the light and dark periods (Steer & Pearson, unpublished) generally fit the pattern of *Vicia faba* whereas in *Allium cepa*, which does not store starch, there is a more simple light/dark effect (Steer & Darbyshire, unpublished).

Maximum rates of ^{14}C export vary markedly between (*e.g.*, Hofstra & Nelson, 1969; Lush & Evans, 1974) and within (Liu *et al.*, 1973; Pearson & Derrick, 1977) species. Rates in the middle of the light period, like photosynthetic rates, depend on the structure and metabolic capacity of the leaf, which in turn are largely determined by the environment in which the leaf has developed. Thus, in *Pennisetum americanum* rates may be very fast or they may be indistinguishable from those of temperate species depending on the temperature during leaf development (Pearson & Derrick, 1977) and the development phase can have more impact on the rates than does the current environment (Table 2).

Table 2. Translocation of ^{14}C–photosynthate, and photosynthetic rates of pearl millet (*Pennisetum americanum*) Tift 71 × Tift 23DB.*

	Temperature during development		
	21/16	33/28	LSD
Rate of ^{14}C loss during first phase % per hr	13.1	45.6	12.2
% of ^{14}C remaining after 20h	14.3	8.7	2.70
Photosynthetic rate ng CO_2 cm^{-2} s^{-1}	21.2	70.9	13.8

*During leaf development the plants were grown in glasshouses at 21/16 and 33/28°C (day/night). Both sets of plants were then grown at 33/28°C for 14 days prior to and during the measurements. From Fussell & Pearson, 1978.

What regulates the daily cycle of translocation? It must reflect many processes in the leaf such as the rates of entry of more recently fixed photosynthate and breakdown of polysaccharides into labile pools, changes in membrane permeability and/or availability of ATP for intracellular transport and vein loading and of ions for enzyme and macromolecule synthesis, together with the rate of utilization of photosynthate in 'sink' organs. Generally the pattern of ^{14}C

or ^{11}C translocation does not follow closely the accumulation of sucrose within the leaf (Pearson, 1974; Pearson & Derrick, 1977; Troughton *et al.* 1977; *cf.* Christy & Swanson, 1976). However, over longer time periods, translocation is in balance with the sucrose gradient between source and 'sink' organs (Ho, 1976; Walker & Ho, 1977a) and it may be manipulated by gross changes in sucrose gradients (Heyser *et al.*, 1975; 1977). The time of maximum photosynthate translocation coincides with that of maximum nitrate uptake, at least in *Capsicum* (Fig. 2): this could reflect the requirement of nitrate uptake for labile photosynthate in the root (Pearson & Steer, 1977) or, perhaps more likely, the dependence of both processes on adenosine phosphate levels (Steer, in this symposium).

DAILY CYCLES IN THE FIELD

There is a circumstantial evidence for a rhythm underlying the daily course of photosynthesis in the field. Data of Chmora (1967) show that the irradiance-response curves of *Zea mays* change with time of day in a way consistent with the crop having an induction period early in the light and a depression period before the start of the dark. I am not aware of anyone else collecting the same kind of data as did Chmora. Nonetheless, the daily course of canopy photosynthesis in the field has been documented by many workers and it does often show marked deviations from the expected single curvilinear correlation with irradiance (*e.g.*, Iwaki *et al.*, 1976). These deviations are sometimes explained by large daily changes in temperature (Takeda, 1976) or by the mid-day depression of photosynthesis. The latter has some parallel with the self-sustaining cycles of small amplitude in controlled environments in that it is associated with transient water deficits and increases in mesophyll temperature which give rise to decreases in both gas-phase and liquid-phase conductance (Mederski *et al.*, 1975). Fluctuations in respiration in the field are positively correlated with changes in net photosynthesis: the ratio of respiration to net photosynthesis decreases with increasing daily irradiance the same way as in a controlled environment (Fig. 4).

Photosynthate translocation in the field has a pattern similar to that in *Vicia faba* in a controlled environment, in that rates of photosynthate flow are fastest in the middle of the day (McNairn, 1972; Sharkey & Pate, 1976) unless there is a mid-day depression in the net photosynthetic rate (Russell & Johnson, 1975). Mid-day water deficits may reduce translocation through reducing the photosynthetic rate (Munns & Pearson, 1974) or by directly reducing vein loading of photosynthate (Wardlaw, 1969). Taking environmental fluctuations aside, field translocation data are consistent with the possibility of underlying cycles which are set up during and constrained by organ development.

REFERENCES

Barta, A.L. & H.F. Hodges. Characterization of photosynthesis in cold hardening winter wheat. *Crop Sci.* 10: *535–538* (1970).

Bomsel, J.L. & A. Pradet. Study of adenosine 5-mono, di- and triphosphates in plant tissues IV. Regulation of the level of nucleotides *in vivo* by adenylate kinase: theoretical and experimental study. *Biochim. Biophys. Acta* 162: *230–242* (1968).

Bravdo, B.A. Oscillatory transpiration and CO_2 exchange of citrus leaves at the CO_2 compensation concentration. *Physiol. Plant.* 41: *36–41* (1977).
Brouwer, R. Root functioning. pp 229–245, in: Landsberg, J. and C.V. Cutting (eds.) Environmental effects on crop physiology. Academic Press, London, New York (1977).
Chapman, E.A. & D. Graham. The effect of light on the tricarboxylic acid cycle in green leaves. 1. Relative rates of the cycle in the dark and light. *Plant Physiol.* 53: *879–885* (1974).
Chmora, S.N. Photosynthetic light curves for a maize stand. pp. 114–119 in: Nichoporovich, A.A. (ed.), Photosynthesis of Productive Systems. Israel Program Sci. Transl., Jerusalem (1967).
Christy, A.L. & C.A. Swanson. Control of translocation by photosynthesis and carbohydrate concentrations of the source leaf. pp. 329–338, in: Wardlaw, I.F. & J.B. Passioura (eds.), Transport and Transfer Processes in Plants. Academic Press, New York, London (1976).
Cowan, I.R. Oscillation in stomatal conductance and plant functioning associated with stomatal conductance: Observations and a model. *Planta* (Berl.), 106: *185–219* (1972).
Cunningham, G.L. & J.P. Syvertsen. The effect of nonstructural carbohydrate levels on dark CO_2 release in creosotebush. *Photosynthetica* 11, *291–295* (1977).
Dittrich, P. & K. Raschke. Uptake and metabolism of carbohydrates by epidermal tissue. *Planta* (Berl.), 134: *83–90* (1977).
Downes, R.W. Effect of light intensity and leaf temperature on photosynthesis and transpiration in wheat and sorghum. *Aust. J. Biol. Sci.* 23: *775–782* (1970).
Flech-Gernatt, Gerlunde von. Investigations on photosynthesis of seneacing leaves 1. Relations between chlorophyll content, nitrogen and intensity of assimilation. *Biol. Zentralblatt.* 90: *479–506* (1971).
Fowler, M.W. Carbohydrate metabolism and differentiation in seedling roots. *New Phytol.* 75: *461–478* (1975).
Fussell, L.K. & C.J. Pearson. Effect of thermal history on photosynthate translocation and photosynthesis. *Austral. J. Plant Physiol.* 5: *547–551* (1978).
Habeshaw, D. The effect of light on the translocation from sugar beet leaves. *J. Expt. Bot.* 20: *64–71* (1969).
Hansen, G.K. Adaptation to photosynthesis and diurnal oscillation of root respiration rates for *Lolium multiflorum*. *Plant Physiol.* 39: *275–279* (1977).
Hartt, C.E. & H.P. Kortschak. Translocation of ^{14}C in the sugar-cane plant during the day and night. *Plant Physiol.* 42: *89–91* (1967).
Hatrick, A.A. & D.J.F. Bowling. A study of the relationship between root and shoot metabolism. *J. Expt. Bot.* 24: *607–613* (1973).
Heichel, G.H. Response of respiration of tobacco leaves in light and darkness and the CO_2 compensation concentration to prior illumination and oxygen. *Plant Physiol.* 48: *178–182* (1971).
Heichel, G.H. Confirming measurements of respiration and photosynthesis with dry matter accumulation. Photosynthetica 5: *93–98* (1972).
Heyser, W., O. Leonard, R. Heyer, E. Fritz & W. Eschrich. The influence of light, darkness and lack of CO_2 on phloem translocation in detached maize leaves. *Planta* (Berl.), 122: *143–154* (1975).
Heyser, W., R. Heyser, W. Eschrich & E. Fritz. The influence of externally supplied sucrose on phloem transport in the maize leaf strip. *Planta* (Berl.). 137: *145–151* (1977).
Ho., L.C. The relationship between the rates of carbon transport and photosynthesis in tomato leaves. *J. Exp. Bot.* 27: *87–97* (1976).
Ho., L.C. & A.F. Shaw. Carbon economy and translocation of ^{14}C in leaflets of the seventh leaf of tomato during leaf expansion. *J. Expt. Bot.* 41: *833–848* (1977).
Hofstra, G. & C.J. Nelson. A comparative study of translocation of assimilated ^{14}C from leaves of different species. *Planta* (Berl.) 88: *103–112* (1969).
Hopkinson, J.M. Studies on the expansion of the leaf surface IV. The carbon and phosphorus economy of the leaf. *J. Expt. Bot.* 15: *125–137* (1964).

Hopmans, P.A.M. Rhythms in stomatal opening in bean leaves. Meded. Landbouwhogeschool Wageningen: *71—73* (86p.) (1971).
Huck, M.G., R.H. Hageman & J.B. Hansen. Diurnal variation in root respiration. *Plant Physiol.*, 37: *371–375* (1962).
Iwaki, H., G. Takeda & U.T. Udagawa. Ecological studies on the photosynthesis of winter cereals. II. Photosynthesis of wheat and rye plants under field conditions. *Proc. Crop Sci. Soc. Japan* 45: *32–39* (1976).
Iyama, J., & Y. Murata & T. Homna. Studies on the photosynthesis of forage crops. III. Influence of the different temperature levels on diurnal changes in the photosynthesis of forage crops under constant conditions. *Proc. Crop Sci. Soc. Japan* 33: *25–28* (1964).
Jackson, W.A. & R.J. Volk. Photorespiration. *Ann. Rev. Plant Physiol.* 21: *385–432* (1970).
King, R.W. & L.T. Evans. Photosynthesis in artifical communities of wheat, lucerne and subterranean clover plants. *Aust. J. Biol. Sci.* 20: *623–635* (1967).
King, R.W., I.F. Wardlaw & L.T. Evans. Effect of assimilate utilization on photosynthetic rate in wheat. *Planta* (Berl.), 77: *261–276* (1967).
Kishitami, S. & Tsunoda. Effect of low and high temperature pretreatment on leaf photosynthesis and transpiration in cultivars of *Oryza sativa. Photosynthetica* 8: *161–167* (1974).
Koh, S. & A. Kumura. Studies on dry-matter production in wheat plant II. Carbon dioxide balance and efficiency of solar energy utilization in wheat stand. *Proc. Crop Sci. Soc. Japan* 44: *335–342* (1975).
Liu, P., D.H. Wallace & J.L. Ozbun. Influence of translocation on photosynthetic efficiency of *Phaseolus vulgaris* L. *Plant Physiol.* 52: *412–415* (1973).
Ludlow, M.M. & G.L. Wilson. Photosynthesis of tropical pasture plants III. Leaf age. *Aust. J. Biol. Sci.*, 24: *1077–1087* (1971).
Ludwig, L.J., D.A. Charles-Edwards & A.C. Withers. Tomato leaf photosynthesis and respiration in various light and carbon dioxide environments. pp 29–36, in: Marcelle, R (ed.) Environment and Biological Control of Photosynthesis. Junk, The Hague (1975).
Lush, W.M. & L.T. Evans. Translocation of photosynthetic assimilate from grass leaves, as influenced by environment and species. *Aust. J. Plant Physiol.* 1: *417–431* (1974).
Martin, J.K. Factors influencing the loss of organic carbon from wheat roots. Soil Biol. Biochem. 9: *1–7* (1977).
Mahon, J.D, S.B. Lowe & L.A. Hunt. Photosynthesis and assimilate distribution in relation to yield of cassava grown in controlled environments. *Can. J. Bot.* 54: *1322–1331* (1976).
Mahon, J.D., S.B. Lowe, L.A. Hunt & M. Thingarajah. Environmental effects on photosynthesis and transpiration in attached leaves of cassava. (*Manihot esculanta* Crantz). *Photosynthetica 11: 121–130* (1977).
McCree, K.J. & J.H. Troughton. Prediction of growth rate at different light levels from measured photosynthesis and respiration rates. *Plant Physiol.* 41: *559–566* (1966).
McNairn, R.G. Phloem translocation and heat-induced callose formation in field-grown *Gossypium hirsutum L. Plant Physiol.* 50: *366–370* (1972).
Mederski, H.Q., R.B. Curry & L.H. Chen. Effects of irradiance and leaf water deficit on net carbon dioxide assimilation and mesophyll and transport processes. *Plant Physiol.* 55: *594–597* (1975).
McPherson, H.G. & R.O. Slatyer. Mechanisms regulating photosynthesis in *Pennisetum typhoides. Aust. J. Biol. Sci.* 26: *329–339* (1973).
Minchin, F.R., C.P. Neves, R.J. Summerfield & A.C. Richardson. A chamber designed for continuous, long-term monitoring of legume root respiration. *J. Expt. Bot.* 28: *507–514* (1977).
Minchin, F.R. & J.S. Pate. Diurnal functioning of the legume root nodule. *J. Expt. Bot.* 25: *295–308* (1974).
Moorby, J. & P.D. Jarman. The use of compartmental analysis in the study of the movement of carbon through leaves. *Planta* (Berl.) 122: *155–168* (1975).
Moorby, J., J.H. Troughton & B.G. Currie. Investigations of carbon transport in plants II. The effect of light and darkness and sink activity on translocation. *J. Expt. Bot.* 25: *937–944* (1974).

Munns, R. & C.J. Pearson. Effect of water deficit on translocation of carbohydrate in *Solanum tuberosum. Aust. J. Plant Physiol.* 1: *529–537* (1974).
Neales, T.F. & J.A. Davies. The effect of photoperiod duration upon the respiratory activity of the roots of wheat seedlings. *Aust. J. Biol. Sci.* 19: *471–480* (1966).
Osman, A.M. Root respiration of wheat plants as influenced by age, temperature and irradiation of shoots. *Photosynthetica* 5: *107–112* (1971).
Osman, A.M. & F.L. Milthorpe. Photosynthesis of wheat leaves in relation to age, illuminance and nutrient supply 1. Techniques. *Photosynthetica* 5: *55–60*.
Ozbun, J.L., R.J. Volk & W.A. Jackson. Effect of potassium deficiency on photosynthesis, respiration and the utilization of photosynthetic reductant by immature bean leaves. Crop Sci. 5: *69–75* (1965).
Pallas, J.E., Y.B. Samish & C.M. Wilmer. Endogenous rhythmic activity in photosynthesis, transpiration, dark respiration, and carbon dioxide compensation point of peanut leaves. *Plant Physiol.* 53: *907–911* (1974).
Pearson, C.J. Daily changes in carbon dioxide exchange and photosynthate translocation of leaves of *Vicia faba. Planta* (Berl.) 119: *59–70* (1974).
Pearson, C.J. & G.A. Derrick. Thermal adaptation of *Pennisetum:* leaf photosynthesis and photosynthate translocation. *Aust. J. Plant Physiol.* 4: *763–769* (1977).
Pearson. C.J. & L.A. Hunt. Studies on the daily course of carbon exchange in alfalfa plants. *Can. J. Bot.* 50: *1377-1384* (1972).
Pearson, C.J. & B.T. Steer. Daily changes in nitrate uptake and metabolism in *Capsicum annuum. Planta* (Berl.), 137: *107–112* (1977).
Penning de Vries, F.W.T. Use of assimilates in higher plants. pp 459–480, in: Cooper, J.P. (ed.), Photosynthesis and Productivity in Different Environments. Cambridge University Press, London (1975).
Pharis, R.P., H. Hellmers & S.E. Schuurman. Kinetics of the daily rate of photosynthesis at low temperatures for two conifers. *Plant Physiol.* 42: *525–531* (1967).
Plaut, Z. & L. Reinhold. Concommitant photosynthesis implicated in the light effect on translocation in bean plants. *Aust. J. Biol. Sci.* 22: *1105–1111* (1969).
Raghavendra, A.S., I.M. Rao, V.S.R. Das, Adenosinetriphosphatase in epidermal tissue of *Commelina benghalensis. Pl. Science Letters* 7: *391–396* (1976).
Rawson, H.M., R.M. Gifford & P.M. Bremner. Carbon dioxide exchange in relation to sink demand in wheat. *Planta* (Berl.) 132: *19–23* (1976).
Rawson, H.M. & C. Hackett. An exploration of the carbon economy of the tobacco plant III. Gas exchange in leaves in relation to position on the stem, ontogeny and N content. *Aust. J. Plant Physiol.* 1: *551–560* (1974).
Robson, M.J. The growth and development of simulated swords of perennial ryegrass II. Carbon assimilation and respiration in a seedling sward. *Ann. Bot.* 37: *501–518* (1973).
Russell, W.J. & D.R. Johnson. Translocation patterns of soybeans exposed to $^{14}CO_2$ at four different time periods of the day. *Crop Sci.* 15: *75–77* (1975).
Scheibe, A., Z. Meyer & H. Drewer. Comparative investigations of the respiration rate in roots of different genotypes in cereals. *Z. Acker – Pflanz.* 108: *225–252* (1959).
Sharkey, P.J. & J.S. Pate. Translocation from leaves to fruits of a legume, studied by a phloem bleeding technique: diurnal changes and effects of continuous darkness. *Planta* (Berl.) 128, *63–72* (1976).
Shirazi, G.A., J.F. Stone & G.W. Todd. Oscillatory transpiration in a cotton plant. I. Experimental characterization. *J. Expt. Bot.* 27: *608–618* (1976).
Steer, B.T. & C.J. Pearson. Photosynthate translocation in *Capsicum annum. Planta* (Berl.) 128: *155–162* (1976).
Takeda, G. Ecological studies in the photosynthesis of winter cereals. I. Diurnal changes in the photosynthesis of two-rowed barley under field conditions in winter. *Proc. Crop Sci. Soc. Japan* 45: *17–24* (1976).
Troughton, J.H. Plant water status and carbon dioxide exchange of cotton leaves. *Aust. J. Biol. Sci.* 22: *289–302* (1969).
Troughton, J.H., B.G. Currie & F.H. Chang. Relations between light level, sucrose concentrations and translocation of carbon 11 in *Zea mays* leaves. *Plant Phsyiol.* 59: *808–820* (1977).

Upmeyer, D.J. & H.R. Koller. Diurnal trends in net photosynthetic rate and carbohydrate levels of soybean leaves. *Plant Physiol.* 51: *871–874* (1973).
Walker, A.T. & L.C. Ho. Carbon translocation in the tomato: carbon import and fruit growth. *Ann. Bot.* 41: *813–823* (1977).
Walker, A.T. & L.C. Ho. Carbon translocation in the tomato: effects of fruit temperature on carbon metabolism and the rate of translocation. *Ann. Bot.* 41: *825–832* (1977a).
Walker, D.A. CO_2 fixation by intact chloroplasts: Photosynthetic induction and its relation to transport phenomena and control mechanisms. pp. 235–278, in: Barker, J. (ed.), The Intact Chloroplast. Elsevier, Amsterdam, New York (1976).
Wardlaw, I.F. The effect of water stress on translocation in relation to photosynthesis and growth. II. Effect during leaf development in *Lolium temulentum* L. *Aust. J. Biol. Sci.* 22, *1–16* (1969).
Wardlaw, I.F. & L. Moncur. Source, sink and hormonal control of translocation in wheat. *Planta* (Berl.) 122: *93–100* (1976).
Webb, W.L. & J.B. Zaerr. Carbon dioxide efflux of douglas-fir seedlings in light and dark. *Photosynthetica* 10: *388–893* (1976).

Author's address:
Department of Agronomy and Horticultural Science
University of Sydney
Sydney, N.S.W. 2006
Australia.

PHOTOSYNTHESIS AND FLOWERING

G. BERNIER & R.M. SACHS
(Liège, Belgium and Davis, California, U.S.A.)

The following is a relatively loose account of a 90 min. tape-recorded discussion on the general topic of the relationship, if any, between photosynthesis and flowering. In some instances the order in which comments were made, particularly where direct question and answer patterns were not evident, has been somewhat altered to give greater coherence to the report. Direct quotes of interviewers were often difficult, incomplete sentences were frequent, and many used the blackboard to illustrate points, substituting diagrams for words.

Bernier (playing the devil's advocate): Is the role of photosynthesis in plant development (flowering) a trivial one? Of course photosynthesis is required, carbohydrates are required, energy is required, for plant growth, but is there a more specific role that can be identified?

Friend: Photosynthesis, *per se*, is not required – plants flower in darkness when they are provided suitable substrates. Nor can one say that the action of high light on flowering is merely to produce more substrate, nor can one say that the action is necessarily a direct one. High photon flux densities affect directly ATP and NADH synthesis, energy charge, *etc*., all of which in turn may affect membrane associated phenomena, including release of hormones which would then have another effect on flowering. Thus, it is dangerous to conclude that high photon flux requirements for flowering are necessarily through synthesis of carbon skeletons (or even to think of carbohydrates serving as an energy source). Reid's work with the Sn gene in peas is particularly interesting. He has some data on a high energy activated gene acting as a timer. May we ask him to comment on that?

Reid: A high intensity light reaction apparently prevents the action of Sn gene in peas; this gene is responsible for the production of an inhibition of flowering while the plant is maintained in darkness. It becomes active about 2 hours after the beginning of darkness. Thus, there is an inverse relationship between the quantity of inhibitor and the number of hours of light to which the plant is exposed. The activity of a similar gene in *Brassica* would account well for the data presented earlier by Friend.

Habeshaw: Dr. Friend, in your paper this morning you made no mention of the importance of photosynthesis for translocation. I became interested in translocation of assimilates from sugar beet leaves and found that translocation itself was photosynthesis dependent, not necessarily for carbon skeleton formation but for some other light activated reactions. Thus we may not be talking about carbohydrate synthesis but transmission of carbohydrate.

Tamas: How did you distinguish between light reactions *per se* as opposed to translocation of photosynthetic products?

Habeshaw: Basically I found that even if the leaves were in CO_2-free atmosphere, translocation was still stimulated – the amount of new carbohydrate formed was insignificant.

Tamas: Translocation of what?

Habeshaw: This was ^{14}C. I was looking at bulk flow of sugar not of specific compounds.

Tamas: Did you analyze what kind of compound the ^{14}C was in?

Habeshaw: ^{14}C was largely in sucrose. I determined the specific activity of the ^{14}C sucrose and, knowing that specific activity, I was able to compute sucrose moving through the plant. Very little of the ^{14}C in the petiole was in forms other than sucrose.

Friend: In several LD plants CO_2 appears to be necessary for the effect of light which would perhaps rule out translocation effects in these cases if lack of CO_2 is preventing fixation. (Note: Subsequent exchange between Habeshaw and Friend not clear on tape.)

Bernier: The effect of light may not be only on the source but on the sink, too. Prof. Wareing can you tell us more about some of the studies that you mentioned this morning.

Wareing: There are two – Galston's studies on the effect of light on sucrose translocation in etiolated pea epicotyls (with an apical bud). Sucrose is applied to the base. If the apex is exposed to light, this greatly increases sucrose accumulation in the bud.

Halevy has shown that light is active in maintaining normal apical dominance in roses. Light is required directly on the bud for its development. If the apical bud, for example, is covered with foil, it will not develop whereas other buds below will begin growth. He can show this effect in CO_2-free air; hence, it is not likely that one is dealing with a normal effect of photosynthesis.

Lurie: I would like to invite speculation about the data that Bodson & Friend presented about the minus CO_2 effect on the induction of flowering.

Bodson: Removing CO_2 depletes carbohydrate production – that is the first thing that we measured – there is a sharp decrease in the just expanded leaf while in the terminal bud, sugar concentration first increases in the same way as in presence of CO_2.

Friend: We see the effect of minus CO_2 even when we are using far red light at levels below the compensation point. In this case the carbohydrate being added is so small that one or two starch grains ought to take care of it. Presumably we are dealing with some other effect, what that is I don't know. Presumably cyclic photophosphorylation will take place in far red light. We have CO_2 fixation as shown with $^{14}CO_2$ – where this is going I don't know – into the PEP carboxylase system or RUBP carboxylase system? Both of these systems will use up ATP, and I suspect it is altering the balance of ATP or energy charge that may be more important than any effect on carbohydrate supply, since there is enough of the latter in the plant to more than compensate for the small change caused by far red.

Tamas: Did you check what was made of that CO_2?
Friend: No.
Tamas: It is very important. If you feed $^{14}CO_2$ at low CO_2 levels very often you don't get any sucrose but products of the glycolate pathway – lot of glycine and serine – perhaps there is some effect there.
Friend: That is possible and we intend to look at the products.
Steer: In contrast to Tamas's comments, nitrate reductase (NR) activity falls very quickly, NR is a very rapid turnover protein. At low CO_2 concentrations there is quite a lot of evidence that NR levels are very severely reduced. So, if you were thinking of the importance of a supply of reduced nitrogen, the minus CO_2 effect may be through this path. That would be in contrast to Tamas's proposal on increased amino acids.
Friend: How does low CO_2 affect NR levels?
Steer: I don't know; I don't think it is very clear. The assumption, or interpretation, usually made is that protein synthesis is dependent upon recently synthesized carbon – the actual synthesis of the enzyme is reduced. But that is not very clear.
Friend: This NR has to be in the light? You do not get activity in the dark?
Steer: No, in the dark. You can extract NR from dark-grown plants but I don't think that *in vivo* much is going on in terms of NO_3 reduction.
Tamas: I think that the important point to make is that whatever the answer you must realize that in removing CO_2 you make major changes in metabolic processes. It is very dangerous to assume that you are just making less sucrose.

NOTE: There was a short discussion on the problem of generalizations made across several species, particularly when one is mainly concerned with environmental controls of flowering without information on evocation. Bernier then made the following comment:

Bernier: Perhaps a good way to cope with the complexity of the discussion, the whole problem of the initiation of flowering, would be by considering: 1) induction in the leaf, 2) floral stimulus (i) loading in the phloem, 3) transport to the target meristem, 4) unloading at the meristem, 5) evocation, and 6) morphogenesis and development of the flower. I would like to know at what part of this complex series of processes photosynthesis is active. Do you think it is important at all steps or at a particular step?
Friend: In *Brassica* and *Anagallis* we have no information on particular steps – no separation. Have you done this in *Sinapis?*
Bodson: Not really. It is very difficult to separate the action of photosynthesis on induction in the leaf from that on evocation at the meristem, because we have no independent measure of induction apart from evocation. We do not know at all what induction in the leaf means. Hence we can only compare something else going on with photosynthetic activity in the leaf. It is just a correlation, but we can't exactly decide whether what we are seeing occurred earlier in the leaf or if it is a direct effect of light at the meristem.
Wareing: I would like to suggest a generalization which you can then refute. Photosynthesis in the leaf is necessary for the synthesis of some sort of flowering promoter, particularly in SD plants. If one remembers the classical work of

Hamner where he placed plants in cycles of 3 min. incandescent light (2000 lux), 3 hrs. of darkness, *etc*., for 36 hours (Hamner, *K. Bot. Gaz. 101*: 658, 1940) and then placed plants in darkness there was no induction; he had to give a period of photosynthesis of relatively high light intensity for about 2 hours (Reporter's note – there was continued enhancement of the rate of inflorescence development with up to 8 hours of sunlight); other workers found that CO_2 was required during this period and then Bonner & co-workers (Liverman, J.L. & J. Bonner. *Bot. Gaz. 115*: 121. 1963) found that sucrose, applied to leaves or to cuttings, could replace in part this high intensity light requirement. I would suggest that as far as the leaves are concerned it may be a question of substrate levels. I think that on the other hand there are other effects of sucrose which are important at the shoot apex – and some of Dr. Bodson's data presented this morning suggest that this may be significant, although we're not sure what is cause and effect in her observations . . . but certainly the levels of sucrose are much higher in the induced plants – I would agree with Dr. Friend that the role of sucrose may not be as a substrate but as an energy source – for unloading, etc.
Bernier: I should like to add a comment, Dr. Wareing, to what you said about the flashing light experiments. Mrs. Lance-Nougarède (*Ann. Sci. Nat., Bot*., IIe série, 18: 91, 1957) happened to look at the meristems of *Xanthium* plants receiving the so-called 'Hamner' treatment and she was very surprised to find the apical meristems in a state of 'semi-dormancy,' no longer could she find mitotic activity, there were very small nucleoli, large vacuols, *etc.* They looked like more differentiated cells (according to her way of expressing such findings). So it could be that the flashing light treatment does more than simply change the carbohydrate level – and perhaps the meristems are unable to respond to other stimuli coming from the leaf.
Landsberg: There is some very good data from a man called Huet working with pears and using selected branches, showing that flowering is inversely related to fruit load during the season in which the flowers are initiated. Thus, since there are good carbohydrate supply conditions when there are few fruit, the data suggest a substrate involvement in flower initiation.
Lenz: I should like to make an opposite comment. In small, two leaved citrus plants with the roots at 12° C, there is not much growth but heavy flowering, whereas with root temperature at 28° C there is much growth but they never will flower. The leaves of the plant at 12° C are full of carbohydrates, have a low photosynthetic rate and relatively high dark respiration and relatively high photorespiration; at 22° C the plants have high photosynthetic rate, high dark respiration rate, and low photorespiration rate and relatively low concentration of carbohydrates in the leaves – because they are making a lot of shoots. By taking off the shoots one can have high carbohydrate content in the leaves of the 22° C plants and still they will not flower.
Friend: I would like to argue against an essential involvement of the immediate products of photosynthesis in accounting for the flowering behavior of *Pharbitis*. One can grow *Pharbitis* for about 4 days in darkness and obtain a miserable etiolated plant with folded cotyledons. If you now give 5 seconds red light at a phytochrome light intensity (this is hardly a photosynthesis effect), the photomorphogenetic effects will occur, *i.e.*, the cotyledons will open. Now, if one waits another 16 hours in darkness and gives another 5 seconds red light,

then, after a subsequent 16 hour dark period, the plants will flower. The first red light jab opens the cotyledons and the second induces an endogenous rhythm. The plants are getting rather miserable; they have exhausted rather a lot of their seed reserves, not totally exhausted certainly.
Wareing: You have answered your own question. Their reserves are not totally exhausted so they are not dependent upon current photosynthesis.
Friend: Right, but we have to make this distinction when we're talking about photosynthates – certainly we have to have photosynthates for any metabolism whatever.
Wareing: But how do you know there are not adequate substrate levels within these cotyledons to respond to red light?
Friend: There must be – they grow – but you do not need an immediate involvement of photosynthesis in the photoperiodic reaction, apparently.
Wareing: Oh, no! I was not necessarily arguing that.
Friend: Even in the flashing light experiments the plants were not completely depleted of reserves.
Kandeler: I shall give an example for the action of ATP on flowering – surely there is more than one effect of photosynthesis on flowering. We have more than one effect on one plant and different effects on different plants, compare LD and SD *Lemna* species, *etc*. Especially in the case of ATP. In *Lemna gibba*, a LDP, we can inhibit flowering by addition of uncouplers of photophosphorylation. And in this case we have a specific effect on flowering because these inhibitors block flowering but not vegetative growth.
Bernier: How do you measure growth rate in this system.
Kandeler: By frond multiplication.
Mousseau: I think that if I have learned a lesson at Gif* it is that one must not think that photosynthesis is necessary for flowering – there are a number of cases reported for flowering of etiolated seedling – of course you need carbon, the plant would not exist if you had no carbon, but there is a certain minimum level required. But photosynthesis itself can be ignored – certain examples show that plants without chlorophyll will flower; so I think it is much more complicated than we think. The lesson I was taught is that we must look not at carbon metabolism but at energy metabolism. Photosynthesis is not only producing sucrose it is also producing ATP. I think it is the ATP produced that is important. If I have retained something from the meeting in Gif, it is this message.
Charles-Edwards: I should like to make one or two observations. If you are modelling flowering, to observe whether flowering occurs tells you very little – it tells you only whether the event has happened, it doesn't tell you how much of the event has happened. Hence, firstly, you have to define a physiological state that corresponds to flowering and observe the rate at which the meristem is passing through that state and approaching flowering. That is generally useful for modelling.

In *Chrysanthemum* we can do that. We show that the rate at which the chrysanthemum meristem approaches the floral state is directly related to the

* Ref. to the 'International Meeting on the Physiology of Flowering', Gif-sur-Yvette, 17–21 July 1978.

growth rate of the apex which is directly related to the photosynthetic environment of the plant. We can predict, on a photosynthesis-based model, almost all of the qualitative behavior of flowering in *Chrysanthemum*. That is *Chrysanthemum* and nothing else. In *Chrysanthemum* we can define the transition to flowering by a critical size of the apical dome. We find that as the dome grows it reaches the critical size and, under whatever conditions, it always flowers at that critical size. We can do experiments where we bisect the dome, reduce that size of it, and take it back down to the exact stage of development that corresponds to the reduced size. Certainly in *Chrysanthemum* you can predict flowering, and flowering behavior, almost entirely in terms of photosynthesis plus a morphogen (an inhibitor that controls not particularly flowering but primordial initiation, both leaf and inflorescence primordia). I think there are systems in which you can explain flowering in very simple terms; perhaps we shouldn't generalize.

Kinet: I believe that the changes that you observe in the size of the meristem are after evocation – an effect of evocation. The criterion of flowering that you use is after the first events have occurred in the leaf and/or meristem.

Charles-Edwards: I don't believe that this is so. Size in *Chrysanthemum* is causal to flowering and not consequent to it. The bisection shows that. You can take it back exactly to a defined vegetative stage that corresponds to the size of the related apex; there's good theoretical support from things like Thornley's phyllotaxis model, in which size of the apex, quite subtle changes in size, apparently lead to discontinuities in the pattern of development. There's a tendency perhaps to dismiss very simple ideas. At Gif the data I saw for clover, and Lloyd Evans's data on wheat, certainly suggest that where you have an inflorescence it's apical dome size that may be the critical factor.

Bernier: About what you said on bisecting the apical meristem and reversion to vegetative function, we have considered in our group at Gif, and it was agreed, that the point at which the irreversible commitment of the meristem happens depends upon the strength of the reversion treatment applied. For instance, if you excise an apical meristem at different times during the transformation, depending upon the nature of the medium on which you put the meristem, you will get the point of no return at different times. That is, reversion will occur late in the transformation if the medium is 'poor' and relatively early if it is 'rich' – the difference may be large. The bisection of the apex is a very strong treatment – the apex suffers a lot; you would expect to come back to the vegatative state very late during the transformation.

Charles-Edwards: In *Chrysanthemum* the irreversible step came, as Cockshull found, after the last floret primordia are initiated. You can cut the apex up until the time the last floret is initiated and still come back.

Antosewski: I am going to insult you a little bit. The problem you have fallen into is a little bit old fashioned. The most important problem is the flow of information in the system and not just the flow of energy or any materials. In order to employ that information you must have concrete things like carbohydrates; there is nothing strange that the plant will not flower, without sufficient carbohydrates. It is not happy enough – to flower you have to be happy. So photosynthesis is just a source of the materials to construct what is directed by the flow of information. What concrete signals are passed or pushed from one part to another to get the concrete physiological result – for which photosynthates are necessary. The second thing is I was very much impressed with the question posed by Fulford during a visit to East Malling. Why does a plant not flower? All of the information is there – there is no reason to ask what information we should supply. Let's ask why the plant remains in the vegetative state.

Sachs: To respond to that question. It has been widely discussed and apparently accepted that inhibition of reproductive development is part of evolutionary strategy to increase size. That is, a basic vegetative phase is often a key to success – and that's true in all organisms; the delay in onset of sexuality is part of the question of increased size. However, one thing that you do not seem to approach – that part about the plant being happy – is not quite explained by the plant remaining vegetative. That is, at least insofar as we understand the problem mechanistically, it requires as much energy to make a leaf primordium as it does a sepal or petal, *etc*. Why does a plant continue to make leaves and not flowers at a certain low photosynthetic level – we need a mechanism to explain this and that is what Charles-Edwards suggests. If the meristem keys to size, and we can make it grow by photosynthetic input, then we have a mechanism (for that plant) for understanding the transition of the meristem as a function of photosynthetic input. We need more of the kind of quantitative relationships drawn by Charles-Edwards and Cockshull. (Then, Sachs presented a possible scheme on blackboard for analyzing morphogenetic change as a function of change in dry weight – after work of Gifford, R. *et al*. (*Austral. J. Agri. Res.* 24: *297*, 1973) on source-sink limitations in crop plants).

Charles-Edwards: That is interesting (referring to graphs on blackboard) because some years ago Cockshull & Rees did something similar for *Chrysanthemum* – they plotted developmental stage against ratio of flower number and weight to total dry weight and across all their treatments they found a nice constant relationship.

Kandeler: I think that the statement of Antosewski is also very old fashioned – the meristem needs information in any case for the production of vegetative or reproductive structures. Ninety-five per cent of the genetic material is blocked – it is not only a case of inhibiting flowering. There are promoting and inhibiting factors for many programs. It is a question of a balance of the two. The plant receives information from the environment, from all the organs, and very important information as well from photosynthesis. It is very important for the plant to know whether photosynthesis is proceeding and at what rate. The plant will flower only when photosynthesis is progressing at a good rate, hence, photosynthesis is a signal – it is only a question of which part of photosynthesis is the signal. I believe it is more than one signal that comes from photosynthesis.

Sachs: One of the problems that we face is that it is very difficult to distinguish (using our experimental techniques) between an hypothesis that proposes loading of carbohydrates, 'energy compounds' or compounds that alter transport of assimilates from one that proposes loading of a specific stimulus (or morphogen). We have been doing combat with people who believe that there is only one interpretation of the translocation/grafting studies. I think that you understand that it is very difficult to distinguish experimentally between the specific hormonal hypotheses for the control of flowering from the more general view of altering the energy status of the leaf or meristem. Until we know more about evocation, what is required to change the genetic program of the meristem (even in a reduced system), we will be unable to test the hypotheses.

Wareing: I think it is important to make a distinction between certain major groups of plants – those that have a chilling requirement, or photoperiodic requirement, and those that are day neutral. An apple tree is day neutral – whereas *Xanthium* is strongly photoperiodic lacking something in long days. I would suggest that the apple tree is not limited by a floral stimulus (lacking in *Xanthium*) but flowering is limited by non-specific factors and that carbohydrates are among these factors. It is well known to growers that carbohydrates are enormously important to flowering. So I think it is well to make the distinction between photoperiodic and cold requiring plants apparently lacking some specific flower stimulus – this is not the problem in day neutral plants. Now, having said that I wonder whether it wouldn't be useful (without getting involved in too many factors other than photosynthesis) to consider the role of photosynthesis in a) day neutral plants and b) plants which are photoperiodic.

Bernier: Why not consider *all* limiting factors as floral stimuli? Is there any reason to make a distinction between cytokinin in the case of the day neutral tobacco of Sachs, which does not flower without the application of cytokinin, and the so-called florigen in *Xanthium* because we do not yet know the factors required in this plant?

Wareing: There is no reason; the only justification I would make is that it is probably worth making the distinction between factors which are acting as signals and factors which are acting as sources of energy, substrates, or environmental factors – wherever.

Antosewski: It is interesting to make the distinction between food and signal. The signal to have any effect must amplify very much – the amplification factor, which can be measured in certain instances, is an indication of the power of the given signal. On the other hand the flow of nutrients may function as a kind of signal, *e.g.*, in any country with a good supply of food it means at the same time, besides that you are happy, that the economy works.

I agree that in order to cause any reaction you must have the proper cocktail for growth – not only in proportions but in time – so that the message must come to you in the proper sequence – as seen in insect behavior. Limiting factors vary in time.

Kandeler: There are some nutritional factors which can release signals; for example in my paper this morning I think that I have given an example. NO_3 withdrawal perhaps gives rise to higher ABA content. At first you have a nutritional factor but this factor induces a signal – perhaps, ABA in this case. Also, in the case where energy is modified, this energy may be used for hormonal transport. Therefore the energy status is correlated with hormonal factors which may be the 'signal'.

Bernier: It is a great problem to distinguish between specific and non-specific signals.

Friend: If you call every non-specific signal 'florigen', you may get into trouble or water will become 'florigen' under some conditions. If you put some strange *Lemna* in water it will flower.

Sachs: But I believe that a very sensible signal for a plant to have is change in water content – or water potential. If I were thinking of the onset of reproductive development being keyed to any environmental signal in terrestrial plants, I would say that reduced water potential would be ideal – if it could be linked to other systems. That would be a danger signal – 'Get on with it now, there's not much time, water, left'. That makes, unfortunately, teleogical sense only – nothing about mechanism – but I feel that we ought to reason like that with our systems. Certainly, the water stress input for flowering is there. Horticulturists have, for decades if not centuries, known that earlier and more prolific flowering can be induced in many species by withholding water – we have evidence for a positive response in *Bougainvillea*.

Farquhar: Prof. Kandeler, does *Lemna* have stomata on it?

Kandeler: Yes, but they are not functional and always open.

Farquhar: When you give ABA do they close?

Kandeler: No – there is no functional cytoplasm in the guard cells.

Farquhar: The reason why I ask is that perhaps malate is made when reducing power is, not used for something else (*e.g.*, NO_3 reduction). In higher plants if you give ABA, you stop photosynthesis, and in high light you can do something with the NADPH, *e.g.*, you can make malate. But if the stomates don't close with ABA that spoils that interpretation.

REPORTER'S NOTE: There was a more extensive exchange among Kandeler, Löppert, and Farquhar dealing with whether NO_3 withdrawal is linked to malate synthesis, not necessarily ABA synthesis, and whether ABA causes increased malate synthesis, but through a different mechanism than nitrate withdrawal. In both cases malate accumulation is affected. Farquhar's final comment was that: 'What needs to be shown is that if you take NO_3 away do you get an increase in ABA?' and Kandeler's reply: 'We are working on this and I hope next week to have the answer.'

Bernier: Prof. Kandeler, do you have an effect of root removal on flowering in *Lemna*?

Kandeler: We have no experimental data on this point although I think one might be able to remove the roots. The roots in *Lemna*, although not like those in terrestrial plants, may be a source of cytokinins – if one makes starch determinations in the frond one finds that starch is accumulated at the point of insertion of the root.

NOTE: A discussion followed on meristematic anatomy of *Lemna* with exchanges between Kandeler and Bernier. It was emphasized that there are two apical meristems in each meristematic pocket – one meristem is specific for frond initiation and the other for flower initiation – thus in *Lemna* one finds frequent examples of inhibition of flowering without concommitant reduction in vegetative growth – there is in a sense a specialized reproductive axis as in *Bougainvillea*, *Vitis*, *etc*.

Bernier: Perhaps now we can turn to another problem which is also close to our topic. That is, the question of distinguishing between quantitative and qualitative photoperiodic or thermoperiodic plants. As far as I know these distinctions are breaking down; *i.e.*, under appropriate conditions one can demonstrate that qualitative (obligate) photoperiodic plants can respond quantitatively and *vice versa*. May I have your comments? Do you still believe in these distinctions?

Wareing: Well, it depends upon whether one is speaking as an ecologist or physiologist. In practice in the field there are biennial plants which remain completely vegetative unless they are chilled although there may be a few that do bolt and flower in the first year. I guess the distinction is an operational one.

Friend: I believe that the definition of quantitative and qualitative depends upon how long one is prepared to wait. In *Anagallis* after 3 years one or two plants flower.

Bernier: So we all agree that the classical categories are operational. I believe it is quite important to recognize this, because if you recall the arguments of people who do not believe that the role of photosynthesis in flowering may be significant is . . . 'How can you explain that a *Xanthium* plant with one leaf exposed to SD and many in LD flowers, while the same plant, with all leaves in LD, does not flower.' They have just taken one extreme example, but we all know that in *Xanthium* you can induce flowering in other conditions – also, there are many strains of *Xanthium* having various photoperiodic requirements.

Wareing: I think it really depends on what you're interested in. If you're really interested in the overall problem of flowering, then you shall not give great weight to cocklebur, or shall we say the Chicago strain, but if you happen to be a biochemist who would like to get the Nobel prize for getting out a new type hormone you will be very interested in this situation. I think that there will be very considerable interest if you can extract a flowering substance from *Xanthium*: *e.g.*, Chaylakhyan claims to have extracted a substance from a SD tobacco that makes *Chenopodium* flower. I believe that this is an important advance.

Bernier: In connection with this perhaps it is important to come back to *Bougainvillea* where there is always a reproductive axis present. Are the treatments necessary to get flowering in this plant mostly acting on morphogenesis and development? Roy, you said clearly that there is no difference between this kind of plant and *Xanthium* on the other hand – plants which do not have an apparent reproductive axis. Is that correct?

Sachs: We do have two kinds of problems; morphologically, there is perhaps something to quarrel about, but as far as induction is concerned I think not. That is, *Bougainvillea* is a SD plant and so is *Xanthium*. Basically we pose issues of induction and morphogenesis positively for purposes of debate to engage people who use *Xanthium* as the model SD plant. I do not believe that SD promoted processes in the leaf are necessarily the same in all plants, and yet, owing to limitations in our ability to recognize induction apart from evocation and morphogenetic change in the meristem, we have no experimental method to distinguish between SD induction in *Bougainvillea* and *Xanthium*. The morphogenetic question interests me as it does you – what does the reproductive axis mean? In *Bougainvillea* it appears to be a peduncle with two bracts on either side of a vegetative meristem. If conditions are favorable for part of inflorescence development (SD, low photon flux) one goes a long way toward formation of recognizable reproductive parts without arriving at anthesis or fertile structures. If conditions are totally unfavorable (LD, low photon flux) the apical dome aborts, or desiccates, without development beyond the bract stage.

If conditions are entirely favorable (SD or LD, high photon flux) development proceeds to anthesis. If our goal were anthesis, many of us would fall long short of that in our present studies. If we were interested in the physiology of meiosis, we would find ourselves in some difficulty. Certainly in *Xanthium* we find that the kind of inflorescence formed, male or female, and the numbers of each kind formed, are a function of the 'strength' of induction – including photon flux.

If we talk about fertile structures then I think keying to photosynthesis is very attractive as part of the plant's strategy. High rates of current photosynthesis provide the plant, where reserves may be minor, with a way of feeding into the large developing reproductive sink (particularly after ovary initiation) which still may not compete well with 'vegetative' sinks. Studies with cytokinin and SD-induced ^{14}C flux into developing reproductive axes in *Bougainvillea* provided us with the first clue that increased assimilate flow preceded any observable morphogenetic change, and might be causal to, not a consequence of, morphogenesis – as proposed by Charles-Edwards for *Chrysanthemum*. All of this research led to the formulation of a 'nutrient diversion' hypothesis to account for environmental and chemical enhancement of reproductive development. Thus, to make the comparison between *Xanthium* and *Bougainvillea*, we suggested that SD in both plants might cause the production or redistribution of factors (such as cytokinins) that would make the presumptive reproductive meristems better sinks, thereby attracting more assimilates, which in turn would be the 'signals' for reproductive development (again as proposed for *Chrysanthemum*).

Friend: There are well-known cases, though, where a soybean will develop flowers under continuous light and will only continue the development to fertile anthers if they get 1 SD; it is hard to think of this as concerned with carbohydrate flux.

Sachs: Right, I do tend to generalize too much – no matter what I say about generalizations. For example, if SD in soybean caused redistribution of cytokinins. . . .

Lenz: Today, we heard that dark respiration may be changed during induction. I wonder if photorespiration is involved. Has this been studied?

Bodson: Yes, but we did not find any changes in *Sinapis*; CO_2 exchange (in a CO_2-free atmosphere) during the light period of a LD or displaced SD was not different of that during a SD.

Mousseau: I looked for that but the only paper I could find was that by Steward *et al.* (*Can. J. Bot.*, 1971) showing that there was a difference in the release of CO_2 in the light when the plant was growing in SD compared to LD inductive conditions. The ratio of photorespiration to photosynthesis was changed.

Farquhar: Dr. Lenz, the citrus plants that you talked about this morning were said to have a different ratio of oxygenase to carboxylase. How was that measured?

Lenz: We measure CO_2 fixation in 21% O_2 for 2 hours and then in 2% O_2 for 2 hours. The CO_2 (2%) – CO_2 (21%) we call photorespiration.

Farquhar: It is complicated because when you go from 21% to 2% O_2 you are not only reducing the oxygenase but enhancing the carboxylase; the main change is releasing the inhibition of the carboxylase. Do you think that this is conclusive evidence that the ratio of oxygenase to carboxylase was changed?

Lenz: Well, this is difficult to say.
Tamas: I think that the way to do it is with a double isotope technique. That way you do not change an important component, *i.e.*, O_2, and you can distinguish between recently absorbed CO_2 and release of CO_2.
Lenz: If we compare the data in the literature, ours is very comparable. Photorespiration is between 20% and more than 50% of photosynthesis.
Mousseau: I want to add something about dark respiration and vegetative and flowering plants. I examined the literature on this subject. It seems that usually induced leaves have a higher dark respiration than non-induced leaves. It means that there is some activation of enzymes in induced compared to non-induced leaves. That is all that you can say.
Bernier: Perhaps we can come to the last point of the discussion – this idea put forward by Prof. Wareing that flowering could be controlled by a linear sequence of gates. Later on he accepted that this could also be a complex interaction of a number of gates. We have some evidence in *Sinapis* for the latter view. In *Sinapis* we are able to produce an early rise in mitotic activity, totally identical to that produced in plants by LD, simply by treating the meristem with a single application of cytokinin but the plants do not flower. Another early change in the meristem, when *Sinapis* is induced to flower by LD, is an increase in the number of mitochondria per cell and increased respiratory activity. This increase can be induced perhaps by an increase in soluble sugar. Apparently these two typical changes occurring in meristems are entirely independent because we are unable to induce increased mitotic activity by increasing the soluble sugars. Probably we will be unable to increase mitochondrial number by a single application of cytokinin because of timing considerations. (Although we have not yet done this experiment.) We thus think of evocation of the meristem in terms of a number of sequences each controlled (induced) by one or several compounds. At least they must be independent at the start, but they probably interact later on. We think the *interaction* of processes is necessary for the completion of evocation. We believe that interactions of other processes may hold true for later flower development because there is no reason to believe that things are different during flower development than during flower initiation. Any comments about this?
Wareing: I wouldn't disagree with this view. I suggested it was a sequence of gates but it need not be a single sequence. However, I will stick to the main point, that I think we have all agreed about, that the overall flowering process involves not a single signal but requires a number of conditions such that if any one of these is limiting then flowering will not occur. I think that is about as general as you can get!

Authors' address:
Department of Botany
University of Liège, Sart Tilman
B–4000 Liège
Belgium.

SESSION III

PHOTOSYNTHESIS AND MINERAL NUTRITION AND GROWTH REGULATORS

THE USE OF MINERAL NUTRIENT STRESS IN THE STUDY OF LIMITING FACTORS IN PHOTOSYNTHESIS*

NORMAN TERRY
(Berkeley, California, U.S.A.)

ABSTRACT

Mineral nutrient stress has been investigated as a possible tool in the study of limiting factors in photosynthesis. Evidence has been obtained which suggests that Fe stress may be used to progressively diminish photochemical capacity, *i.e.*, the capacity of a leaf to convert absorbed light to chemical energy. The reduction in photochemical capacity with Fe stress is due to a decrease in the number of photosynthetic units per unit leaf area, there being no effect on the size of photosynthetic units. The effect of Fe stress is remarkably specific, other photosynthetic and growth processes being relatively unaffected.

The effect of Fe stress on the formation of photosynthetic units within a growing leaf was explored by following changes in the chlorophyll content and leaf area after Fe is withheld from the plant. The results indicate that there was a failure to form photosynthetic units in the absence of an adequate supply of Fe and that leaf expansion continued at a constant rate which was independent of the changes in photosynthetic unit production. The effect of Fe stress on photosynthetic unit formation was rapidly and almost completely reversible. Furthermore, Fe stress was shown to have no effect on leaf thickness, fresh weight per unit area, or on leaf tissue volume. The data support the view that the effect of Fe stress is sufficiently specific for it to be used as an experimental means of manipulating photochemical capacity in an intact plant system.

INTRODUCTION

If the steps which regulate or control the rate of photosynthesis could be identified it should be possible to genetically or physiologically manipulate crop plants to improve their productivity. The research program presently underway in our laboratory is concerned with the identification of the photosynthetic processes which potentially limit photosynthesis. There are four main processes which are involved in the photosynthetic conversion of light energy and CO_2 into carbohydrates. These four processes are: 1) the absorption of light energy; 2) the conversion of absorbed light energy into chemical energy; 3) the diffusion and transport of CO_2 to the chloroplasts; and 4) the enzymatic fixation and reduction of CO_2 to carbohydrate. The objective of our research program is to find ways of experimentally varying the capacity of the photosynthetic apparatus to carry out one of these four processes independently of other factors

* This work was supported by a grant from the Beet Sugar Development Foundation. I thank Gary Low for excellent technical assistance during the course of this project.

so as to assess the quantitative effect of that individual capacity on the rate of photosynthesis.

Mineral nutrient stress was investigated as a possible experimental technique to vary specific photosynthetic capacities. After an extensive survey, Fe stress was discovered to have a major effect on the capacity of a leaf to convert absorbed light into chemical energy. This is referred to as 'photochemical capacity' and is defined as the maximum rate per unit area at which a leaf can convert a nonlimiting supply of light energy to chemical products (such as ATP or NADPH). Iron stress was subsequently found to reduce photochemical capacity by decreasing the number of photosynthetic units per unit leaf area (Spiller & Terry, 1978). The sugar beet photosynthetic unit contained about 450 chlorophyll molecules per molecule of P_{700} and per molecule of cytochrome f. The molar ratio of chlorophyll to P_{700}, which is a measure of the size of the photosynthetic unit, was not changed by Fe stress. The molar ratios of chlorophyll a to chlorophyll b and chlorophyll to β-carotene also remained reasonably constant during Fe chlorosis. Assuming that chlorophyll a, chlorophyll b, β-carotene, P_{700} and cytochrome f constituted an integral part of the sugar beet photosynthetic unit, it is evident that the size of the photosynthetic unit was not altered by Fe stress. Thus, Fe chlorosis and the accompanying reduction in photochemical capacity was due to there being fewer photosynthetic units per leaf, and not to a change in the size of the photosynthetic unit.

Iron stress had relatively small effects on other photosynthetic attributes (Terry, in press). The maximum rate of photosynthesis per unit chlorophyll (*i.e.*, the rate of CO_2 uptake attained under optimal environmental conditions) was about 450 μmoles of CO_2 fixed per mg chlorophyll per hour. Iron stress did not decrease this rate of photosynthesis and also had very little effect on stomatal conductance. The quantum yield of attached leaves irradiated with red light was 0.10 mole CO_2 per absorbed einstein; it was decreased only when Fe stress was severe. The absorptance of Fe stressed leaves was decreased relatively little by a large reduction in chlorophyll and carotenoid content with Fe stress; Fe stressed leaves containing only 5 μg chlorophyll cm^{-2} absorbed 53% of the incident white light while control leaves with 10 times as much chlorophyll absorbed 84%. Iron stress had no effect on the leaf contents of soluble protein and chloroplast protein and only slightly decreased ribulose bisphosphate carboxylase activity per unit soluble protein. The principal and dominating effect of Fe stress was to decrease the photosynthetic electron transport capacity *in vivo*: rates of noncyclic ATP formation were not decreased by Fe stress when expressed per unit chlorophyll, but when expressed per unit leaf area they decreased linearly with leaf chlorophyll content. This was because the Fe stressed leaves had fewer photosynthetic units and therefore fewer reaction centers and electron carriers to carry out photosynthetic electron transport.

Iron stress had remarkably little effect on many attributes of leaf growth, including the number of cells per unit leaf area and average leaf cell volume (Terry, 1978). There were no differences in the numbers of chloroplasts per unit area (or per cell) or in average chloroplast volume between Fe stressed and control leaves; the main effect of Fe stress was to reduce the amount of chlorophyll per chloroplast. Visual observation of Fe stressed chloroplasts by electron microscopy indicates that there was less chlorophyll per chloroplast due to a reduction in the number of chloroplast lamellae. Iron stressed chloroplasts of sugar beet (Spiller & Terry, unpublished), corn (Stocking, 1975), tomato, spinach and maize (Vesk *et al.*, 1966), and Xanthium (Bogorad *et al.*, 1958) have been shown to have reduced numbers of granal and stromal lamellae within a relatively large stroma. Furthermore, Machold (1971) observed that the decrease in chlorophyll in Fe stressed leaves was associated with a substantial reduction in the amounts of chlorophyll-protein complexes, *i.e.*, the light harvesting chlorophyll a-chlorophyll b protein complex and the P_{700}-chlorophyll a protein complex. In our work the reduction in chlorophyll per unit area with Fe stress was accompanied by a coordinate decrease in the amount of carotenoid pigments; similar coordinate decreases of carotenoid with chlorophyll have been observed in potato (Bolle-Jones & Norton, 1953) and barley (Mackinney, 1935). These observations suggest that Fe stress may retard the formation of the complete light harvesting apparatus, including the chloroplast membranes and the chlorophyll-protein complexes, carotenoids, reaction centers, and electron carriers associated with them.

In this presentation, the effect of Fe stress on the formation of the light harvesting apparatus within a growing leaf is further explored by following the changes in the chlorophyll content and area of individual leaves after Fe is withheld from the plant. The precise objective was to obtain more information on the specificity of the Fe stress effect on the production of photosynthetic units relative to the effect of Fe stress on leaf growth. Such information is valuable in evaluating the validity of using Fe stress to determine the quantitative effect of photochemical capacity on the rate of photosynthesis. The measurement of chlorophyll provided a direct index of the number of photosynthetic units owing to the fact that the amount of chlorophyll per photosynthetic unit is not changed by Fe stress. The interpretation of the data was facilitated by measurement of additional parameters including the numbers and volumes of cells and chloroplasts, fresh weight per unit leaf area, leaf thickness and leaf tissue volume.

METHODS

Plant culture

Sugar beets (*Beta vulgaris* L. cv. F58–554H1) were cultured hydroponically at 25° C, and illuminated at 40,000 lux over a 16-hour day. The plants were cultured for 2 weeks following planting in vermiculite with half-Hoagland solution. They were then transplanted into a culture solution containing (in mmol/l), 2.5 $Ca(NO_3)_2$, 1.0 KH_2PO_4, 2.5 KNO_3, 1.0 Mg SO_4, and 0.5 NaCl, and (in mg/l) 0.25 B, 0.25 Mn, 0.025 Zn, 0.01 Cu, 0.005 Mo, and 0.5 Fe (as ferric-sodium ethylene diamine tetraacetate complex). After two weeks, plants which were to be Fe stressed were transferred to solutions with the same composition as above except for Fe, which was withheld; 2 M NaOH and solid $CaCO_3$ were added to raise the pH to about 8.5.

Leaf expansion and chlorophyll content

At the time that Fe was withheld, a young growing leaf was selected with an area of 25 to 40 cm^2. Seven plants were cultured with Fe and seven without for a period of 8 days. Iron (2.5 mg/l) was resupplied to the Fe stressed plants on the eighth day and the experiment terminated on the 12th, day. At 0, 2, 4, 6, 8, 10 and 12 days leaf tracings were made and 3 leaf discs taken for determination of leaf area and chlorophyll content (Terry & Huston, 1975), respectively.

Leaf thickness and tissue volume

The measurements described below were carried out on young rapidly expanding leaves which were about 150 to 200 cm^2 in area. The leaves were selected from plants which were about 2 to 3 weeks old (from transplanting). Three sets of 10 discs were removed from each leaf at random: one set was used for determination of chlorophyll content and one for determination of fresh weight per unit leaf area. The remaining set of 10 leaf discs was placed in a dry 10 ml volumetric flask and distilled water added to make a total volume of 10 ml. The volume of leaf discs was determined by subtracting the volume of water added from the total volume of leaf discs and water. Leaf thickness was calculated by dividing the leaf disc volume by disc area. Leaf tissue volume was estimated from the fresh weight of leaf discs assuming a density of 1.03 g fresh weight per cm^3 of leaf tissue.

RESULTS

Effect of withholding the Fe supply on leaf expansion and chlorophyll (photosynthetic unit) production

In this experiment, the Fe supply was withheld from and then restored to the plant and the changes in leaf area and chlorophyll (photosynthetic unit) content monitored during the life of an individual leaf. The experiment was repeated seven times and the changes which occurred for one of these leaves are shown in Figs. 1–3. The leaf from the treated plant was a little larger than the control and had a slightly greater chlorophyll content per unit area (40 compared to 30 μg chlorophyll (a + b) per cm^2). The chlorophyll content per unit area decreased significantly between 2 and 4 days after Fe was withheld and reached a minimum value of 17 $\mu g\ cm^{-2}$ which did not change from 6 to 8 days (Fig. 1). Over the same 8 day period chlorophyll/area in the control leaf increased curvilinearly with time and then remained constant at about 70 $\mu g\ cm^{-2}$.

After 8 days of withholding, the Fe supply was restored and chlorophyll production rapidly resumed. Within four days the chlorophyll content per unit area had reached a value of 40 $\mu g\ cm^{-2}$ which was the value held by the leaf when the experiment began. In another experiment which was continued over a period of 16 days after Fe was resupplied, values of 48.0 and 53.4 μg chlorophyll cm^{-2} were attained compared to about 59 $\mu g\ cm^{-2}$ in the two control leaves. Thus, the effect of withholding the Fe supply on chlorophyll production could be rapidly and almost completely reversed by restoring the Fe supply to the plant.

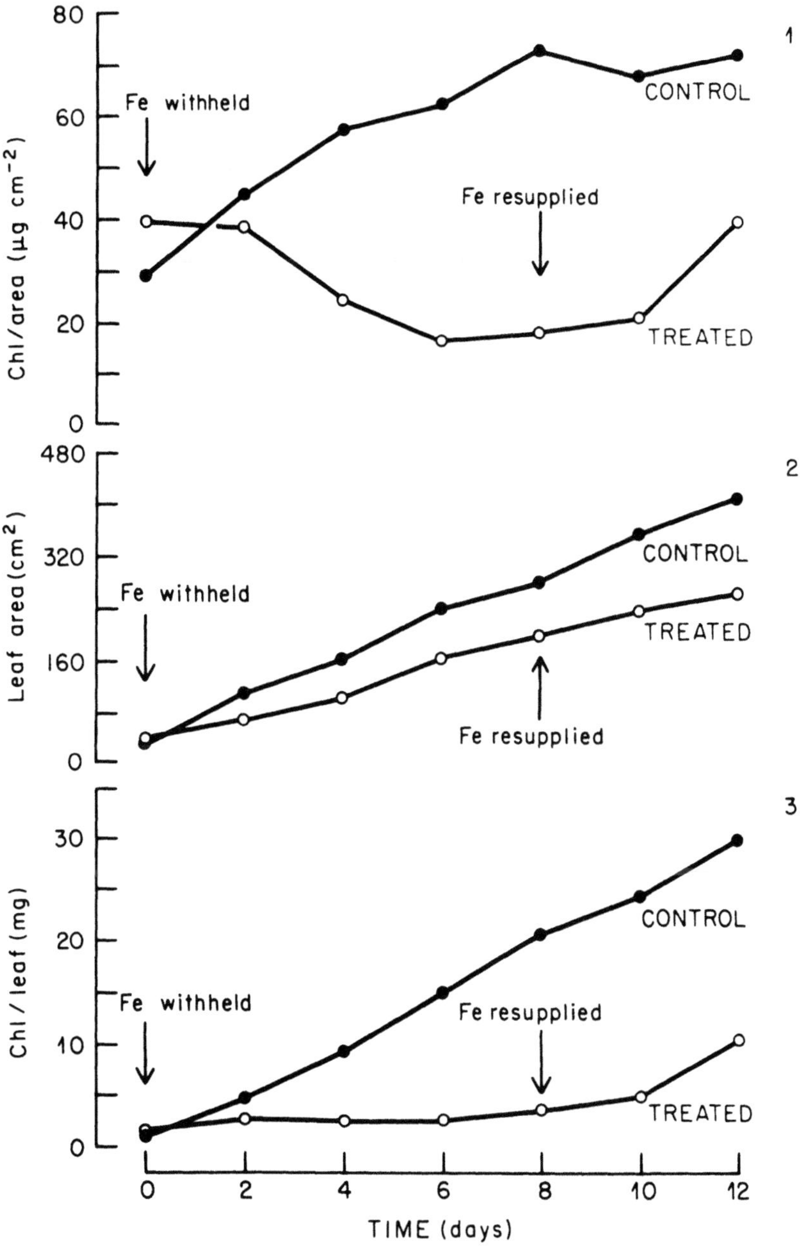

Figs. 1–3 Effects of withholding and restoring the Fe supply on chlorophyll content and leaf expansion. Changes with time are shown for a single leaf of the treated plant (○) and for a single leaf of the control (●).

1. Chlorophyll per unit leaf area.
2. Area per leaf.
3. Chlorophyll content per leaf.

The production of chlorophyll by the growing leaf quickly ceased when Fe was withheld (Fig. 3). There was perhaps a slight increase in the chlorophyll content per leaf during the first 2 days but after that the chlorophyll content did not change until the Fe supply was restored. In contrast, the leaf continued to expand at a constant rate (Fig. 2). Although the Fe stressed leaf expanded at a slower rate than the control, its rate of expansion appeared to be independent of the changes in chlorophyll production which occurred when Fe was withheld and restored. These data suggest that changing the Fe supply had a rapid and specific effect on leaf chlorophyll production which was largely independent of the effect of Fe on leaf expansion. Furthermore, since the amount of chlorophyll/-leaf did not decline after Fe was withheld, there was no evidence of chlorophyll destruction. The data suggest that the chlorophyll content per unit area decreased because no new photosynthetic units were formed in the absence of Fe and because the existing chlorophyll within the leaf was diluted as the leaf increased in size. However, it is also possible that some turnover of chlorophyll could have occurred in Fe stressed leaves with the rate of chlorophyll synthesis within a photosynthetic unit equalling the rate of chlorophyll breakdown.

Development of chlorosis and its relation to chloroplast production and leaf cell growth

The growth which continued after chlorophyll production ceased most likely occurred by cell enlargement, although some growth may have occurred as the result of new cell production. As the leaf expanded from 30 or 40 cm^2 to about 100 to 200 cm^2, average leaf cell volume could have increased by as much as 5-fold while the number of cells per leaf most likely would have increased by about 50% (Table 1). Over the same phase of leaf growth, there was almost certainly a very large production of chloroplasts. (The method of measurement for the numbers and volumes of cells and chloroplasts is described by Terry (in press).) The number of chloroplasts per unit area could have increased almost 2-fold as the leaf expanded from 40 to 200 cm^2 and there may have been an accompanying increase in chloroplast volume (Table 2). The earlier work showed that Fe stress did not affect the numbers and volumes of chloroplasts or cells when leaves of similar size and stage of development were compared (Terry, in press).

These observations suggest that Fe stress may have induced chlorosis in the following way. As the leaves expanded for 8 days without Fe, there was little

Table 1. Number of cells per leaf and average leaf cell volume for leaves at different nodal positions.

Leaf position (in order of formation)	12	13	14	15	16
Leaf area (cm^2)	198	159	123	71	30
Number of cells/ leaf ($\times 10^8$)	2.48	2.81	2.68	2.02	1.85
Average leaf cell volume (10^{-8} cm^3)	2.27	1.52	1.15	0.88	0.43

increase in the amount of chlorophyll per leaf. Over the same period it seems likely that there was some increase in the number of cells, but leaf growth probably proceeded mainly by cell enlargement. Accompanying cell enlargement was a large increase in the number of chloroplasts, perhaps as much as 10-fold increase in the number per leaf. These data suggest that chlorosis occurred because chloroplasts replicated several times during leaf growth dividing a fixed amount of chlorophyll (number of photosynthetic units) equally among them. This conclusion received additional support from electron microscopy carried out by Spiller & Terry (unpublished) who showed that the number of chloroplast lamellae per chloroplast was equally reduced in all of the chloroplasts examined in a given Fe stressed leaf.

Table 2. Number of chloroplasts per unit leaf area and average chloroplast volume for leaves at different nodal positions.

Leaf position (in order of formation)	16	17	18	19	20
Leaf area (cm^2)	178	125	68	43	41
Number chloroplasts/cm^2 (10^8)	1.02	0.68	0.46	0.55	0.60
Average chloroplast volume (μ^3)	23.7	29.8	15.5	14.7	20.6

Validity of expressing chlorophyll on a per unit leaf area basis

Since Fe stress could conceivably change the thickness or the tissue volume of the leaf, expression of chlorophyll content as mass per unit area might not be a valid basis for comparison. To clarify this point, leaf thickness was estimated from measurements of the volume and area of leaf discs (Table 3). Leaf thickness was about 0.50 ± 0.05 mm and did not vary with chlorophyll content. In these same leaves the average fresh weight per unit leaf area was 33.0 ± 3.1 mgs/cm^2 and the average proportion of the total leaf volume which was actually leaf tissue (not intercellular air space) was 63.6 ± 6.4%. Since leaf thickness, fresh weight/area, and tissue volume did not differ significantly in leaves of

Table 3. Fresh weight per unit leaf area, leaf thickness and tissue volume for leaves with different chlorophyll contents.

	Chlorophyll ($\mu g\ cm^{-2}$)	Fresh weight/ area ($mg\ cm^{-2}$)	Leaf thickness (mm)	Tissue volume (%of leaf vol.)
Control	65.6	32.3	0.446	69.9
	52.5	28.1	0.496	54.6
	50.7	32.5	0.446	70.3
	45.9	34.5	0.556	59.9
	44.1	31.3	0.516	58.5
Fe stressed	21.4	32.5	0.476	65.6
	18.8	30.6	0.476	62.0
	17.2	39.1	0.511	73.3
	16.4	35.9	0.595	58.3

different chlorophyll contents, it is concluded that it is reasonable to express chlorophyll on a mass per unit area basis. Furthermore, these data support the view that the effect of Fe stress on chlorophyll content is specific, having no effect on these other attributes of leaf growth.

DISCUSSION

The results obtained in these experiments show that Fe stress exerted an effect on chlorophyll (photosynthetic unit) production which was largely independent of the effect on growth. When Fe was withheld from the plant, chlorophyll production by a young growing leaf virtually ceased. Nevertheless, the young leaf continued to expand at a constant rate increasing 5-fold in area over a period of 8 days. The decrease in the amount of chlorophyll per unit leaf area which occurred during Fe stress is attributed to the dilution of chlorophyll as the leaf grew in volume. The chloroplasts replicated several times during leaf cell enlargement apparently dividing the number of photosynthetic units present in the leaf at the time of withholding more or less equally among them. These data suggest that there was a failure to form photosynthetic units in the absence of Fe. When Fe was resupplied, chlorophyll (photosynthetic unit) production rapidly resumed and the Fe stress effect on chlorophyll production was almost completely reversed. Also, when Fe stressed low chlorophyll leaves were compared with high chlorophyll control leaves of comparable age and size, there were no differences between them in leaf thickness, fresh weight per unit leaf area or tissue volume, *i.e.* volume exclusive of the leaf intercellular air space. These data support the view (Terry, in press; Spiller & Terry, unpublished) that Fe stress has a specific effect on the formation of photosynthetic units and very little effect on the rate of leaf expansion and other attributes of leaf growth.

It is postulated therefore, that in the case of Fe, mineral nutrient stress presents a useful means of studying limiting factors in photosynthesis. Iron stress can be used to experimentally vary the photochemical capacity of leaves independently of other factors. This is because Fe stress has a reasonably specific effect on the formation of photosynthetic units. The number of photosynthetic units affects photochemical capacity because it determines the number of reaction centers and electron carriers available to convert absorbed light to chemical energy. The quantitative effect of photochemical capacity has since been explored by Fe stressing plants to obtain leaves with different numbers of photosynthetic units per unit leaf area and then measuring the rate of photosynthesis of these leaves in 25 light/CO_2 environments. The chlorophyll content per unit area, which is a direct measure of the number of photosynthetic units, was used as the index of photochemical capacity.

This study shows that at light saturation the rate of photosynthesis in an environment approximating field conditions, *i.e.,* 30° C and an ambient CO_2 concentration of 300 μl CO_2 l^{-1} air, was curvilinearly related to chlorophyll-/cm^2 from 5 to 65 μg cm^{-2}. This means that photosynthesis was limited by photochemical capacity over the entire range of leaf chlorophyll contents, the increase in the rate of photosynthesis with chlorophyll becoming progressively less as chlorophyll content increased. At low irradiances of about 5% of full sunlight, the rate of photosynthesis was limited by photochemical capacity only up to 30 μg chlorophyll cm^{-2}; beyond this chlorophyll content (actually photosynthetic unit content) the photochemical capacity was more than sufficient to convert this low level of radiation (100 μeinstein [400–700nm] m^{-2} sec^{-1}) to chemical energy. However, at irradiances of 500 μeinstein m^{-2} sec^{-1} or more, the rate of photosynthesis was limited by photochemical capacity over the entire range of leaf chlorophyll contents.

It is concluded that the rate of photosynthesis of sugar beets growing under field conditions may be simultaneously colimited by at least three factors; internal photochemical capacity, CO_2 supply and irradiance, provided that the latter is less than 1500 μeinstein cm^{-2} sec^{-1}. I consider that photosynthesis is 'colimited' because an increase in any one of these three factors independently of the other two could potentially result in an increase in the rate of photosynthesis. These data confirm observations made by Šesták (1966) and Buttery & Buzzell (1977) who showed that the rate of photosynthesis per unit leaf area for crop plants under field conditions was linearly related to chlorophyll/cm^2. It is clear that the empirical relation obtained by these workers was not between photosynthesis and chlorophyll *per se* but between photosynthesis and the number of photosynthetic units per unit leaf area. Photosynthesis is not increased by high chlorophyll due to improved light absorption, but rather to an increased photochemical capacity resulting from the greater number of photosynthetic units associated with high chlorophyll. An increased photochemical capacity is particularly desirable at high irradiances where quanta are absorbed faster then they can be utilized in photochemistry; this explains why Šesták (1966) found a greater dependence of photosynthesis on chlorophyll content at high than at low irradiances.

REFERENCES

Bogorad, L., G. Pires, H. Swift & W.J. McIlbrath. The structure of chloroplasts in leaf tissue of iron deficient *Xanthium*. pp. 132–137, in: Brookhaven Symposia in Biology II. The Photochemical Apparatus: Its Structure and Function (1958).

Bolle-Jones, E.W. & B.A. Norton. The relative proportions of the chloroplast pigments as influenced by different levels of iron and potassium supply. *Pl. Soil* 5: *87–100* (1953).

Buttery, B.R. & R.I. Buzzell. The relationship between chlorophyll content and rate of photosynthesis in soybeans. *Can. J. Pl. Sci.* 57: *1–5* (1977).

Machold, O. Lamellar proteins of green and chlorotic chloroplasts as affected by iron deficiency and antibiotics. *Biochim. biophys. Acta* 238: *324–331* (1971).

MacKinney, G. Development of the chlorophyll and carotenoid pigments in barley seedlings. *Pl. Physiol.*, Lancaster 10: *365–373* (1935).

Šesták, Z. Limitations for finding a linear relationship between chlorophyll content and photosynthetic activity. *Biologia Pl.* 8: *336–346* (1966).

Spiller, S. & N. Terry. Iron deficiency reduces photochemical capacity via an effect on the number but not size of photosynthetic units. *Pl. Physiol.*, Lancaster 61: *S–87* (1978).

Stocking, C.R. Iron deficiency and the structure and physiology of maize chloroplasts. *Pl. Physiol.*, Lancaster 55: *626–631* (1975).

Terry, N. Photochemical capacity as a limiting factor in photosynthesis. *Pl. Physiol.*, Lancaster 61: *S–87* (1978).

Terry, N. Limiting factors in photosynthesis. I. Use of iron stress to control photochemical capacity *in vivo*. *Pl. Physiol.*, Lancaster (in press).

Terry, N. & R.P. Huston. Effects of calcium on the photosynthesis of intact leaves and isolated chloroplasts of sugar beets. *Pl. Physiol.*, Lancaster 55: *923–927* (1975).

Vesk, M., J.V. Possingham & F.V. Mercer. The effect of mineral nutrient deficiencies on the structure of the leaf cells of tomato, spinach, and maize. *Aust. J. Bot.* 14: *1–18* (1966).

Author's address:
Department of Soils and Plant Nutrition
University of California
Berkeley, California 94720
U.S.A.

PHOTOSYNTHESIS, CARBONIC ANHYDRASE AND AMINO ACIDS RELATED TO ZINC STATUS IN COTTON

K. OHKI & C.T. YOUNG
(Experiment, Georgia, U.S.A.)

ABSTRACT

The relationship between Zn status and photosynthesis, carbonic anhydrase and amino acid concentrations in cotton (*Gossypium hirsutum* L.) leaves was determined. Plant tissue ranging in Zn status from very severe deficiency to adequacy was obtained from plants grown in nutrient solution with seven Zn rates from 10 to 2000 μg/liter Zn as $ZnSO_4$. With very severe Zn deficiency in blades containing 10 μg/g Zn, photosynthesis was 31% of controls. The critical deficiency level related to photosynthesis was determined to be 14 μg/g. Carbonic anhydrase activity was not detectable in very severely deficient blades. As Zn concentration in the blades approached the critical deficiency level of Zn for photosynthesis and growth, carbonic anhydrase activity increased sharply to 200 EU/g. Carbonic anhydrase activity continued to increase as Zn concentration in the blades increased. Data suggest that carbonic anhydrase was not the limiting factor in photosynthesis.

Zinc deficiency decreased total glycine, alanine, valine, leucine, isoleucine, serine, threonine, glutamic acid, lysine, cystine, methionine, proline, phenylalanine, tyrosine and tryptophan concentrations. Critical Zn concentration was 11 to 15 μg/g in recently matured blades for these amino acids. Total histidine was not influenced by Zn status. Total aspartic acid and arginine concentrations were greater in Zn-deficient blades than in blades with adequate Zn for plant growth.

INTRODUCTION

The influence of mineral nutrient status on photosynthesis, carbonic anhydrase and amino acid levels in higher plants has been reported from numerous studies. The studies have been conducted primarily on effects of nutrient deficiencies compared to non-deficient plants. Botrill *et al.* (1970) reported that photosynthesis by spinach (*Spinacea oleracea* L.) on fresh weight basis was depressed by all mineral nutrients except Ca with Mn, Cu, P and K deficient plants showing the greatest depression. From nutrient withdrawal studies, low N decreased CO_2 assimilation in maize (*Zea mays* L.), cotton (*Gossypium hirsutum* L.), and bean (*Phaseolus vulgaris* L.) (Ryle & Hesketh, 1969). Terry & Ulrich (1973a) reported that net photosynthetic CO_2 uptake by sugar beets (*Beta vulgaris* L.) decreased to about 33% of controls 30 days after P cut-off. Similar depletion studies indicated that K deficient bean leaves reduced photosynthetic CO_2 fixation (Ozbun *et al.*, 1965) and K deficient sugar beet leaves attained CO_2 fixation rates of approximately one-third of the control rate (Terry & Ulrich, 1973b). Reduction in photosynthetic rates has also been reported for Mn (Terry & Ulrich, 1974b; Botrill *et al.*, 1970) and other essential micronutrients (Botrill *et al.*, 1970).

Utilization of the critical nutrient concept (Ulrich & Hills, 1969) relating nutrient status and photosynthetic CO_2 assimilation is a means whereby rates of photosynthesis can be associated with the extent of nutrient deficiencies. Moreover, if critical deficiency levels are established from nutrition experiments, the information on minimal nutrient levels required for near maximum rates of photosynthesis and other physiological processes would be valuable in assessing nutrient factors associated with reduced growth. Critical deficiency levels of K related to photosynthetic CO_2 assimilation (Peaslee & Moss, 1966; Estes *et al.*, 1973) and Mg (Peaslee & Moss, 1966) in maize leaves have been reported. Recently, critical deficiency levels for photosynthesis on sugar beet leaves have been reported for the macroelements, Mg, Ca and S (Terry & Ulrich, 1974a; Terry & Huston, 1975; Terry, 1976). In case of microelements, Ohki (1978) reported critical Zn deficiency level for photosynthesis by soybean (*Glycine max* L.) and Terry (1977) reported critical Cl deficiency levels for sugar beets. In most instances, the critical deficiency levels for plant growth were approximately equal to the critical deficiency levels for photosynthesis.

Inhibition of carbonic anhydrase activity has been related to Zn deficiency in tomato (Wood & Sibly, 1951), bean (Edwards & Mohamed, 1973), soybean (Ohki, 1978), and citrus (Bar-Akiva & Lavon, 1969). The role of carbonic anhydrase in photosynthesis in higher plants has not been resolved. However, Randall & Bouma (1973) have reported that carbonic anhydrase does not have a role in facilitating CO_2 to the carboxylation sites.

Mineral nutrient deficiencies have been reported to influence amino acid contents in several crop plants. Haghiri (1966) reported that N deficiency reduced total aspartic acid, arginine and histidine contents in soybean tops while P deficiency increased arginine and aspartic and decreased glutamic proline, glycine, alanine, valine and leucine contents. Calcium deficiency had no influence on amino acid contents. In corn, free glutamic acid and aspartic acid contents were decreased by K deficiency while asparagine increased (Barber & Bradfield, 1963). In soil pot culture, total lysine and arginine contents of bromegrass (*Bromus inermis* L.) decreased as the substrate Ca and K concentrations increased (Reed *et al.*, 1960). Sulfur deficiency in alfalfa (*Medicago sativa* L.) reduced the 16 common amino acids and increased aspartic and arginine contents (Mertz *et al.*, 1952). They suggested that asparagine and arginine acts as a primary and secondary, respectively, storage forms for nitrogen in S deficient alfalfa. Microelement studies indicated that Mn deficiency in cauliflower (*Brassica oleracea* L.) induced increases in free arginine content (Hewitt *et al.*, 1949) and that B deficiency in peanut (*Arachis hypogaea* L.) induced increases in free arginine, aspartic, glutamic, proline and serine (Shiralipour *et al.*, 1969). Studies with Zn have indicated that Zn deficiency in citrus induced increases in free arginine, serine, threonine and proline contents with decreases in alanine content (Stewart, 1962), no effect on pear (*Pyrus communis* L.) (Lewis *et al.*, 1963) and induced decreases in free glycine and isoleucine contents in beans (*Phaseolus vulgaris* L.) (Schütte & Schendel, 1958).

The objectives of this study were to determine critical Zn deficiency levels for growth, photosynthesis and total amino acid by relating activity and Zn concentration in recently matured blades of cotton, and to determine the influence of Zn nutrition on carbonic anhydrase activity.

MATERIAL AND METHODS

Seeds of upland cotton (*Gossypium hirsutum* L. cv. Coker 310) were planted in white quartz sand washed with deionized water. Seedlings were transplanted in the early cotyledonary stage and grown in the greenhouse with solution culture provided with seven discrete Zn levels from 10 to 2000 μg Zn/l as described previously (Ohki, 1976). Plants were grown for 35 days following transplanting. Recently matured blades, usually blade three from the top, were used for photosynthesis measurements. These blades were sampled and freeze-dried for subsequent Zn and amino acid analyses. The residue was oven dried at 70°C and weighed.

Zinc was determined in plant material by atomic absorption spectrophotometry after wet ashing with a nitric-perchloric acid mixture. Net photosynthesis (P_N) was measured on intact recently matured leaves by a flow through system using a differential IR Beckman 215A analyzer. The air temperature in a water jacketed chamber was 25° C with leaf irradiation of 1100 $\mu E\ m^{-2}s^{-1}$ in the 400 to 700 nm range. Carbonic anhydrase activity on freeze dried material was assayed by the CO_2-Veranol buffer procedure described by Wilbur & Anderson (1948). Enzyme units are described earlier (Ohki, 1976).

Total amino acid analyses with the exception of tryptophan were performed by the ion exchange chromatography technique of Spackman *et al.* (1958) with a Durrum Model D-500 amino acid analyzer using a 1.75 internal diameter and 48 cm length column, packed with the Durrum high resolution cation exchanger (bead diameter 8 ± 2 microns). Running time was 70 min. including regeneration of column. Duplicate samples (50 mg) were hydrolized in 20 ml of 6N HCl and flushed with high purity nitrogen for 2 hr at 145°C. The sample pH was adjusted to 2.1–2.2 with 12 N NaOH and the volume diluted to 50 ml. A portion of the sample was centrifuged on a Beckman microfuge and a 20 μl aliquot was analyzed.

Duplicate samples (100 mg) were prepared for tryptophan analysis by the method of Hugli & Moore (1972) with the following modifications. Teflon faced screw cap – glass tubes were used for flushing with nitrogen at greater than 5 ppm. The pH of the cooled sample was adjusted to 4.50 with 6N HCl and diluted to 50 ml with water. After centrifuging, 40 μl of supernatant were analyzed for tryptophan on the Durrum analyzer using a 9.80 pH buffer for elution. The run time was 40 minutes.

RESULTS

The critical Zn concentration for growth was determined from the relationship of Zn concentration in recently matured blades and top dry weight (Fig. 1). Plant weights ranged from 43 g under very severe deficiency to an average maximum of 93 g with adequate Zn supply. Zinc concentration in blades ranged from 10 μg/g dry weight for the very severely deficient leaves to 48 μg/g for non-deficient leaves. From the Zn nutrient calibration for growth the critical deficiency level which is within the transition level zone was determined as 13 μg/g dry weight. The critical deficiency level is defined as the nutrient concentration in blade tissue associated with a 10% reduction in growth from maximum. The critical deficient level is considered to be a narrow range of values that are less than or greater than the determined level rather than a definitive single value. Plants with recently matured blades containing less Zn than the critical deficiency level are considered Zn deficient, whereas plants with blades containing more Zn than the critical deficiency level are considered Zn sufficient. Similar interpretations can be made for other physiological processes such as photosynthetic activity as indicated in Fig. 2. Photosynthesis was inhibited to 31% of control in recently matured blades containing 10 μg Zn/g dry weight. Maximum P_N was attained when blades contained approximately 17 μg/g and maximum P_N continued to 48 μg/g. A critical Zn concentration of 13 μg/g for photosynthesis was determined from Fig. 2.

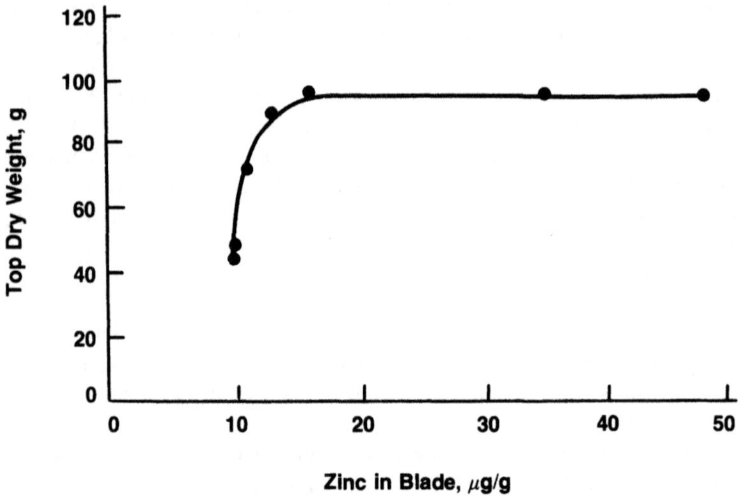

Fig. 1. Nutrient calibration relating Zn concentration in recently matured blade and top dry weight. Critical deficiency level is 13 μg Zn/g dry weight.

Very severe Zn deficiency reduced carbonic anhydrase activity to a non-detectable level (Fig. 3). Zinc concentration of 10 μg/g in blades was associated with non-detectable levels of carbonic anhydrase activity. As Zn concentration in blades was increased to 16 μg/g where maximum growth was attained, carbonic anhydrase activity increased sharply to 400 EU/g. The enzyme activity continued to rise sharply in a curvilinear manner as Zn concentration in blades increased to 48 μg/g. Apparently Zn is one of the limiting factors to carbonic anhydrase activity since no maximum was attained.

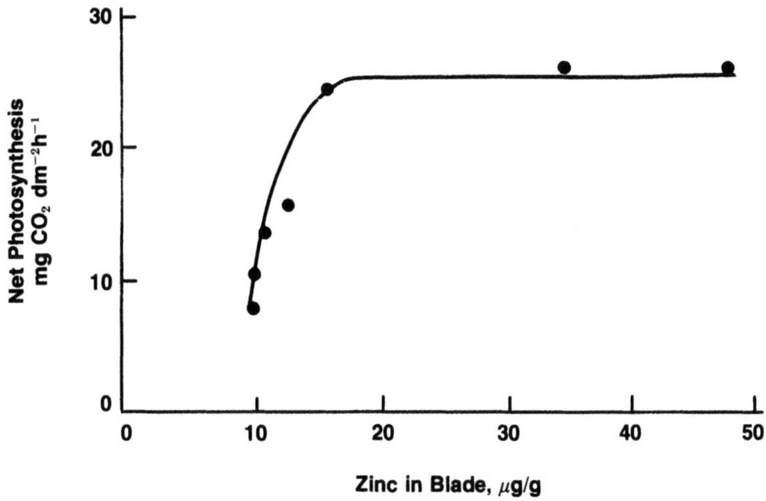

Fig. 2. Relationship of Zn status in recently matured blade and net photosynthesis. Zinc critical deficiency level is 14μg Zn/g dry weight.

The influence of Zn deficiency on amino acid concentrations in recently matured leaves falls into three general categories (Table 1). In group 1, Zn deficiency reduced amino acid concentration; group II, Zn deficiency had no effect on amino acid concentration; and group III, Zn deficiency increased amino acid concentration in leaf blades. The group I amino acids were glycine, alanine, valine, leucine, isoleucine, serine, threonine, glutamic acid, lysine, cystine, methionine, proline, phenylalanine, tyrosine and tryptophan. The relationship of amino acid concentrations and Zn concentration in blades of group I are typified by the glutamic acid and tryptophan responses indicated in Figs. 4 and 5. The response curve indicates a deficient, transition and adequate zones of Zn nutrition. The critical Zn concentration of 14 μg/g is in the transition zone and is the Zn concentration associated with a 10% reduction from maximum glutamic acid in blades. The critical Zn concentration for other group I amino acids range from 11 to 15 μg/g Zn (Table 1). All critical values are essentially alike in that approximately 11 to 14 μg/g Zn and greater are required for maximum amino acid concentrations in the group I category.

Histidine was the only amino acid measured that was not influenced by Zn nutrition from severe deficiency to adequacy (Fig. 6).

Aspartic acid and arginine were the two amino acids in group III category in which Zn deficiency induced an increase in amino acid concentration (Fig. 7 and Table 1). Aspartic acid increased approximately 38% due to very severe Zn deficiency. As Zn concentration in blades increased from 10 to 16 μg/g, aspartic acid concentration decreased from 35.2 mg/g to a minimum of 25.6 mg/g with a critical Zn concentration of 15 μg/g (Table 1). Arginine accumulated approximately 3.4-fold under severe Zn deficiency as compared to adequate Zn nutrition (Table 1).

Table 1. Effect of Zn nutrition on amino acid concentration and critical Zn concentration for amino acids in cotton leaf blades.

Amino Acid	Zn rates, μg/l 10	20	50	100	500	1,000	2,000	Critical Zn Concentration
	mg/g dry wt.							μg/g
Glycine	7.74e*	9.01d	10.76c	12.05b	14.45a	14.45a	15.13a	13
Alanine	7.73d	8.84d	11.07c	12.77b	16.06a	15.91a	16.62a	13
Valine	8.11d	9.21d	11.52c	12.98b	15.83a	15.92a	16.49a	14
Leucine	13.1d	14.4d	17.7c	20.4b	24.9a	25.0a	25.9a	14
Isoleucine	7.02d	7.74d	9.34c	10.41b	12.31a	12.32a	12.86a	14
Serine	7.62e	9.20d	10.09c	10.48bc	11.58a	10.87abc	11.46ab	11
Threonine	6.48e	7.29d	8.65c	9.73b	11.95a	12.01a	12.36a	13
Glutamic Acid	16.4c	17.6c	20.6b	22.4b	26.2a	26.1a	26.7a	14
Lysine	10.6d	11.5d	13.7c	15.3b	17.9a	17.9a	18.7a	14
Cystine	0.81b	0.84b	1.06a	1.00ab	1.05a	1.21a	1.05a	11
Methionine	5.09c	5.29bc	5.48bc	5.70b	6.16a	6.22a	6.27a	12
Proline	10.9b	11.2b	12.0b	12.9b	16.6a	16.8a	17.7a	14
Phenylalanine	7.80e	9.06d	11.11c	12.77b	15.78a	15.72a	16.33a	14
Tyrosine	4.70e	5.74d	6.78c	8.01b	10.26a	10.34a	10.81a	14
Tryptophan	2.46c	2.61c	3.15b	3.32b	3.87a	3.91a	3.97a	13
Histidine	6.00a	5.86a	6.27a	6.63a	6.65a	5.94a	6.02a	–
Aspartic Acid	35.2a	33.6ab	33.4ab	28.9bc	25.6c	25.2c	26.1c	13
Arginine	56.0a	43.4b	40.3bc	32.4c	17.5d	15.8d	16.5d	15
Total Amino Acid	219d	218d	239c	243bc	257ab	255ab	265a	

* Means followed by the same letter within rows are not significantly different at the 5% level.

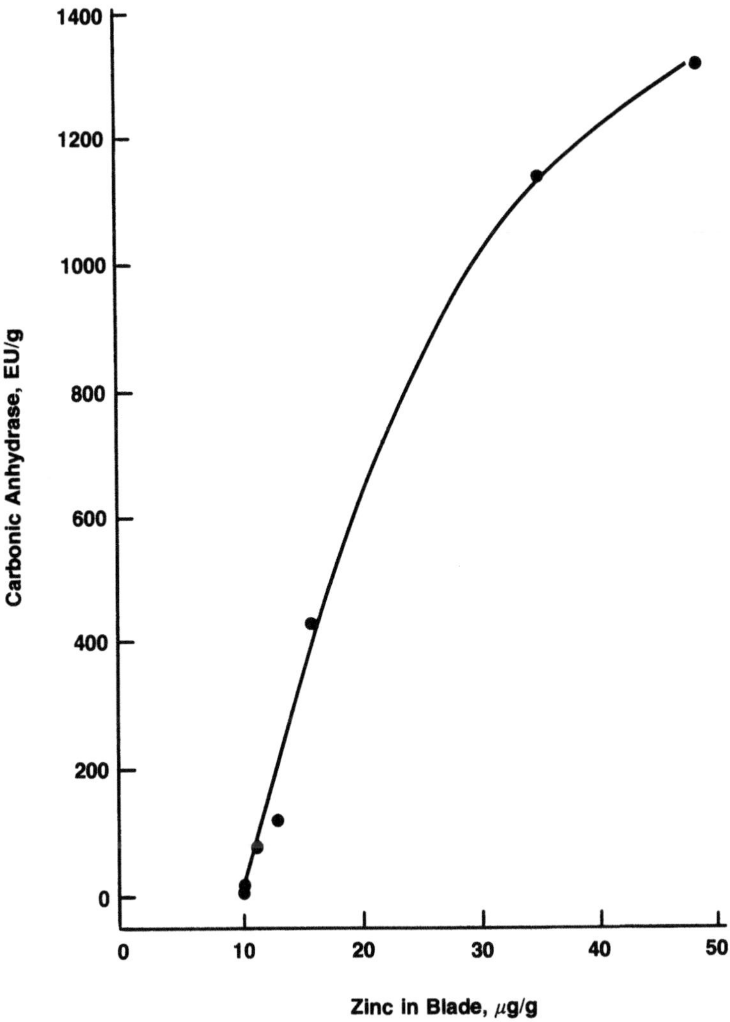

Fig. 3. Influence of Zn status in recently matured blades on carbonic anhydrase activity.

DISCUSSION

The level of zinc nutrition influenced greatly the plant growth processes related to plant dry matter production, photosynthesis and total amino acid concentrations in recently matured cotton leaves. The concept of critical deficiency levels is significant in this study to relate these processes to zinc status. The critical deficiency level as proposed by Ulrich & Hills (1969) is defined as the nutrient concentration in the tissue associated with an approximate 10% reduction from maximum activity. The critical nutrient concentration is used as

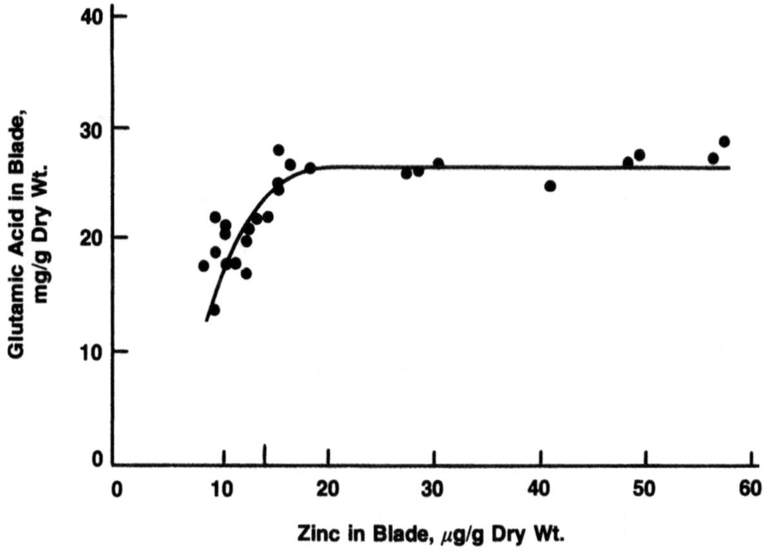

Fig. 4. Glutamic acid concentration as influenced by Zn status in recently matured blades. Critical deficiency level is 14 μg Zn/g dry weight.

a guideline for separating nutrient deficient plants and nutrient sufficient plants. In our study, the critical deficiency level of Zn for plant growth as dry matter production was 13 μg/g dry weight in recently matured blades. This value agrees closely with the Zn critical deficiency level reported by Ohki (1975) in an earlier study on critical deficiency and toxicity levels related to cotton growth. Critical Zn deficiency levels of similar magnitude have been reported for soybean (*Glycine max* L.) (Melsted *et al.*, 1969; Jones, 1967; Ohki, 1978) and corn (*Zea mays* L.) (Viets *et al.*, 1953). Zinc critical deficiency level of 9μg/g has been reported by Rosell & Ulrich (1964) in sugar beets grown in nutrient solution. The data suggest that for these plant species, Zn concentrations in recently matured blades ranged from 9 to 15 μg/g to achieve near maximum yield.

Inadequate Zn supply greatly reduced the photosynthetic activity in cotton blades. Photosynthesis in very severe Zn deficient blades was 31% of controls that were provided with adequate Zn blades. Randall & Bouma (1973) reported that photosynthesis in low Zn leaves of spinach (*Spinacia oleracea* L.) was 70% of controls on a leaf area basis. On a chlorophyll basis, Botrill *et al.* (1970) concluded that all nutrient deficiencies except Fe and Mo reduced carbon assimilation by spinach. However, these studies did not report the critical nutrient concentration in blade tissues associated with near maximum photosynthetic activity. In our study the critical deficiency level of Zn for photosynthesis in cotton was 14 μg/g as indicated from the data relating Zn concentration in blades and CO_2 fixation rates. The critical deficiency level reported in this study agrees with the critical deficiency level of 12 μg/g for soybean (*Glycine max* L.) reported by Ohki (1978). These studies indicate that

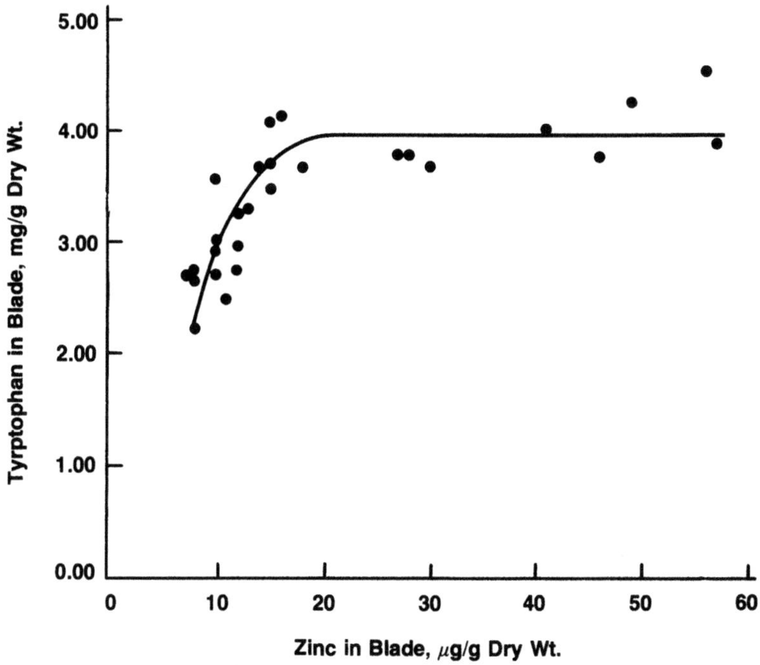

Fig. 5. Relationship of tryptophan concentration and Zn in recently matured blade. Critical deficiency level is 13 μg Zn/g dry weight.

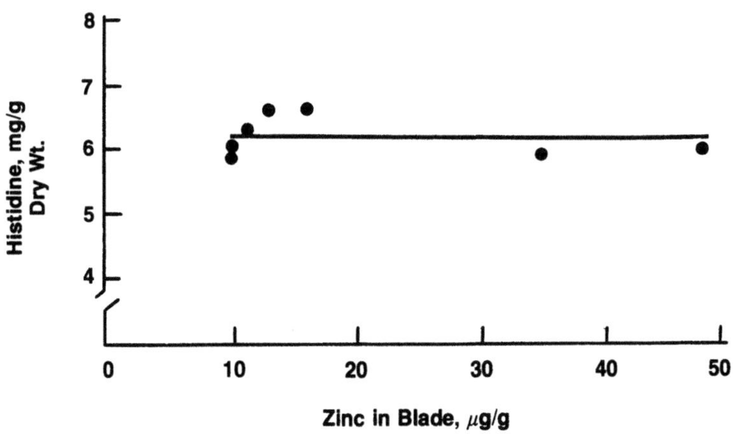

Fig. 6. Non-influential effect of Zn status on histidine concentration in recently matured blade.

a minimal range of 12 to 14 μg Zn/g is required in recently matured blades for near maximum photosynthetic activity. Critical nutrient levels for K (Estes *et al.*, 1973; Peaslee & Moss, 1966), Mg (Peaslee & Moss, 1966; Terry & Ulrich, 1973b) and sulfate-S (Terry 1976) have been reported. In most cases, the critical deficiency levels of nutrients for photosynthesis and growth were approximately the same. This was also the case for Zn in our present study on cotton and for another study on soybeans (Ohki, 1978).

Carbonic anhydrase activity in cotton leaves was dependent on Zn status. No activity was detected in blades from plants grown under very severe deficient conditions. These results that Zn deficiency reduces carbonic anhydrase agree with results of other workers (Randall & Bouma, 1973; Bar-Akiva & Lavon, 1969; Wood & Sibly, 1952). From the two relationships between Zn concentration in blades and carbonic anhydrase and Zn concentration in blades and photosynthesis, carbonic anhydrase does not limit photosynthesis when plants are supplied with adequate Zn. Maximum photosynthesis is attained at about 14 to 48 μg/g Zn while carbonic anhydrase continues to increase. Randall & Bouma

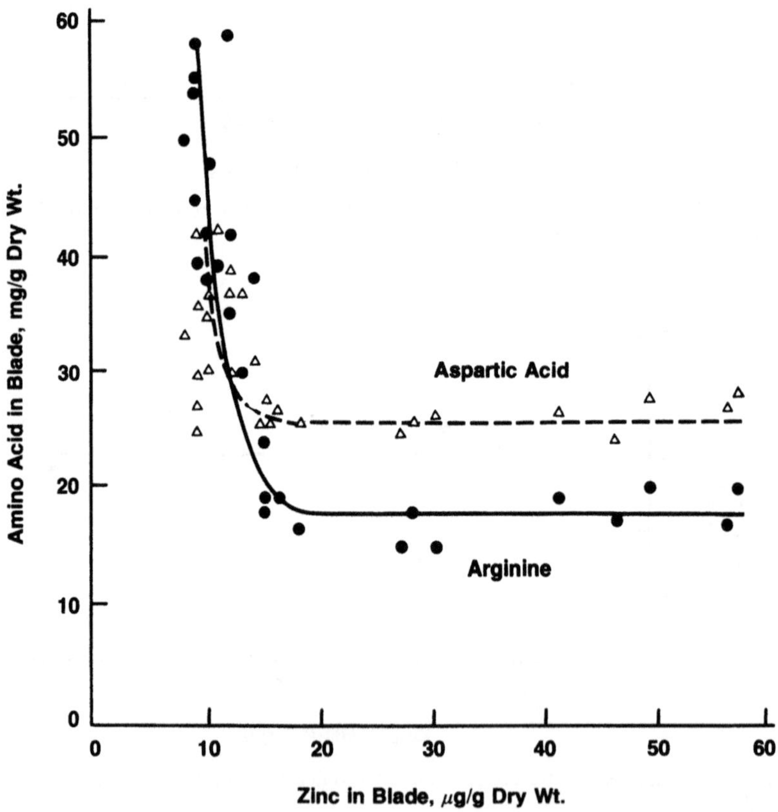

Fig. 7. Aspartic acid and arginine concentrations related to Zn status in recently matured blades. Critical deficiency levels are 13 and 15 μg Zn/g dry weight.

(1971) concluded that carbonic anhydrase does not have the role of facilitating the diffusion of CO_2 to carboxylation sites within the chloroplast as proposed by other workers (Hatch & Slack, 1970).

Zinc deficiency influences amino acid metabolism as evidenced by a decrease in total concentration of fifteen amino acids, an increase in aspartic acid and arginine, and no effect on histidine. In another preliminary study on bean (*Phaseolus vulgaris* L.) zinc deficiency decreased the extracted amino acid glycine and isoleucine levels and had no effect on serine, valine, phenylalanine, glutamic acid, alanine, methionine leucine and threonine (Schütte & Schendel, 1958). Our results agree for only the decrease in glycine and isoleucine. The lack of decreases in other amino acids reported by Schütte and Schendel (1958) is attributed to the extracted rather than total amino acids.

Stewart (1962) reported that Zn deficiency in orange (*Citrus* sp.) decreased free amino acid content of alanine and increased serine, threonine, proline and arginine in deficient leaves. Increases in free amino acid in tomato (*Lycopersicon esculentum* L.) plants by Zn, Cu and Mn deficiency have been reported by Possingham (1956). Aspartic acid and arginine in Zn deficient leaves increased about 33 and 340%, respectively, of control leaves. Our results on arginine accumulation induced by Zn deficiency agree with those reported by Stewart (1962) where free arginine accumulated 120-fold in Zn deficient orange leaves. Arginine accumulation due to nutrient deficiencies has been reported for Fe in macadamia (*Macadamia tetraphylla* L.) (Gilfillan & Jones, 1968), Mn in cauliflower (Hewitt *et al.*, 1949), and B in peanut (*Arachis hypogaea* L.), (Shiralipour *et al.*, 1969). It is concluded that the level of micronutrient nutrition greatly influences the amino acid concentration in leaves.

The influence of Zn status on tryptophan in leaves is of interest. Zinc in plant nutrition is commonly associated with auxin (Rains, 1976). Skoog (1940) reported that Zn deficiency resulted in the excessive destruction of auxin probably through oxidation. Later, Tsui (1948) concluded that Zn is required for tryptophan synthesis and indirectly for auxin synthesis. However, Takaki and Kushizaki (1970) reported conflicting data that tryptophan levels in maize increased under very severe Zn deficiency. This may be a pathological response from extreme zinc deficiency.

The continuous relationship between growth processes and Zn status ranging from very severe deficiency to more than adequate Zn levels was accomplished by the determination of critical deficiency level for Zn through plant analysis. Zinc critical deficiency levels for growth, photosynthesis and amino acids were about 14 μg Zn/g dry matter in recently matured blades. These data suggest that a minimum Zn level of about 14 μg/g is required for maximum dry matter production, photosynthesis and all amino acids except aspartic acid, arginine and histidine. Although aspartic acid and arginine approached a minimum of about 14 μg/g, the minimum concentrations were considered as near normal levels of these amino acids. Apparently Zn deficiency does not inhibit the formation of precursors of these two amino acids and the data suggest that Zn may be involved in the degradation of two amino acids.

Zinc concentrations less than the critical deficiency level inhibited growth and photosynthesis through inhibition of metabolic processes requiring zinc.

Inhibition of amino acid synthesis of 15 amino acids and the accumulation of aspartic acid and arginine were attributed to the dearrangement of amino acid metabolism. It is significant that the normalcy of amino acid accumulation was dependent on maintaining Zn concentrations equal to or greater than the critical deficiency level determined by plant growth.

REFERENCES

Bar-Akiva, A. & R. Lavon. Carbonic anhydrase activity as an indicator of zinc deficiency in citrus leaves. *J. Hort. Sci.* 44: *359–362* (1969).

Barber, A.V. & R. Bradfield. Effect of potassium and nitrogen on the free amino acid content on corn plants. *Agron. J.* 55: *465–470* (1963).

Botrill, D.E., J.V. Possingham & P.E. Kriedemann. The effect of nutrient deficiencies on photosynthesis and respiration in spinach. *Plant and Soil* 32: *424–438* (1970).

Edwards, G.E. & A.K. Mohamed. Reduction in carbonic anhydrase activity in zinc deficient leaves of *Phaseolus vulgaris* L. *Crop Sci.* 13: *351–354* (1973).

Estes, G.O., D.W. Koch & T.F. Bruetsch. Influence of potassium nutrition on net CO_2 uptake and growth in maize (*Zea mays* L.) *Agron. J.* 65:972-975 (1973).

Gilfillan, I.M. & W.W. Jones. Effect of iron and manganese deficiency on the chlorophyll amino acid and organic acid status of leaves of Macademia. *Proc. Am. Soc. Hort. Sci.* 93: *210–214* (1968).

Haghiri, F. Influence of macronutrient elements on the amino acid composition of soybean plants. *Agron. J.* 58: *609–612* (1966).

Hatch, M.D. & C.R. Slack. Photosynthetic CO_2-fixation pathways. *Ann. Rev. Plant Physiol.* 21: *141–162* (1970).

Hewitt, E.S., E.W. Jones & A.H. Williams. Relation of molybdenum and manganese to the free amino-acid content of the cauliflower. *Nature* 163: *681–682* (1949).

Huglie, T.E. & S. Moore. Determination of the tryptophan content of proteins by ion exchange chromatography of alkaline hydrolysates. *J. Biol. Chem.* 247: *2828–2834* (1972).

Jones, J.B. Jr. Interpretation of plant analysis for several agronomic crops, p. 49 to 58, in: Soil Testing and Plant Analysis, Part II, Plant Analysis, Soil Sci. Soc. of Am. Inc., Madison, WI (1967).

Lewis, L.N., N.E. Tolbert & A.C. Kenworthy. Influence of mineral nutrition on the amino acid composition of Bartlett pear trees. *Proc. Am. Soc. Hort. Sci.* 83: *185–192* (1963).

Melsted, S.W., H.L. Motto & T.R. Peck. Critical plant nutrient composition values useful in interpreting plant analysis data. *Agron. J.* 61: *17–20* (1969).

Mertz, E.T., V.L. Singleton & C.L. Garey. The effect of sulfur deficiency on the amino acids of alfalfa. *Arch. Biochem. Biophy.* 38: *139–145* (1952).

Ohki, K. Lower and upper critical zinc levels in relation to cotton growth and development. *Physiol. Plant.* 35: *92–100* (1975).

Ohki, K. Effect of zinc nutrition on photosynthesis and carbonic anhydrase activity in cotton. *Physiol. Plant.* 38: *300–304* (1976).

Ohki, K. Zinc concentration in soybean as related to growth, photosynthesis, and carbonic anhydrase activity. *Crop Sci.* 18: *79–82* (1978).

Ozbun, J.L., R.T. Volk & W.A. Jackson. Effect of potassium deficiency on photosynthesis, respiration and the utilization of photosynthetic reductant by immature bean leaves. *Crop Sci.* 5: *69–75* (1965).

Peaslee, D.E. & D.N. Moss. Photosynthesis in K- and Mg- deficient maize (*Zea mays* L.) leaves. *Soil Sci. Am. Proc.* 30: *220–223* (1966).

Possingham, J.V. The effect of mineral nutrition on the content of free amino acids and amides in tomato plants. *Aust. J. Biol. Sci.* 9: *539–551* (1956).

Rains, D.W. Mineral metabolism, p. 561 to 597, in: J. Bonner and J.E. Varner (ed.). Plant Biochemistry. Academic Press, New York (1976).

Randall, P.J. & D. Bouma. Zinc deficiency, carbonic anhydrase, and photosynthesis in leaves of spinach. *Plant Physiol.* 52: *229–232* (1973).

Reed, L.W., V.L. Sheldon & W.A. Albrecht. Soil Fertility and the organic composition of plants lysine arginine and aspartic acid variations. *Agron. J.* 52: *523–526* (1960).
Rosell, R.A. & Ulrich. Critical zinc concentrations and leaf minerals of sugar beet plants. *Soil Sci.* 97: *152–167* (1964).
Ryle, G.J.A. & T.D. Hesketh. Carbon dioxide uptake in nitrogen-deficient plants. *Crop Sci.* 9: *451–454* (1969).
Schütte, K.H. & H.E. Schendel. Influence of trace elements upon plant protein composition. *Nature* 182: *958–959* (1958).
Shiralipour, A., H.C. Harris & S.H. West. Boron deficiency and amino acid and protein contents of peanut leaves. *Crop. Sci.* 9: *455–456* (1969).
Skoog, F. Relationships between zinc and auxin in growth of higher plants. *Am. J. Bot.* 27: *939–951* (1940).
Spackman, D.H., W.H. Stein & S. Moore. Automatic recording apparatus for use in the chromatography of amino acids. *Anal. Chem.* 30: *1190–1206* (1958).
Stewart, I. The effect of minor element deficiencies on free amino acids in citrus leaves. *Proc. Am. Soc. Hort. Sci.* 81: *244–249* (1962).
Takaki, H. & M. Kushizaki. Accumulation of free tryptophan and tryptamine in zinc deficient maize seedling. *Plant Cell Physiol.* 11: *793–804* (1970).
Terry, N. Effects of sulfur on the photosynthesis of intact leaves and isolated chloroplasts of sugar beets. *Plant Physiol.* 57: *477–479* (1976).
Terry, N. Photosynthesis, growth, and the role of chloride. *Plant Physiol.* 60: *69–75* (1977).
Terry, N. & R.P. Huston. Effects of calcium on the photosynthesis of intact leaves and isolated chloroplasts of sugar beets. *Plant Physiol.* 55: *923–927* (1975).
Terry, N. & A. Ulrich. Effects of phosphorus deficiency on the photosynthesis and respiration of leaves of sugar beet. *Plant Physiol.* 51: *43–47* (1973a).
Terry, N. & A. Ulrich. Effects of potassium deficiency on the photosynthesis and respiration of leaves of sugar beet. *Plant Physiol.* 51: *783–786* (1973b).
Terry, N. & A. Ulrich. Effects of magnesium deficiency on the photosynthesis and respiration of leaves of sugar beet. *Plant Physiol.* 54: *379–381* (1974a).
Terry, N. & A. Ulrich. Photosynthetic and respiratory CO_2 exchange of sugar beet leaves as influenced by manganese deficiency. *Crop Sci.* 14: *502–504* (1974b).
Tsui, C. The role of zinc in auxin synthesis in the tomato plant. *Am. J. Bot.* 35: *172–178* (1948).
Ulrich, A. & F.J. Hills. Sugar beet nutrient deficiency symptoms. 36 pp. University of California, Division of Agricultural Sciences (1969).
Viets, F.G. Jr., L.C. Boawn, C.L. Crawford & C.E. Nelson. Zinc deficiency in corn in central Washington. *Agron. J.* 45: *559–565* (1953).
Wilbur, K.M. & N.G. Anderson. Electrometric and colorimetric anhydrase. *J. Biol. Chem.* 176: *147–154* (1948).
Wood, J.G. & P.M. Silby. Carbonic anhydrase activity in plants in relation to zinc content. *Aust. J. Sci. Res.* B. 5: *244–255* (1952).

Authors' address:
Departments of Agronomy and Food Science
Georgia Experiment Station
University of Georgia
Experiment, 6A 30212
U.S.A.

PHOTOSYNTHESIS IN *PHASEOLUS VULGARIS* L., AS INFLUENCED BY SUPRA-OPTIMAL ZINC NUTRITION*

F. VAN ASSCHE, H. CLIJSTERS & R. MARCELLE
(Diepenbeek and Sint-Truiden, Belgium)

ABSTRACT

Uptake from the root medium of supra-optimal amounts of zinc decreases growth, dry matter production and total leaf chlorophyll content. Apparent photosynthesis is significantly reduced, dark respiration is not affected.

Increase in total leaf diffusive resistance can account for this effect on photosynthesis, but other mechanisms of action are possible.

Non-cyclic electron transport in chloroplast fragments was found to be reduced by zinc concentrations of about 0.5 mM, supplied to the reaction medium. This inhibitory effect could not be confirmed in chloroplasts from plants, grown under conditions of zinc stress. These plants show significant inhibition of different enzymes, related to the photosynthetic and photorespiratory CO_2-exchange.

INTRODUCTION

When supplied to the root medium, several heavy metals (*e.g.*, thallium, cadmium, nickel, ...) inhibit net or apparent photosynthesis of the plant (Carlson *et al.*, 1975).

After addition to the reaction medium some of them affect the activity of isolated chloroplasts. Reduction of CO_2-fixation by lead (Hampp *et al.*, 1973a), aluminium (Hampp & Schnabl, 1975), cadmium and zinc (Hampp *et al.*, 1976), of non-cyclic electron transport by cadmium and zinc (Van Duyvendijke-Matteoli & Desmet, 1975; Hampp *et al.*, 1976), of ATP-production (Hampp *et al.*, 1973a) and activities of different Calvin-cycle enzymes (Hampp *et al.*, 1973b) by lead, are described. In several cases the inhibitory effect shown on isolated chloroplasts *in vitro* was taken as explanation of the effect observed in the living plant.

As a consequence of industrial pollution the zinc content of the soil can increase considerably, inducing severe zinc uptake by the roots. In these conditions plant growth and dry matter production are inhibited, and leaf pigment content is reduced (Cottenie *et al.*, 1976).

In the present paper we investigate to what extent uptake of supra-optimal but sublethal amounts of zinc by the roots can interfere with plant photosynthesis.

* Skilful technical assistance of Mrs. C. Bogaert-Vanherle and Mr. W. Brugmans is gratefully acknowledged.

This study was supported by the Instituut tot Aanmoediging van het Wetenschappelijk Onderzoek in Nijverheid en Landbouw (I.W.O.N.L.), Brussels.

MATERIAL AND METHODS

Plant material

Dwarf beans, *Phaseolus vulgaris* L. c.v. Limburgse Vroege, were cultivated in a growth chamber at 25° C, 80% relative humidity and a light intensity of 140 microeinsteins per m^2 per sec (PAR) during the 12 hr day period, at 21° C and 80% relative humidity during the night. Seeds were soaked for 4 hr in polyethylene pots of 11 cm diameter, containing 400 ml vermiculite, 100 ml distilled water and 50 ml nutrient solution (5 g NH_4NO_3; 2.5 g $Ca(H_2PO_4)_2$; 2.5 g K_2SO_4; 2.5 g $MgSO_4$ per liter). Zinc sulphate was applied before sowing as aqueous solutions of different concentration, resulting in a final metal concentration in the pot of 0, 50, 100 or 200 ppm (w/v). During plant growth distilled water was added if necessary. Except for the estimation of leaf diffusive resistance, primary leaves were utilized 17 to 23 days after sowing for the different analyses and activity measurements.

Leaf content

Chlorophyll analyses were performed according to MacKinney (1941). The method of Lowry *et al.* (1951) was utilized for protein analysis. Zinc was measured by atomic absorption.

Gas exchange

CO_2- and O_2-exchange of leaves attached to the plant was measured in an open system, as described by Marcelle (1975). Light intensity on the leaves was 250 microeinsteins per m^2 per sec (PAR), and was measured with a LI-COR quantum photometer LI 170, provided with a LI-190S quantum sensor.

Leaf resistance

Diffusive resistance for water vapor of the abaxial leaf surface was examined with a LI-COR diffusive resistance meter LI 60, provided with an horizontal sensor LI-155. As *Phaseolus vulgaris* has hypostomatous leaves, the resistance of the adaxial leaf surface is 4 to 6 times higher. Calibration of the apparatus and measurements were performed in the growth chamber. Plants were abundantly watered 16 hrs before measuring.

Photosynthetic electron transport

As the method of Cockburn *et al.* (1968) for isolating intact chloroplasts gave very poor yields on bean leaves, non-cyclic electron transport was measured on chloroplast fragments, prepared according to Oben & Marcelle (1975). Reduction of methylviologen with water or DCIP/HAsc as electron donor was estimated as already described by these authors. Reduction of ferricyanide was followed by measuring O_2 production in a Rank oxygen electrode at 25° C with light saturation. The reaction medium was saturated with N_2 and contained 0.3

ml Sorbitol 0.66 M, 0.2 ml Tricine pH 7.5 0.5 M, 0.3 ml $MgCl_2$ 0.3 M, 0.08 ml NaCl 0.5 M, 0.2 ml K_2HPO_4 0.1 M, 0.2 ml EDTA (2 Na) 0.02 M, 0.2 ml $K_3Fe(CN)_6$ 0.025 M, 0.59 ml H_2O. Measurement of the effects of $ZnSO_4$ in the reaction medium were corrected for change of pH.

Enzyme activities

Measurement of enzyme activities was performed on a crude extract, prepared by grinding 2 g leaf tissue in a chilled mortar with 20 ml tris-HCl buffer 0.1 M pH 8, containing 1 mM EDTA and 5 mM dithiothreitol. The slurry was filtered by 8 layers of cheese cloth and 1 layer of nylon, and the filtrate obtained was used for estimating enzyme activities.

RuDPC was measured according to Lilley *et al.* (1974). For NADPH-glyoxylate reductase, NADH-hydroxypyruvate reductase and glycolate oxidase, the methods, described by Murray *et al.* (1973) were applied. Catalase was measured according to Bergmeyer *et al.* (1974) and serine-glyoxylate-aminotransferase according to Rehfeld *et al.* (1972). For this enzyme test endogenous pyridoxal-5-phosphate and glyoxylate reductase concentrations were found to be in excess; therefore they were not added to the reaction medium. Carbonic anhydrase was estimated following the method of Ohki (1976).

RESULTS

Zinc nutrition and plant growth

Total plant growth and dry matter production were optimal when the zinc concentration in the primary leaf was about 200 ppm on a dry matter basis (Fig. 1). Lower zinc content resulted in deficiency and growth inhibition. When supra-optimal amounts of zinc were applied, leaf content increased up to 400 ppm and more, and plant growth and total dry matter production progressively declined. In these conditions less total chlorophyll was found in the primary leaf, but dry matter and protein content per fresh weight unit were significantly higher, as compared to plants receiving optimal zinc nutrition (Table 1).

Table 1. Content of the primary leaf in relation to zinc nutrition.

ppm Zn in nutrient solution[1]	ppm Zn in dry matter	total chlorophylls (mg) per cm²	total chlorophylls (mg) per g fresh weight	% dry matter in fresh weight	mg protein per g fresh weight
0	38.1	49.5 ± 3.3[2]	2.85 ± 0.23	9.20 ± 0.68	15.28 ± 4.18
100	257.6	48 ± 2.2	2.91 ± 0.23	9.91 ± 1.05	15.89 ± 3.37
200	343.3	46.5 ± 6.8	2.54 ± 0.45[3]	10.40 ± 0.86[3]	19.96 ± 4.39[3]

[1] For details *cfr.* "Material and Methods"
[2] Means ± standard errors
[3] Significant difference ($p = 0.05$; t-test)

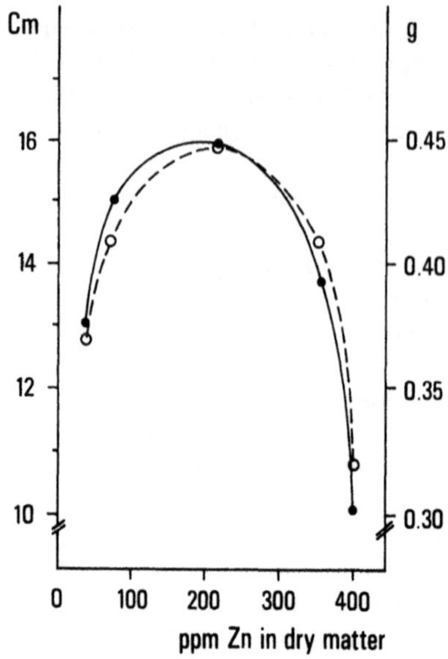

Fig. 1. Mean shoot length (●) and dry matter per plant (○) in relation to the Zn-content of the primary leaf.

Effect of zinc on apparent photosynthesis and dark respiration

Apparent photosynthesis, measured as CO_2-fixation, was significantly reduced in the primary leaves of plants, grown with supra-optimal amounts of zinc (Table 2). This result is confirmed by data on photosynthetic O_2-production. Dark respiration was not affected by heavy zinc nutrition; a slight decrease of this activity (significant on a fresh weight basis only) could be observed in control plants with zinc deficiency.

Table 2. Apparent photosynthesis and dark respiration (μmoles CO_2 per hr) in primary leaves as a function of zinc nutrition.

		control without zinc	zinc nutrition optimal	supra-optimal
apparent photosynthesis	per g fresh weight	97	103	72.5[2]
	per g dry matter	1,046	1,054	705[2]
	per mg total chlorophyll	34	35.5	29[1]
	per mg protein	6.8	6.8	3.9[2]
	per leaf area (dm^2)	151	147	116.5[2]
dark respiration	per g fresh weight	6.4[1]	8.3	8.5
	per g dry matter	69	83	80
	per mg protein	0.46	0.55	0.44
	per leaf area (dm^2)	10	12	13.8

[1] significant difference ($p = 0.05$; t = test).
[2] significant difference ($p = 0.01$; t-test)

Up to 15 days after sowing the diffusive resistance of the primary leaf was independent of the zinc uptake by the plant (Fig. 2). Later on a sharp increase of this resistance was observed in plants, receiving supra-optimal amounts of zinc. When grown in deficient or optimal conditions, the primary leaves did not show any appreciable change before day 25, when the first trifoliate leaf was completely developed and the diffusive resistance suddenly rose.

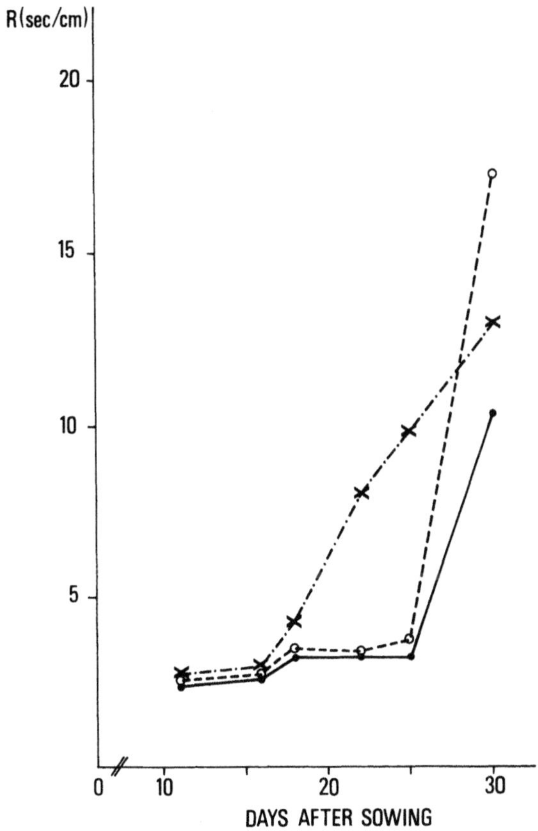

Fig. 2. Diffusive resistance to water vapor of the primary leaf of plants grown with different zinc regimes. Each point represents the mean of 40 measurements on 10 different plants.
(●) control without zinc (deficiency)
(○) optimal zinc nutrition
(×) supra-optimal zinc nutrition

Similar effects of supra-optimal zinc nutrition were observed when leaf diffusive resistance was measured on the subsequently developing trifoliate leaves.

Electron transport from water to methylviologen declined in chloroplasts isolated from primary leaves, when zinc was applied to the reaction medium as sulphate or nitrate at concentrations above 0.5 mM (Fig. 3A). Potassium sulphate had no effect.

Similar inhibition by zinc was observed for the Hill-reaction, measured by ferricyanide reduction (Fig. 3B), and for the reduction of methylviologen with DCIP/HAsc as electron donor (Fig. 3C). Both activities were completely suppressed for zinc concentrations above 1 mM in the reaction medium.

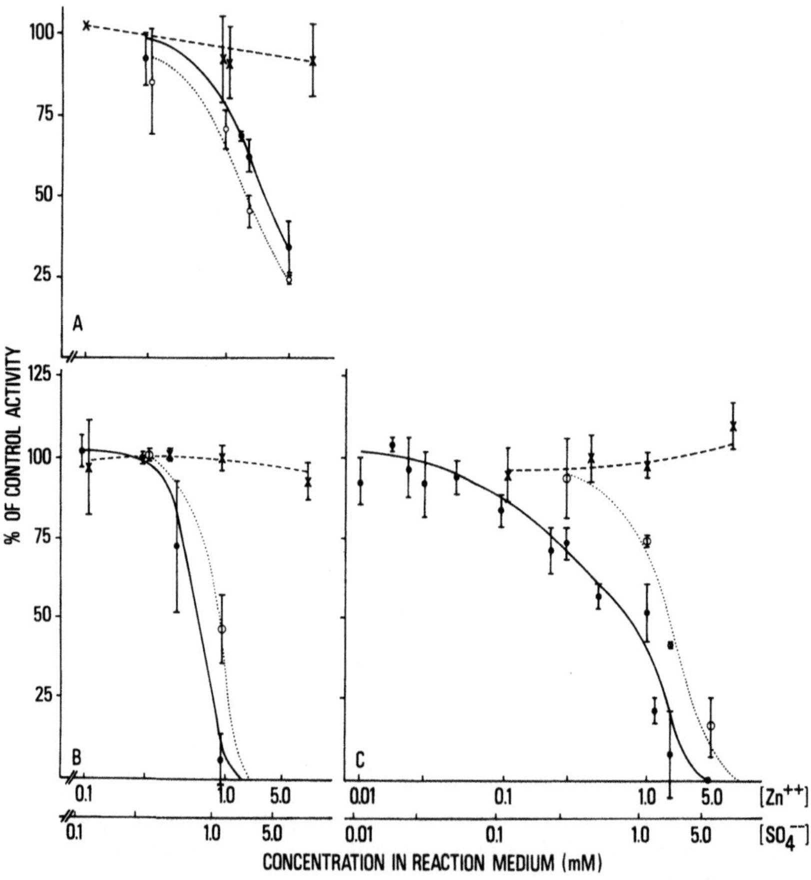

Fig. 3. Photosynthetic electron transport in chloroplasts after addition of different salts to the reaction medium.

A – Reduction of methylviologen with water as electron donor
B – Reduction of ferricyanide
C – Reduction of methylviologen with DCIP/HAsc.
(●) $ZnSO_4$
(○) $Zn(NO_3)_2$
(×) K_2SO_4

However, when chloroplasts were isolated from primary leaves of plants, provided with suboptimal, optimal or supra-optimal amounts of zinc, no significant decrease of the non cyclic electron transport activities could be observed (Table 3). On the contrary, when zinc nutrition was supra-optimal, reduction of methylviologen with DCIP/HAsc significantly increased.

Table 3. Photosynthetic electron transport (microequivalents per mg chlorophyll per hr) in chloroplast fragments, isolated from primary leaves of plants, grown at various zinc concentrations.

	control without zinc	zinc nutrition optimal	zinc nutrition supra-optimal
reduction of ferricyanide	208	224	236
reduction of methylviologen with			
- DCIP/HAsc	252	256	312[1]
- water	156	168	176

[1] significant difference (p = 0.05; t = test).

Enzyme activities

The activity of the zinc containing chloroplast enzyme carbonic anhydrase, expressed per protein, was significantly reduced in control plants without zinc, as compared to plants, receiving optimal zinc nutrition (Table 4). Only a negligible decline in activity could be observed when supra-optimal amounts of zinc were provided. In these conditions however the RUDPC and NADPH-glyoxylate reductase activities were significantly reduced. A similar decrease in activity per protein with supra-optimal zinc nutrition was found for the peroxisomal enzymes glycolate oxidase, serine-glyoxylate-amino-transferase and NADH-hydroxypyruvate reductase. Catalase was not affected by the zinc nutrition applied.

Table 4. Enzyme activities (milli-units per mg protein) in the primary leaf in relation to zinc nutrition.

	control without zinc	zinc nutrition optimal	zinc nutrition supra-optimal
Carbonic anhydrase ($\times 10^3$)	127[1]	257	202
RudP-carboxylase	179	192	133[1]
NADPH-glyoxylate reductase	70	85	59[1]
Glycolate oxidase	112	108	60[1]
Serine-Glyoxylate amino-transferase	53	62	31[1]
NADH-hydroxypyruvate reductase	350	386	222[1]
Catalase ($\times 10^3$)	80	84	75

[1] Significant difference (p = 0.05; t = test).

DISCUSSION

Due to the zinc content of the seeds (about 50 ppm of dry weight) and of the vermiculite used, zinc deficiency was not very pronounced in our experimental conditions. Nevertheless the results on control plants confirm the data of Botrill *et al*. (1970) on spinach, showing that zinc deficiency decreases respiration and photosynthesis, although in our conditions the decline of apparent photosynthesis is not significant.

Supra-optimal amounts of zinc in the root medium change leaf content and decrease total dry matter production of *Phaseolus vulgaris*, as was already shown for other plant species by Cottenie *et al*. (1976). In these conditions apparent photosynthesis is inhibited. This effect is not due to a possible breakdown of pigments by high zinc concentrations in the leaf (*cf* Table 1): even apparent photosynthesis per mg chlorophyll significantly declines (Table 2).

Reduction of apparent photosynthesis was also observed by Carlson *et al*. (1975) on corn and sunflower after application of different other heavy metals to the root medium. Direct inhibition of stomatal opening was considered to be an important factor explaining the effect on apparent photosynthesis. A rise of leaf diffusive resistance (Fig. 2) after zinc stress can partly or completely account for the decrease of photosynthesis, found in our experiments. As the glycolate-glyoxylate cycle is probably involved in the regulation of stomatal opening (Zelitch, 1971), the observed increase in leaf diffusive resistance could be attributed to the inhibitory effect of zinc on glycolate oxidase and NADPH-linked glyoxylate reductase (table 4).

Thallium also interfered with the biophysical and biochemical reactions of photosynthesis (Carlson *et al., l.c.*). Experiments with intact spinach chloroplasts in the presence of different amounts of zinc brought Hampp *et al*. (1976) to the same conclusion. Our data on chloroplast fragments confirm their results, showing that non cyclic photosynthetic electron transport is inhibited by the metal, present in the reaction medium. In our experimental conditions however this inhibition is not a valid explanation for the decline of apparent photosynthesis *in vivo*. The non cyclic electron transport of chloroplasts, isolated from plants with severe zinc inhibition of apparent photosynthesis, is not affected (Table 3). Taking into account that the primary leaves of such plants contain 400 ppm zinc (Fig. 1), and assuming that 10% (Ernst, 1972) to 20% (Polar, 1976) of the metal is associated to the plastids, we calculate that the zinc concentration per unit chlorophyll was 50 to 100 times lower in the leaf than in the reaction medium containing 0.5 mM zinc and inhibiting photosynthetic electron transport in isolated chloroplasts.

The suggestion of Hampp *et al*. that zinc as well as other heavy metals could interact with the biochemical mechanism of photosynthetic CO_2-fixation, is directly supported by the decrease in activity of the RUDPC (Table 4) after supra-optimal zinc nutrition. Carbonic anhydrase only declined with zinc deficiency, as already shown by Ohki (1976).

True photosynthesis is the sum of apparent photosynthesis and photorespiration. The inhibition of different peroxisomal enzymes argues for a decrease of photorespiration by zinc, resulting in a further decline of true photosynthesis. Protein purification of all measured enzymes is necessary to exactly assess the zinc effect on their specific activities.

It is however clear that zinc, assimilated by the roots in supra-optimal amounts, interferes with leaf photosynthesis at different levels from diffusive leaf resistance till biochemical CO_2-fixation.

ABBREVIATIONS

DCIP = 2–6, dichlorophenol indophenol
HAsc = ascorbic acid
mU = milliunit
RUDPC = Ribulose 1,5-diphosphate carboxylase
PAR = Photosynthetic, active radiation.

REFERENCES

Bergmeyer, H.U., K. Gawehn & M. Grassl. Enzymes as biochemical reagents. pp. 425–522, in: Bergmeyer H.U. (ed.), Methods of Enzymatic Analysis, vol. I. Verlag Chemie, Weinheim; Academic Press Inc., New York, London (1974).

Botrill, D.E., J.V. Possingham & P.E. Kriedemann. The effect of nutrient deficiencies on photosynthesis and respiration in spinach. *Plant Soil* 32: *424–438* (1970).

Carlson, R.W., F.A. Bazzaz & G.L. Rolfe. The effect of heavy metals on plants II. Net photosynthesis and transpiration of whole corn and sunflower plants treated with Pb, Cd, Ni and Tl. *Environm. Res.* 10: *113–120* (1975).

Cockburn, W., D.A. Walker & C.W. Baldry. The isolation of spinach chloroplasts in pyrophosphate media. *Plant Physiol.* 43: *1415–1418* (1968).

Cottenie, A., A. Dhaese & R. Camerlynck. Plant quality response to uptake of polluting elements. *Qual. Plant. Pl. Fds. Hum. Nutr.* XXVI: *293–319* (1976).

Ernst, W. Zink- und Cadmium-Immissionen auf Böden und Pflanzen in der Umgebung einer Zinkhütte. *Ber. Deutsch. Bot. Ges.* 85: *295–296* (1972).

Hampp, R., H. Ziegler & I. Ziegler. Die Wirkung von Bleiionen auf die $^{14}CO_2$-Fixierung und die ATP-Bildung von Spinatchloroplasten. *Biochem. Physiol. Pflanzen.* 164: *126–134* (1973a).

Hampp, R., H. Ziegler & I. Ziegler. Der Einfluss von Bleiionen auf Enzyme des reduktiven Pentosephosphatcyclus. *Biochem. Physiol. Pflanzen.* 164: *588–595* (1973b).

Hampp, R. & H. Schnabl. Effect of aluminium ions on $^{14}CO_2$-fixation and membrane system of isolated spinach chloroplasts. *Z. Pflanzenphysiol.* 76: *300–306* (1975).

Hampp, R., K. Beulich & H. Ziegler. Effects of zinc and cadmium on photosynthetic CO_2-fixation and hill activity of spinach chloroplasts. *Z. Pflanzenphysiol.* 77: *336–344* (1976).

Lilley, R. McC & D.A. Walker. An improved spectrophotometric assay for ribulosebiphosphate carboxylase. *Biochem. Biophys. Act.* 358: *226–229* (1974).

Lowry, O.H., N.J. Rosebrough, A.L. Farr and R.J. Randall. Protein measurement with the folin phenol reagent. *J. biol. Chem.* 193: *265–275* (1951).

MacKinney, G. Absorption of light by chlorophyll solutions. *J. Biol. Chem.* 140: *315–322* (1941).

Marcelle, R. Effect of photoperiod on the CO_2 and O_2 exchanges in leaves of *Bryophyllum daigremontianum* (Berger). pp. 349–356, in: Marcelle R. (ed.), Environmental and Biological Control of Photosynthesis. Junk, The Hague (1975).

Murray, D.R., O. Wara-Aswapati, H.M.M. Ireland & J.W. Bradbeer. The Development of activities of some enzymes concerned with glycolate metabolism in greening bean leaves *J. Exp. Bot.* 24: *175–184* (1973).

Oben, G. & R. Marcelle. The effects of CCC and GA on some biochemical and photochemical activities of primary leaves of bean plants. pp. 211–216, in Marcelle R.

(ed.), Environmental and Biological Control of Photosynthesis. Junk, The Hague (1975).

Ohki, K. Effect of zinc nutrition on photosynthesis and carbonic anhydrase activity in cotton. *Physiol. Plant*. 38: *300–304* (1976).

Polar, E. Variations in zinc content of subcellular fractions from young and old roots, stems and leaves of broad beans (*Vicia Faba*). *Physiol. Plant*. 38: *159–165* (1976).

Rehfeld, D.W. & N.E. Tolbert. Aminotransferases in peroxisomes from spinach leaves. *J. Biol. Chem*. 247: *4803–4811* (1972).

Van Duyvendijk-Matteoli, M.A. & G.M. Desmet. On the inhibitory action of cadmium on the donor side of photosystem II in isolated chloroplasts. *Biochim. Biophys. Act.* 408: *164–169* (1975).

Zelitch, I, Photosynthesis, photorespiration and plant productivity. Academic Press Inc., New York, London (1971).

Authors' address:
Limburgs Universitair Centrum
Department SBM
Universitaire Campus
B-3610 Diepenbeek
Belgium.

EFFECTS OF CALCIUM DEPRIVATION ON THE INTRACELLULAR DISTRIBUTION OF THE EARLY PRODUCTS OF PHOTOSYNTHESIS IN TOMATO LEAF CELLS*

P. RANGNEKAR
(Mississauga, Ontario, Canada)

ABSTRACT

Distribution of the early products of photosynthesis between plastids and their cytoplasmic environment in the leaf cells of calcium-deprived and calcium-supplied tomato plants was compared.

Leaves allowed to photoassimilate $^{14}CO_2$ for increasing time intervals ranging from 30 seconds to 1 hour, were separated into chloroplast and non-chloroplast fractions using non-aqueous medium. The labeled compounds in the two fractions were separated and their radioactivities measured.

In +Ca leaf cells, the label appeared outside the chloroplasts very early, increased rapidly and equalled the label within the chloroplasts in less than 5 minutes. In −Ca leaf cells, however, it took much longer for the label to appear outside the chloroplasts and it remained considerably lower than the label within the chloroplasts even after one hour. A decrease in the rate of appearance and accumulation of ^{14}C-products of photosynthesis outside the calcium-deprived chloroplasts was evident.

Significant differences in the radioactivities of starch and sucrose in the chloroplasts of the two leaf tissues were evident. These differences also reflected upon the possibility that the transport of the early products of photosynthesis between the plastids and cytoplasm was restricted in calcium-deprived leaf cells.

INTRODUCTION

A number of publications have appeared on the effects of removal of calcium from the nutrition of plants, in which a multitude of its functions have been inferred at molecular, cellular or whole plant level. Accumulation of various metabolites, mainly carbohydrates, occurs in the leaves accompanied by a cessation of the terminal growth of plant, under calcium starvation (Gauch, 1940; Groom, 1896; Joham, 1957; Joham & Johnson, 1973; Nightingale, 1937). Investigation to determine the cause and effect relationship between these two events, revealed that accumulation of assimilates in calcium-deprived tomato leaves occurred because of their decreased export out of the source leaf and not as a consequence of their decreased demand for plant growth (Rangnekar, 1974). Furthermore, a study of the carbon metabolism in photosynthesis and respiration of calcium-starved tomato plants indicated a decrease in the utilization of ^{14}C-photosynthate in respiration, cell wall synthesis

* Financial support from National Research Council of Canada is gratefully acknowledged. Thanks are also due to Mr. S. Mankikar for technical assistance, and Miss Leena R. for typing the manuscript.

and in other reactions which normally occur outside the chloroplasts (Rangnekar, 1975). A restriction in the movement of the early products of photosynthesis at the chloroplast level was suspected.

The exchange of material between the plastid and its cytoplasmic environment is known more by inference than from direct measurement *in vivo* (Arnon, 1961). The fate of early products of carbon fixation and especially their use in extra-plastid reactions, would depend, among other things, on the ease with which these products move out of the chloroplast. Since calcium ions are known to be essential for the normal functioning of membranes (Kalra, 1956; Marinos, 1962), the removal of calcium from the nutrition of plant could easily upset this transfer between plastids and the rest of the cytoplasm. An attempt was therefore made to study (comparatively) the transfer of early ^{14}C-products of photosynthesis between the chloroplasts and the hyaloplasm of calcium-deprived tomato leaf cells and the findings are reported in this communication.

MATERIAL AND METHODS

Plant material

Seeds of tomato (*Lycopersicon esculentum*, var. Starfire) were germinated in vermiculite. Two-week-old seedlings of uniform size were transferred to half-strength nutrient solution (Hoagland & Arnon, 1938) in 4-liter mason jars provided with adequate aeration and placed on growth benches under controlled environmental conditions with regard to illumination (1200 ft. c.), day length (14hr), relative humidity (80%) and temperature (25° C daytime, 22° C at night). After ten days' growth, about 24 to 36 seedlings selected again for uniform growth were supplied with full strength Hoagland's nutrient solution, half of them deprived of calcium by the omission of calcium nitrate but supplemented by equivalent amounts of potassium nitrate. Minus-calcium (−Ca) plants thus received 100 ppm calcium for 10 days prior to its total omission from their nutrition, whereas control plants (+Ca) received 100 ppm for the first 10 days and 200 ppm thereafter until the time of their harvest. Potassium concentration for the minus-calcium plants was, however, twice (*i.e.* 390 ppm) that available for the control plants during the experimental period. The nutrient solutions were renewed every three days.

Experimental

On the third day of calcium deprivation, a set of three (or five) plants each of −Ca and +Ca treatments were used for feeding experiment, in which a single leaf was exposed to $^{14}CO_2$ for a certain length of time, either 30 seconds, 1 minute, 5 minutes, 15 minutes, 30 minutes or 1 hour. All feedings were carried out at the same time of the day in a fume hood, well-illuminated with banks of cool white fluorescent and incandescent lights (illumination 2,000 ft. c., temperature 26 ±1° C). The third leaf of each plant was enclosed in a plexiglass leaf chamber made especially for the purpose and sealed with plasticene around its petiole. Radioactive carbon dioxide from 100 microcuries of sodium carbonate was introduced in the leaf chamber. In trial runs under similar conditions this amount of $^{14}CO_2$ could be fixed within 20 minutes by the control leaf (+Ca). In experiments exceeding 15 minutes photoassimilation, therefore, the leaf chamber was briefly flushed with $^{12}CO_2$ at the end of 15-minute-exposure to $^{14}CO_2$ and the leaf was allowed to photosynthesize for the remaining length of time under the same conditions.

Immediately at the end of the feeding time, the leaf was plunged into liquid nitrogen and then lyophilized at −18°C until dry. The time lapse between the end of the experiment and the complete freezing of the leaf did not exceed 5 seconds. Extreme care was exercised to prevent thawing of the leaves during lyophilization. The water condensation trap was cooled with a dry ice-acetone mixture.

Chloroplast Isolation

Separation of plastids from non-plastid cytoplasm and cell walls of the leaf tissues was carried out by nonaqueous method using hexane/carbon tetrachloride as the suspending medium (Stocking, 1959). To ensure minimum of plastid rupture, small (0.1g) samples of lyophilized leaf were ground in a hand model tissue grinder using ice cold hexane/carbon tetrachloride mixture (density 1.32). Two or three minutes of gentle grinding was sufficient to reduce the sample to a uniform homogenate. The ground samples per treatment were combined and fractionated into three components, *viz*., chloroplasts, cell walls or cell debris, and supernatant, by following Stocking's method (*loc. cit.*) of particle sedimentation. Mixtures of hexane and carbon tetrachloride in different proportions were made to obtain the required densities of the suspending medium. The density of chloroplasts appropriate for the experimental leaf age in the two treatments of the tomato plants was determined by trial runs (Jagendorf, 1955). Of the three components so obtained, the chloroplast fraction (density 1.32 to 1.38) contained some nuclei, the denser (greater than 1.38) cell wall fraction (cell debris) was not entirely free of unbroken cells with trapped chloroplasts and the supernatant (non-plastid cytoplasm, density less than 1.32) was clear but light green, because of some chlorophyll removal which occurred primarily during the initial grinding period from chloroplasts broken during grinding. There was no evidence of any leaching from the chloroplasts during isolation mainly because the early products of photosynthesis were essentially insoluble in the suspending medium (Stocking *et al*., 1963). The magnitude of chloroplast contamination in the 'non-chloroplast' fractions was determined by chlorophyll analyses and appropriate corrections were made. Aliquots of each fraction were extracted with 80% acetone and optical density readings were taken at 652mμ to determine chlorophyll concentrations by Arnon's method (Arnon, 1949).

Extraction

Each fraction of the leaf tissue was extracted in boiling 80% ethanol three times using a small volume of ethanol each time. Ethanol-insoluble residues were ground to a finer uniform size, if necessary, prior to their extraction in perchloric acid to dissolve and separate starch from the cell wall residue (Pucher *et al*., 1948).

Radioactivity assay

Radioactivity measurements of extracts and residues were carried out in a liquid scintillation counter (Packard Tricarb Model No. 3375) using appropriate aliquots in 20ml of scintillation mixture (6g PPO+0.3g POPOP per liter of toluene).

Radioactivity in each fraction was expressed both as per cent of the initial dose given, as well as, in absolute amounts (microcuries) recovered.

Ethanol and water-soluble compounds in each fraction were separated by paper chromatography. Radioactive compounds were located by radioautography and identified by specific chemical reactions when in adequate amounts otherwise by cochromatography with authentic sample in different solvent systems. Radioactivity measurements of the compounds were done by using an aliquot of the eluate from the paper in the scintillation mixture described above.

RESULTS

The amount of $^{14}CO_2$ photoassimilated by +Ca and −Ca leaves in each time interval, ranging from 30 seconds to 1 hour, is shown in Table 1.

Table 1. Per cent C^{14} fixed by calcium-deficient and the control leaf. (Dose 100 μCi).

Treatment	30s	1m	5m	15m	30m	60m
+Ca	25.0*	40.8	66.3	83.8	94.2	98.7
	29.1	44.1	70.1	79.4	100.5	99.9
	26.7	46.6	70.7	81.7	100.3	100.7
−Ca	12.2	23.9	35.0	60.8	70.6	80.6
	12.1	25.8	41.3	56.9	72.1	77.1
	14.4	28.9	39.6	58.7	70.5	82.3

* microcuries or percent.

Calcium-deficient leaf tissue was less than half as efficient (48%) as the control leaf in short term photosynthesis (30 seconds) but the difference was not as great under longer periods. In 15 minutes (15m), the deficient leaf fixed little more than half of the available $^{14}CO_2$ and attained about 72% efficiency as compared to that of the control leaf. This was consistent with the photosynthetic efficiency of −Ca tomato leaves exposed to non-radioactive carbon dioxide under similar conditions, observed earlier (Rangnekar, 1975). It required less than 30 minutes for the control leaf to fix all the available radiocarbon under the conditions of the experiment whereas the deficient leaf could not fix much more than 80% (by average) of the available ^{14}C even by the end of 1 hour of photosynthesis.

Intracellular distribution of radioactivity in the leaf cells of the control (+Ca) and the deficient (−Ca) tissues for different time intervals is presented in Table 2. The values represent per cent of the total ^{14}C fixed by the leaf tissues.

In thirty seconds (0.5m), from about 14% to 31% of the photoassimilated ^{14}C

Table 2. Intracellular distribution of radioactivity in leaf cells of calcium-deprived (−Ca) and control (+Ca) tomato plants.

	Per cent of total C^{14} fixed					
	In chloroplasts		Outside the chloroplasts			
Exposure			Supernatant		Cell Debri	
to $C^{14}O_2$	+Ca	−Ca	+Ca	−Ca	+Ca	−Ca
0.5m	86.1	91.4	10.5	5.4	3.4	3.2
	82.8	89.1	14.1	6.8	3.1	4.1
	76.8	86.9	20.3	8.3	2.9	4.8
	74.6	85.6	21.5	10.1	3.9	4.3
	70.1	83.6	26.8	12.0	3.1	4.4
1m	71.3	86.3	24.6	9.5	4.1	4.2
	69.6	84.6	26.6	10.2	3.8	5.2
	67.9	83.5	28.2	12.1	3.9	4.4
	62.5	80.8	33.2	14.5	4.3	4.7
	59.8	78.2	36.2	16.7	4.0	5.1
5m	53.5	83.8	40.6	10.4	5.9	5.8
	52.8	80.1	41.9	15.0	5.3	4.9
	46.3	79.7	49.2	14.8	4.5	5.5
15m	51.3	79.8	41.8	14.4	6.9	5.8
	49.2	78.7	44.3	15.8	6.5	5.5
	43.6	76.9	50.1	17.4	6.3	5.7
30m	49.1	77.8	44.4	17.2	6.5	5.0
	46.0	75.3	46.3	18.5	7.7	6.2
	45.7	74.8	47.3	19.2	7.0	6.0
60m	50.2	75.8	42.0	18.7	7.8	5.5
	45.3	74.2	46.8	19.5	7.9	6.3
	44.1	72.6	48.7	19.8	7.2	7.6

had migrated out of the control (+Ca) chloroplasts whereas only about half as much (from 8% to 16%) moved out of the calcium-deficient ones in the same time interval. The difference became greater over longer periods of time. In 5 minutes (5m), the difference was about two- to three-fold with the average migration of about 49% of the fixed radiocarbon out of the control chloroplasts as compared to about 19% from the deficient ones. With additional time interval beyond 5 minutes, the chloroplast and the hyaloplasm (supernatant) of the control leaf tissue reached an equilibrium with regard to the synthesis and the movement of the ^{14}C products of photosynthesis and of the subsequent reactions. The only appreciable increase in the radioactivity here onwards was evident in the cell wall fraction (cell debris) of the control leaf cells, as would be expected. The chloroplasts of the calcium-deficient leaf cells, on the other hand, retained a bulk (75%) of the total fixed ^{14}C even by the end of one hour of photosynthesis. A meagre amount (5 to 7%) of ^{14}C-products moved out of these chloroplasts in the additional 55 minutes and there was no appreciable gain in the radioactivity of the cell wall fraction.

Analysis of the ethanol-soluble chloroplast and nonchloroplast fractions revealed information about the identity of the specific compounds and their relative rates of movement outside the chloroplasts.

Within 30 seconds of exposure of the leaves to $^{14}CO_2$, radioactive sucrose appeared in the chloroplasts of the control leaf cells. Sucrose had the bulk (60–70%) of the soluble radiocarbon in the plastids. Appreciable ^{14}C-sucrose did not appear in the non-chloroplast fraction of these cells until after one minute. This gave credence to the method of isolation of the chloroplasts adopted in the present investigation, in that, there was no detectable leaching from the chloroplasts during the process of their isolation. In the chloroplasts of calcium-deficient leaf cells, radioactive sucrose was detected in trace amounts after 1 minute of photosynthesis and it did not increase appreciably with additional time. No radioactive sucrose was detected outside calcium-deficient chloroplasts even after 15 minutes of photosynthesis. Appearance of ^{14}C-sucrose in extra plastid cytoplasm could be a result of diffusion from the plastid or a result of secondary synthesis occurring outside the plastid, utilizing simpler ^{14}C-products that diffused out earlier. Apparently, neither of these two events occurred to an appreciable extent in calcium-deprived leaf cells.

Analysis of the ethanol-insoluble residues of the chloroplasts presented another difference in the two treatments. A bulk (70–80%) of the radioactivity of the −Ca chloroplasts, from all time periods, was present in insoluble form extractable in perchloric acid and identified as starch. In control chloroplasts, on the other hand, the bulk of the radioactivity was in soluble form, mostly in sucrose (50–60%), with a small proportion (15–20%) in insoluble form, mainly starch.

Secondly, radioactive starch appeared early in both chloroplasts, within 30 seconds. It increased considerably more with time, reaching 90% of the total fixed ^{14}C in the calcium-deficient chloroplasts. In the control chloroplasts, it reached its highest level in one minute and leveled off thereafter acquiring not more than 20% of the total label.

DISCUSSION

The variability of the results reported within one time period is not any greater than would be expected from a leaf tissue which is composed of a wide variety of cells, many of which may not photosynthesize at the same time. Even within the population of photosynthesizing cells a wide difference in rate and amount of $^{14}CO_2$ fixed within the shorter time series undoubtedly exists, which precluded undertaking experiments with less than 30 seconds' duration.

Prior to the $^{14}CO_2$ feeding, the leaves were illuminated in closed leaf chambers without added CO_2 level and would result in the photosynthetic build up of ATP and NADPH in the cells just prior to the $^{14}CO_2$ feeding. It would thus increase the efficiency of the leaf tissues (both +Ca and −Ca) for photosynthetic carbon fixation and the subsequent reactions determining the intracellular fate of these early products. Furthermore, assimilates accumulated previously in the leaf cells, if any, would be lowered, thus decreasing their slowing influence on these reactions mentioned above. All this was intended to ensure enough labeled products in the chloroplasts so that in the subsequent study of their migration out of the plastids it would be easier to detect differences, if any, between the two leaf cells (+Ca and −Ca).

Although calcium-starved tomato leaf cells had 30 to 50% lower efficiency for photosynthetic fixation of $^{14}CO_2$, there was sufficient label in the −Ca chloroplasts, in each time interval, to study the migration of the early products out of the plastids.

As would be expected, the per cent of total fixed carbon present in the chloroplasts at the end of the photosynthesis period decreased with increasing time of photosynthesis in both leaf tissues. The highest, at 30 seconds interval, was 86% label in +Ca and 91% label in −Ca chloroplasts. Decrease in the chloroplast label with concurrent increase in the extraplastid portion of the leaf cells was evidence of the movement of the early products of photosynthesis out of the chloroplasts. In that respect, it was very clear (Table 2), that the migration of ^{14}C-products out of the calcium-deficient chloroplasts was 50% slower in short term exposures to $^{14}CO_2$ and decreased further to about a third of the control, with longer exposures. In five minutes, the level of radioactivity moving outside the control chloroplasts equalled that which was being fixed newly within, and this balance was evident in the extended exposures of the leaf tissues to $^{14}CO_2$. In calcium-deficient leaf cells, on the other hand, such a balance between synthesis of new carbon compounds within plastids and their migration out of the plastids was not attained even by the end of 1 hour exposure of leaf cells to $^{14}CO_2$. Instead, due to a restriction on their movement, there was a considerable build up of the early products within these deficient chloroplasts and their consequential channeling into starch synthesis occurred.

Restriction in the movement of the carbon products out of the chloroplasts was evident in a leaf which predominantly exports to the terminal meristem of the plant at a time when growth in extension is still occuring at the meristem (Rangnekar, 1974). In the control leaf, the ^{14}C-products moved rapidly out of the chloroplasts to the sites of respiratory or secondary synthetic activities in the cytoplasm. As a result of this, radioactive carbon could be detected, on a large scale, in the respired carbon dioxide or in the cell wall polymers, but this was not the case in calcium-deficient leaf cells (Rangnekar, 1975).

Restriction in the movement of early products of photosynthesis in calcium-deprived leaf tissue was inferred from observations made earlier (Rangnekar 1974, 1975). The present findings are in conformity with this inference. The main purpose of the present investigation was to locate the site of this restriction in the leaf tissue. To this end the findings are sufficient to conclude that this restriction occurs at the chloroplast level. Although the exact mechanism of the influence of Ca^{++} ions on the normal functioning of membrane is not known, their deficiency in nutrition causes initially a restriction in the movement of early carbon products out of the chloroplasts. This restriction may lead to an excessive accumulation of assimilates in calcium-deficient leaf tissues reported by a number of workers.

REFERENCES

Arnon, D.I. Changing concepts of photosynthesis. *Bull. Torrey Bot. Club* 88: *215–259* (1961).

Gauch, H.G. Response of the bean plant to calcium deficiency. *Plant Physiol.* 15: *1–21* (1940).
Groom, P. On the relation between calcium and the transportation of carbohydrates in plants. *Ann. Bot.* 10: *91–96* (1896).
Hoagland, D.R. & D.I. Arnon. The water-culture method of growing plants without soil. *Calif. Agr. Expt. Sta. Circ.* No. 374: *32* (1938).
Jagendorf, A.T. Purification of chloroplasts by a density technique. *Plant Physiol.* 30: *138–143* (1955).
Joham, H.E. Carbohydrate distribution as affected by calcium deficiency in cotton. *Plant Physiol.* 32: *113–117* (1957).
Joham, H.E. & L. Johnson. The effect of sodium and calcium on the translocation of ^{14}C-sucrose in excized cotton roots. *Physiol. Plantarum* 28: *121–126* (1973).
Kalra, G.S. Responses of the tomato plant to calcium deficiency. *Bot. Gaz.* 118: *18–37* (1956).
Marinos, N. Studies on submicroscopic aspects of mineral deficiencies. 1. Ca^{++} deficiency in the shoot apex of barley. *Am. J. Botany* 49: *834–841* (1962).
Nightingale, G.T. Potassium and calcium in relation to nitrogen metabolism. *Botan. Gaz.* 98: *725–734* (1937).
Pucher, G.W., A.J. Wakeman & C.S. Leavenworth. Determination of starch in plant tissue. *Anal. Chem.* 20: *850–853* (1948).
Rangnekar, P.V. Effect of calcium deficiency on the translocation of ^{14}C photosynthate in tomato plants. *Plant and Soil* 41: *589–600* (1974).
Rangnekar, P.V. Effect of calcium deficiency on the carbon metabolism in photosynthesis and respiration of tomato leaf. *Plant and Soil* 42: *565–583* (1975).
Stocking, C.R. Chloroplast isolation in nonaqueous media. *Plant Physiol.* 34: *56–61* (1959).
Stocking, C.R., G.R. Williams & A. Ongun. Intracellular distribution of the early products of photosynthesis. *Biochem. Biophys. Res. Commun.* 10(5): *416–421* (1963).

Author's address:
3571 Gallager Drive
Mississauga, Ontario
Canada.

THE INFLUENCE OF KINETIN ON THE BIOSYNTHESIS OF CHLOROPHYLL*

C. BUSCHMANN
(Karlsruhe, Fed. Rep. of Germany)

ABSTRACT

Seedlings of *Raphanus sativus* L. (Saxa Treib) were grown without (control) or with added kinetin (2 ppm = 9.29 μM). During growth in the dark kinetin induces a higher accumulation of total protochlorophyll(ide), especially of its phototransformable form $P_{657-650}$. During a 30 min. dark period after phototransformation by one flash of light less protochlorophyll(ide) is resynthesized in 6-day-old kinetin-treated seedlings. In older plants kinetin causes the Shibata shift to be more accomplished. After longer illumination times the 77° K absorption spectra for cotyledons from kinetin-treated seedlings show a relatively higher absorption for long wavelength chlorophyll forms. This is interpreted as a rise in the amount of chlorophyll directly participating in light energy trapping. Kinetin induces a higher accumulation of P700, the reaction center chlorophyll of pigment system I. The amount of photosynthetic active chlorophylls connected with pigment system II, as measured by the 680 nm fluorescence kinetic, is slightly increased in kinetin-treated plants. It is concluded that kinetin does not only induce a higher accumulation of chlorophyll and protochlorophyll(ide) but also a change in the amount of the different *in vivo* chlorophyll forms, which leads to a more effective photosynthetic apparatus.

INTRODUCTION

The rise in the accumulation of total chlorophyll induced by cytokinins is well documented (Sugiura, 1963; Banerji & Laloraya, 1967; Beevers *et al.*, 1970; Fletcher & McCullagh, 1971a, b; Yakushkina & Pushkina, 1971; Fletcher *et al.*, 1973; Straub & Lichtenthaler, 1973; Woźny & Szweykowska, 1975) and was already proposed as a cytokinin bioassay with cucumber cotyledons (Fletcher & McCullagh, 1971b).

In some cases an increase of the chlorophyll a/b ratio due to the action of kinetin was reported (Straub & Lichtenthaler, 1973; Bengtson *et al.*, 1977).

The influence of kinetin on the synthesis of protochlorophyll(ide), the direct precursor of chlorophyll, was investigated mainly with excised leaves. Kinetin was found to increase the amount of accumulated protochlorophyll(ide) during illumination (Banerji & Laloraya, 1967) and in the dark (Yakushkina & Pushkina, 1971). According to Shlyk (Shlyk & Averina, 1969; Shlyk *et al.*, 1970; Walter *et al.*, 1977) due to kinetin more protochlorophyll(ide) is resynthesized after phototransformation during a 24 hour dark period. However, another working group could not find an effect of kinetin on the

* This work was partly sponsored by a NATO and an EMBO fellowship.

resynthesis of protochlorophyll(ide) during an 8 hour dark period (Bengtson *et al.*, 1977).

Cytokinins are also reported to increase the synthesis of δ-aminolevulinic acid, the essential intermediate of the tetrapyrrole pathway (Fletcher & McCullagh, 1971a; Fletcher *et al.*, 1973). This effect was explained as a stimulation of the δ-aminolevulinic acid synthetase in the presence of light.

In the studies described here the influence of kinetin on the biosynthesis of chlorophyll was investigated with special regard to the different protein and lipid bound *in vivo* pigment forms.

The accumulation of these *in vivo* forms was followed by measuring the absorption spectra of intact leaves (Shibata, 1957; Goedheer, 1967; Sironval *et al.*, 1968). The 680 nm fluorescence kinetic was taken as an indicator for the amount of photosynthetic active chlorophyll connected with pigment system II (Lavorel & Etienne, 1977).

MATERIALS AND METHODS

Seedlings of *Raphanus sativus* L. (Saxa Treib) were grown on tap water. Added kinetin was present in the water during germination and growth (2 ppm = 9.29 μM).

Manipulations in the dark were carried out with a green safety light (540 nm).

Protochlorophyll(ide) was extracted with acetone from the cotyledons ($\varepsilon = 3.56 \times 10^4$ l$\times$Mol$^{-1}\times$cm^{-1} after Koski & Smith, 1948).

The 77° K absorption spectra were measured with a CARY 17R spectrophotometer (see Dujardin, 1976). For the calculation of the absorption ratios the changes of the spectral baseline were corrected. A 1 msec polychromatic flash was used for illumination (Multiblitz Report Proba, electric energy: 125 J, color temperature: 5800° K).

The concentration of P700 was measured according to Marsho & Kok (1971) with a Shimadzu MPS 5000 spectrophotometer ($\varepsilon = 6.4 \times 10^4$ l$\times$Mol$^{-1}\times$cm^{-1} after Hiyama & Ke, 1972). Chloroplasts isolated (according to Nobel, 1967) from radish seedlings grown one day in the dark and 7 days under fluorescent light (fluora, 7500 mw/m^2) were used.

For the 680 nm fluorescence kinetic dark adapted cotyledons were excited with a 441 nm light (He-Cd-Laser, Type 401, Liconix). The fluorescence was measured by a photomultiplier (Type 7265, RCA) through RG 665 (2 mm, Schott u. Gen., Mainz) and IF 680 (Shimadzu) filters. For these experiments seedlings were grown 4, 6, or 8 days in total darkness and then transferred to the light (Osram HQ, 5.6×10^4 mw/m^2).

RESULTS

The content of total protochlorophyll(ide) accumulated in cotyledons of etiolated radish seedlings practically does not change between day 4 and day 8 after germination. This is true both for control and kinetin-treated plants (Fig. 1). Cotyledons from 4-, 6-, and 8-day-old seedlings grown in the presence of kinetin always show higher amounts of total protochlorophyll(ide) than the controls.

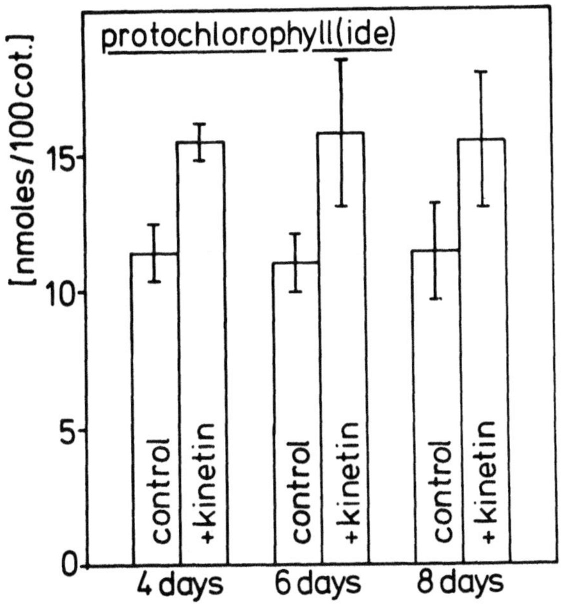

Fig. 1. The amount of total protochlorophyll(ide) (nmoles per 100 cotyledons) extracted from cotyledons of 4-, 6-, and 8-day-old etiolated radish seedlings grown without (control) or with added kinetin.

77° K absorption spectra of etiolated radish cotyledons exhibit peaks at 650 and 637 nm (Fig. 2). The absorption at 650 nm (phototransformable protochlorophyll(ide)) compared to that at 627 nm (non-phototransformable protochlorophyll(ide)) decreases first, between day 4 and day 6 following germination. Afterwards the ratio A650/A627 shows a gradual increase for control seedlings. The etiolated cotyledons from kinetin-treated plants exhibit always a relatively higher absorption at 650 nm; in contrast to the controls the A650/A627 ratio decreases slightly between day 8 and day 13 after germination.

Immediately after phototransformation of protochlorophyll(ide) by one single flash of light the *in vivo* 77° K absorption spectra of radish cotyledons exhibit peaks at 627 and 677 nm (Fig. 3). Spectra measured after a 30 min dark period following phototransformation have the long wavelength peak shifted to 672 nm and new absorption maxima appear at 650 and 637 nm due to protochlorophyll(ide) resynthesized in the dark. With increasing age of the seedlings this absorption at 650 and 637 decreases relatively to that at 672 nm. The A650/A672 ratio decreases most after day 4 and day 6 following germination for kinetin-treated and control plants, respectively. Therefore seedlings grown in the presence of kinetin show smaller amounts of resynthesized protochlorophyll(ide) at day 6. Before or after this developmental stage practically no differences in the A650/A672 ratio can be seen.

30 sec after the phototransformation by one flash of light the 77° K absorption peak is found at 680 nm and is subsequently shifted to 672 nm (Fig. 4). This

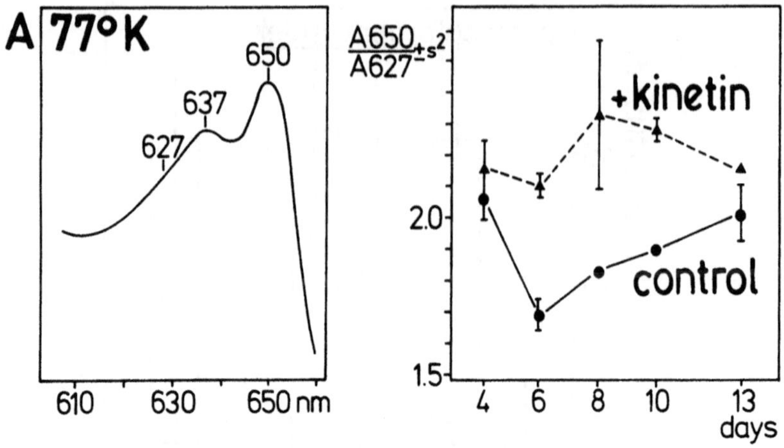

Fig. 2. 77° K absorption spectrum of an etiolated radish cotyledon (left) and the changing amount of phototransformable protochlorophyll(ide) relatively to that of the non-tranformable protochlorophyll(ide) (right) as measured from the absorption spectra (77° K) at 650 and 627 nm, respectively, for cotyledons from 4-, 6-, 8-, 10-, and 13-day-old etiolated radish seedlings grown without (control) or with added kinetin.

shift, called Shibata shift (Shibata, 1957), needs a longer time for its accomplishment with increasing age of the seedlings. The 77° K absorption at 672 nm compared to that at 680 nm rises up to day 8 following germination. The A672/A680 ratio, which indicates the accomplishment of the Shibata shift, decreases rapidly for control plants after day 8 whereas it maintains its value for cotyledons of seedlings grown in the presence of kinetin. After 140 flashes of

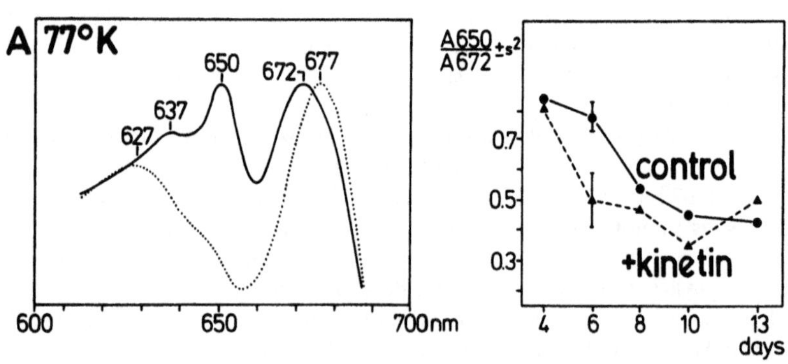

Fig. 3. 77° K absorption spectrum of an etiolated radish cotyledon after one flash of light with a subsequent dark period of 0 (. . . .) or 30 min. (———) (left) and the changing amount of protochlorophyll(ide) resynthesized in a 30 min. dark period after one flash of light (right) as measured from the absorption (77° K) at 650 nm relatively to that at 672 nm for cotyledons from 4-, 6-, 8-, 10-, and 13-day-old etiolated radish seedlings grown without (control) or with added kinetin.

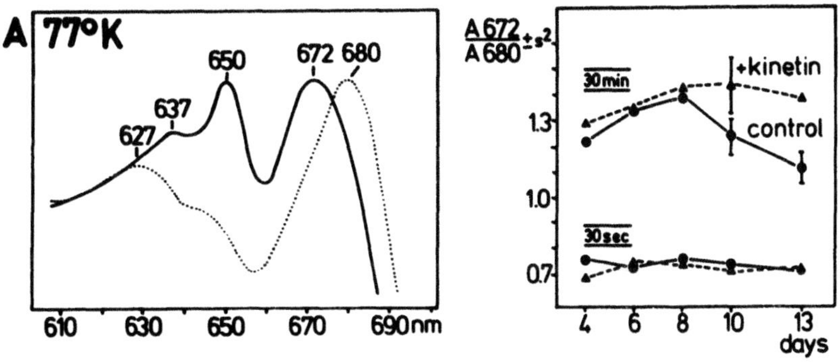

Fig. 4. 77° K absorption spectrum of an etiolated radish cotyledon after one flash of light with a subsequent dark period of 30 sec. (·····) or 30 min. (———) (left) and the changing in the accomplishment of the Shibata shift during a 30 sec. and a 30 min. dark period after one flash of light (right) as measured from the absorption (77° K) at 672 nm relatively to that at 680 nm for cotyledons from 4-, 6-, 8-, 10-, and 13-day-old etiolated radish seedlings grown without (control) or with added kinetin.

light (one flash followed by a dark period of 10 min) the 77° K absorption spectrum shows peaks at 672, 647, 620, and 580 nm (Fig. 5). In comparison with spectra measured after a shorter illumination time the maxima are broader and higher. Cotyledons from seedlings grown in the presence of kinetin, when compared with the controls, exhibit relatively higher absorption before and after the 672 nm maximum. The amount of the reaction center chlorophyll of pigment system I (P700) was measured with chloroplasts from fully green radish cotyledons (7 days continuous light). Kinetin, when added to the growing plant, induces higher amounts of P700 (on a chlorophyll basis) compared with the controls (Table 1). During the development in the light, the variable part of the

Table 1. The concentration of P700 per chlorophyll in chloroplasts isolated from radish seedlings grown without (control) or with added kinetin.*

	10^{-4} P700/chlorophyll	P700/chlorophyll
control	44.4	1/225
+ kinetin	62.3	1/161

* 8-day-old radish seedlings grown for one day in the dark and afterwards for seven days in the light (fluora, 7500 mw/m²), standard deviation 6%.

680 nm fluorescence kinetic curve (F_v) increases up to a certain level (Fig. 6). Cotyledons from 6- and 8-day-old seedlings in contrast to 4- day-old seedlings clearly show a lag phase between 2 and 6 hours' illumination. Kinetin induces a slightly higher variable fluorescence, compared with the controls.

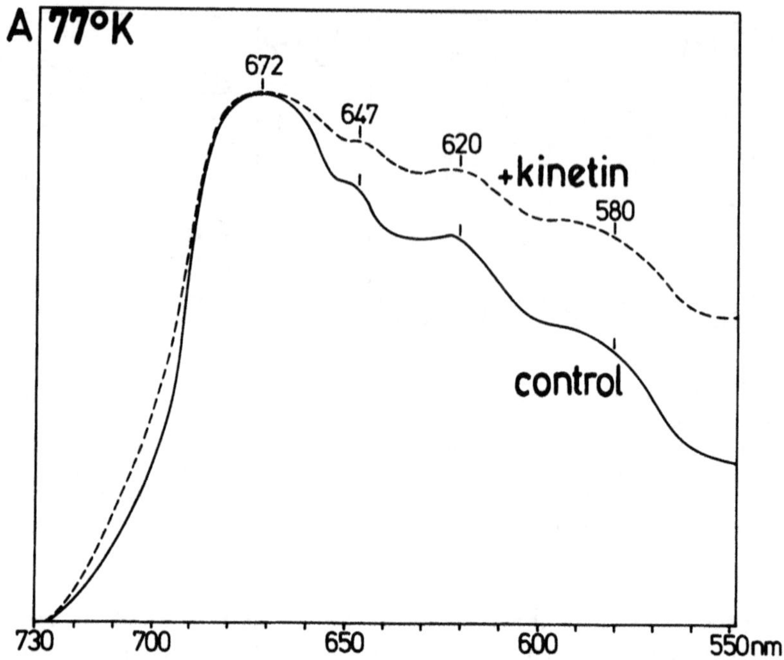

Fig. 5. 77° K absorption spectrum of a radish cotyledon taken from a 6-day-old etiolated seedling grown without (control) or with added kinetin measured after 140 flashes of light (one 1 msec. flash followed by 10 min. darkness).

DISCUSSION

The rise in the amount of total protochlorophyll(ide) in etiolated radish cotyledons due to kinetin treatment (Fig. 1) is in agreement with findings for maize (Yakushkina & Pushkina, 1971). It indicates that enzymes of the chlorophyll biosynthesis are already stimulated in total darkness. No changes of the protochlorophyll(ide) content can be seen both in control and kinetin-treated plants between day 4 and day 8 after germination. That means that in radish cotyledons most of the protochlorophyll(ide) is synthesized before day 4, a much faster accumulation than in bean leaves (Akoyunoglou & Siegelman 1968).

The main binding site of cytokinins was found to be located on cytoplasmic ribosomes (Berridge *et al.*, 1970; Fox & Erion, 1957). As the synthesis of enzymes leading to the formation of δ-aminolevulinic acid is expected to take place in the cytoplasm (Beisenherz & Schneider, 1974) the enzyme(s) stimulated by kinetin should be catalyzing a process which participates in the formation of δ-aminolevulinic acid.

In the dark kinetin especially stimulates the formation of the phototransformable protochlorophyll(ide) (Boardman, 1966; Boardman *et al.*, 1970; Kahn *et al.*, 1970; Bovey *et al.*, 1974) $P_{657-650}$ (Fig. 2). Phototransformable protochlorophyll(ide), which is accumulated during the development in the dark,

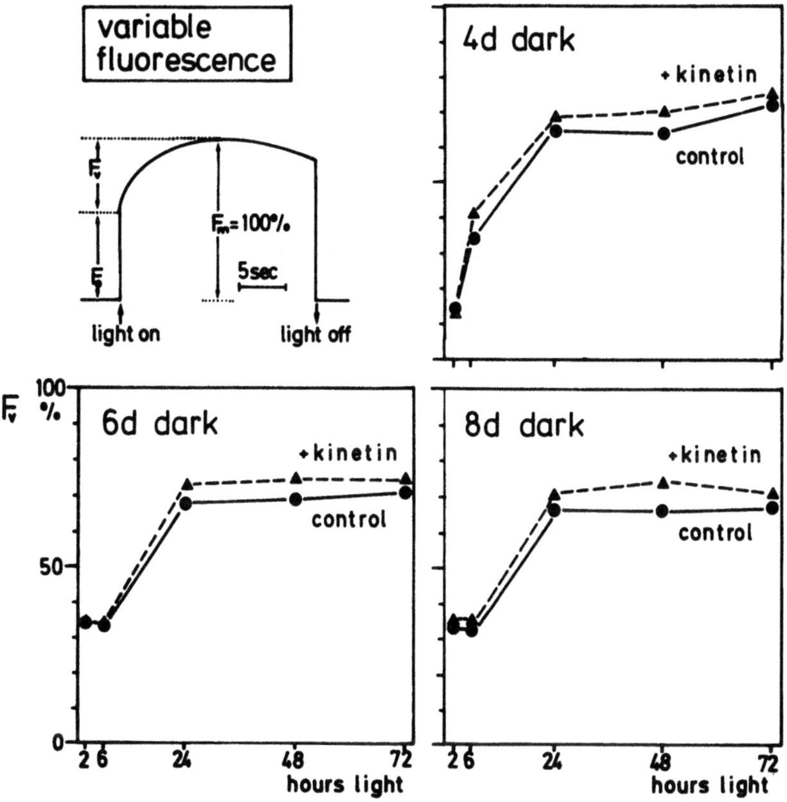

Fig. 6. The variable fluorescence as measured from the 680 nm fluorescence kinetic (maximal fluorescence = 100%) for cotyledons from 4-, 6-, and 8-day-old etiolated radish seedlings grown without (control) or with added kinetin after 2, 6, 24, 48, and 72 hours' illumination (Osram HQ 5.6 × 10^4 mw/m^2).

(Boardman *et al.*, 1970; Thorne, 1971; Klein & Schiff, 1972) *in vitro* can be formed from non-phototransformable protochlorophyll(ide) by the addition of ATP (Horton & Leech, 1972, 1975) or NADPH (Griffiths, 1974). The transformation of the non-phototransformable protochlorophyll(ide) to the phototransformable form was explained in terms of a firmer binding to proteins (Dujardin & Sironval, 1970). Thus the action of kinetin on the formation of $P_{657-650}$ may be explained both by a higher ATP and NADPH level and/or by an increased protein binding.

The resynthesis of protochlorophyll(ide) after one flash of light was decreased by kinetin treatment only in 6-day-old seedlings; at earlier or later stages of the development no effect could be seen (Fig. 3). A particular endogenous cytokinin content must be expected for these 6-day-old radish seedlings. At day 4, 8, 10, and 13 following germination no effect was visible as found by Bengtson *et al.* (1977). Shlyk (Shlyk & Averina, 1969; Shlyk *et al.*, 1970; Walter *et al.*, 1977) described a higher amount of resynthesized protochlorophyll(ide) in excised barley leaves. Both the illumination time (24 hr) and the subsequent dark period (24 hr), however, were longer in the experiments of Shlyk; this could be an explanation for the different results.

With increasing age of the seedlings the Shibata shift (Shibata, 1957) needs longer times to be accomplished (Akoyunoglou & Siegelman, 1968; Akoyunoglou & Michalopoulos, 1971). Kinetin suppresses here (Fig. 4) the general ageing process as it has been demonstrated for several other cases (Richmond & Lang, 1957; Mothes, 1961). The molecular change during the Shibata shift is not yet fully understood. Phytylation of chlorophyllide and changes in the pigment-protein binding together with ultrastructural changes in the plastids are expected (Sironval *et al.*, 1965; Butler & Briggs, 1966; Akoyunoglou & Michalopoulos, 1971; Mathis & Sauer, 1973; Virgin & French, 1973).

The 77° K absorption spectra of radish cotyledons illuminated for longer periods show higher absorption before and after the main peak at 672 nm when seedlings were grown in the presence of kinetin (Fig. 5). Part of this rise in absorption, especially at wavelength < 672 nm, can be interpreted as a possible rise in light scattering for kinetin-treated leaves. However, the higher absorption in the long wavelength region (> 672 nm) must be explained as an increase in the concentration of the long wavelength chlorophyll forms induced by kinetin. These long wavelength forms of chlorophyll play an important role in the light trapping pigment systems of the photosynthesis (Brown, 1972). Because of the excitation energy transfer in the photosynthetic apparatus (Borisov & Godik, 1973) it is expected that chlorophyll forms with long wavelength absorption maximum are in close relation to the reaction center chlorophylls. The long wavelength chlorophyll forms are supposed to be more firmly bound to proteins than the short wavelength forms (see protochlorophyll(ide)). Protein synthesis, which is known to be particularly increased by cytokinins (Maaß & Klämbt, 1977) should play an important role in the increase of long wavelength chlorophyll forms. Thus kinetin seems to lead to a more effective light energy trapping. The increase of P700 (per chlorophyll) in kinetin-treated plants (Table 1) also indicates a more effective photosynthetic apparatus than in control plants. The rise in P700 as well as in the accumulation of the long wavelength chlorophyll forms explains the increased Hill activity found in chloroplasts isolated from radish seedlings grown in the presence of kinetin (Buschmann & Lichtenthaler, 1977). The relative amount of the variable part of the 680 nm fluorescence kinetic curve increases up to an illumination time of 24 hours corresponding with the maximum in photosynthetic activity (Fig. 6). In kinetin-treated plants the variable fluorescence is slightly increased. This is a further indication for the higher amount of photosynthetic active chlorophylls, especially those connected with pigment system II (Lavorel & Etienne, 1977), due to kinetin action. To sum up kinetin does not only increase the total amount of protochlorophyll(ide) and chlorophyll but also gives rise to changes in the content of the different *in vivo* pigment forms, leading to a more effective photosynthetic apparatus.

REFERENCES

Akoyunoglou, G.A. & G. Michalopoulos. The relation between the phytylation and the 682→672 nm shift *in vivo* of chlorophyll a. *Physiol. Plant.* 25: *324–329* (1971).

Akoyunoglou, G.A. & H.W. Siegelman. Protochlorophyllide resynthesis in dark grown bean leaves. *Plant Physiol.* 43: *66–68* (1968).

Banerji, D. & M.M. Laloraya. Chlorophyll formation in isolated pumpkin cotyledons in the presence of kinetin and chloramphenicol. *Plant & Cell Physiol.* 8: *263–268* (1967).

Beevers, L., B. Loveys, J.A. Pearson & P.F. Wareing. Phytochrome and hormonal control of expansion and greening of etiolated wheat leaves. *Planta* (Berl.), 90: *286–294* (1970).

Beisenherz, W.W. & H.A.W. Schneider. Syntheseort von Enzymen der Chlorophyllbiosynthese. *Ber. Deutsch. Bot. Ges.* 87: *161–166* (1974).

Bengtson, C., B. Klockare, S. Larsson & C. Sundquist. The effect of phytohormones on chlorophyll (ide), protochlorophyll(ide), and carotenoid formation in greening dark grown wheat leaves. *Physiol. Plant.* 40: *198–204* (1977).

Berridge, M.V., R.K. Ralph & D.S. Letham. The binding of kinetin to plant ribosomes. *Biochem. J.* 119: *75–84* (1970).

Boardman, N.K. Protochlorophyll. pp. 437–479, in: Vernon, L.P. & G.R. Seely (eds.), The chlorophylls. Academic Press, New York (1966).

Boardman, N.K., J.M. Anderson, A. Kahn, S.W. Thorne & T.E. Treffry. Formation of photosynthetic membranes during chloroplast development. pp. 70–84, in: Boardman, N.K., A.W. Linnane & R.M. Smillie (eds.), Autonomy and biogenesis of mitochondria and chloroplasts. North Holland, Amsterdam (1970).

Borisov, A.Y. & V.I. Godik. Excitation energy transfer in photosynthesis. *Biochim. Biophys. Acta* 301: *227–248* (1973).

Bovey, F., T. Ogawa & K. Shibata. Photoconvertible and non-photoconvertible forms of protochlorophyll(ide) in etiolated bean leaves. *Plant & Cell Physiol.* 15: *1133–1137* (1974).

Brown, J.S. Forms of chlorophyll *in vivo*. *Ann. Rev. Plant Physiol.* 23: *73–86* (1972).

Buschmann, C. & H.K. Lichtenthaler. Hill-activity and P700 concentration of chloroplasts isolated from radish seedlings treated with β-indoleacetic acid, kinetin or gibberellic acid. *Z. Naturforsch.* 32c: *798–802* (1977).

Butler, W.L. & W.R. Briggs. The relation between structure and pigments during the first stages of proplastid greening. *Biochim. Biophys. Acta* 112: *45–53* (1966).

Dujardin, E. Reversible transformation of $P_{657-650}$ form into $P_{633-628}$ in etiolated bean leaves. *Plant Sci. Lett.* 7: *91–94* (1976).

Dujardin, E. & C. Sironval. The reduction of protochlorophyllide into chlorophyllide III. The phototransformability of the forms of the protochlorophyllide-lipoprotein complex found in darkness. *Photosynthetica* 4: *129–138* (1970).

Fletcher, R.A. & D. McCullagh. Benzyladenine as regulator of chlorophyll synthesis in cucumber cotyledons. *Can. J. Bot.* 49: *2197–2201* (1971a).

Fletcher, R.A. & D. McCullagh. Cytokinin-induced chlorophyll formation in cucumber cotyledons. *Planta* (Berl.), 101: *88–90* (1971b).

Fletcher, R.A., C. Teo & A. Ali. Stimulation of chlorophyll synthesis in cucumber cotyledons by benzyladenine. *Can. J. Bot.* 51: *937–939* (1973).

Fox, J.E. & J.L. Erion. A cytokinin binding protein from higher plant ribosomes. *Biochem. Biophys. Res. Commun.* 64: *694–700* (1975).

Goedheer, J.C. Les changements du spectre d'absorption et de fluorescence au cours du verdissement et du vieillissement des plastes. pp. 77–85, in: Sironval, C. (ed.), Le chloroplaste. Masson, Paris (1967).

Griffiths, W.T. Protochlorophyll and protochlorophyllide as precursors for chlorophyll synthesis *in vitro*. *FEBS Letters* 49: *196–200* (1974).

Hiyama, T. & B. Ke. Difference spectra and extinction coefficients of P700. *Biochim. Biophys. Acta* 267: *160–171* (1972).

Horton, P. & R.M. Leach. The effect of ATP on photoconversion of protochlorophyllide into chlorophyllide in isolated etioplasts. *FEBS Letters* 26: *277–280* (1972).

Horton, P. & R.M. Leech. The effect of ATP on the photoconversion of protochlorphyllide on isolated etioplasts of *Zea mays*. *Plant Physiol.* 56: *113–120* (1975).
Kahn, A., N.K. Boardman & S.W. Thorne. Energy transfer between protochlorophyllide molecules: evidence for multiple chromophores in the photoactive protochlorophyllide-protein complex *in vivo* and *in vitro*. *J. Mol. Biol.* 48: *85–101* (1970).
Klein, S. & J.A. Schiff. The correlated appearance of prolamellar bodies, protochlorophyll from etiolated barley seedlings. *J. Am. Chem. Soc.* 70: *3558–3562* in the dark. *Plant Physiol.* 49: *619–626* (1972).
Koski, V.M. & J.H.C. Smith. The isolation and spectral absorption properties of protochlorophyll from etiolated barley seedlings. *J. Am. Chem. Soc.* 70: *3558–3562* (1948).
Lavorel, J. & A.-L. Etienne. *In vivo* chlorophyll fluorescence pp. 203–268, in: Barber, J. (ed.), Primary processes in photosynthesis. Elsevier, Amsterdam (1977).
Maaß, H. & D. Klämbt. Cytokinin effect on protein synthesis *in vivo* in higher plants. *Planta* (Berl.) 133: *117–120* (1977).
Marsho, T.V. & B. Kok. Detection and isolation of P700. pp. 515–522, in: San Pietro, A. (ed.), Methods in enzymology (vol. 23A). Academic Press, New York (1971).
Mathis, P. & K. Sauer. Chlorophyll formation in greening bean leaves during the early stages. *Plant Physiol.* 51: *115–119* (1973).
Mothes, K. Der Beitrag der Kinetinforschung zum Verständnis pflanzlicher Korrelationen. *Ber. Deutsch. Bot. Ges.* 74: (24)–(41). (1961).
Nobel, P.S. A rapid technique for isolating chloroplasts with high rates of endogenous photophosphorylation. *Plant Physiol.* 42: *1389–1394* (1967).
Richmond, A. & A. Lang. Effect of kinetin on protein content and survival of detached *Xanthium* leaves. *Science* 125: *650–651* (1957).
Shibata, K. Spectroscopic studies on chlorophyll formation in intact leaves. *J. Biochem.* 44: *147–173* (1957).
Shlyk, A.A. & N.G. Averina. Stimulation by kinetin of the accumulation of protochlorophyllide in normal green leaves in the dark. *Dokl. Akad. Nauk SSSR* 186: *1209–1212* (1969).
Shlyk, A.A., G. Walter, N.G. Averina & G.E. Savchenko. Influence of kinetin on the formation of active protochlorophyllide in green and post-etiolated leaves of wheat. *Dokl. Akad. Nauk SSSR* 193: *1429–1432* (1970).
Sironval, C., M. Brouers, J.-M. Michel & Y. Kuiper. The reduction of protochlorophyllide into chlorophyllide I. The kinetics of $P_{657-647} \rightarrow P_{688-676}$ phototransformation. *Photosynthetica* 2: *268–287* (1968).
Sironval, C., M.R. Michel-Wolwertz & A. Madsen. On the nature and possible functions of the 673- and 684-mμ forms *in vivo* of chlorophyll. *Biochim. Biophys. Acta* 94: *344–354* (1965).
Straub, V. & H.K. Lichtenthaler. Die Wirkung von Gibberellinsäure A_3 und Kinetin auf die Bildung der Photosynthesepigmente, Lipochinone und Anthocyane in *Raphanus*-Keimlingen. *Z. Pflanzenphysiol.* 70: *308–321* (1973).
Sugiura, M. Promotion of chlorophyll synthesis by kinetin. *Bot. Mag. Tokyo* 76: *309–310* (1963).
Thorne, S.W. The greening of etiolated bean leaves, I. The initial photoconversion process. *Biochim. Biophys. Acta* 226: *113–127* (1971).
Virgin, H.I. & C.S. French. The light induced protochlorophyll-chlorophyll a –transformation and the succeeding interconversions of the different forms of chlorophyll. *Physiol. Plant.* 28: *350–357* (1973).
Walter, G., N.G. Averina & A. Meister. Protochlorophyllid-Resynthese unter dem Einfluß von Kinetin. Spektrophotometrische Untersuchungen an Weizenkeimpflanzen (*Triticum aestivum* L.) *in vivo*. *Biochem. Physiol. Pflanzen* 171: *409–417* (1977).
Woźny, A. & A. Szweykowska. Effect of cytokinins and antibiotics on chloroplast development in cotyledons of *Cucumis sativus*. *Biochem. Physiol. Pflanzen* 168: *195–209* (1975).

Yakushkina, N.I. & G.P. Pushkina. Effects of gibberellin and kinetin on chlorophyll content and photophosphorylation in maize seedlings. *Fiziol. rast.* 18: *898–903* (1971).

Author's address:
Botanical Institute II
University of Karlsruhe
D–7500 Karlsruhe
Federal Republic of Germany.

CHLOROPHYLL-DEPENDENT OXIDATION OF INDOLEACETIC ACID AND ITS ANALOGS. FACTORS AFFECTING DECARBOXYLATION AND OXYGEN UPTAKE*

IMRE A. TAMAS, DEBRA B. SHERMAN,**
JOSEPH D. BECKER*** &
ROBERT M. OBERLANDER
(Ithaca, New York, U.S.A.)

ABSTRACT

Light-dependent oxidation of indoleacetic acid and its analogs by Triton X-100-solubilized chloroplasts was demonstrated in this work. Solubilized chloroplasts showed high rates of light-dependent oxygen uptake when indoleacetic acid was present. No oxygen was utilized by chloroplasts in the absence of Triton X-100. Oxygen uptake was also observed using a chlorophyll solution. This fact indicates that the reaction is mediated by chlorophyll and does not require the participation of enzymes. Blue and red light were most efficient in promoting oxygen uptake. The reaction rate was negligible under green light. No oxygen uptake was observed in darkness or in the absence of chlorophyll.

Incubation of carboxyl-labeled [^{14}C] indoleacetic acid with Triton X-100-solubilized chloroplasts resulted in $^{14}CO_2$ evolution. The reaction was dependent on light and molecular oxygen.

The chlorophyll-dependent oxidation of various indoles and other nonindolic analogs of indoleacetic acid was also tested. Indoleacetic acid and other carboxylated indoles showed high rates of oxidation, whereas oxidation of indoles without a free carboxyl group was slow. No reaction occurred with 2,4-dichlorophenoxyacetic acid or naphthaleneacetic acid.

INTRODUCTION

The plant hormone indoleacetic acid (IAA) is known to be readily decomposed by tissues of many plant species. The current status of the problem has recently been reviewed by Schneider & Wightman (1974). A key role in IAA degradation is played by plant peroxidases which catalyze the oxidation of the molecule. According to a reaction scheme proposed by Hinman & Lang (1965), IAA is oxidized by peroxidases utilizing molecular oxygen. In subsequent steps the reaction product is decarboxylated and finally converted to 3-methyleneoxindole. Thus, the initial steps of IAA degradation by peroxidases involve the utilization of molecular oxygen and release of CO_2.

* We are grateful to Drs. A. Eberhard and W. R. Bergmark, Chemistry Department, Ithaca College, for helpful discussions.

A gift of Alaska pea seeds from the Ferry Morse Seed Co. is gratefully acknowledged.

** Undergraduate Research Participant, National Science Foundation Grant SMI 76–83170.

*** Undergraduate Research Participant, National Science Foundation Grant GY-11175.

IAA has been shown to undergo photodegradation in the absence of enzymes. The reaction can be catalyzed by various pigments, such as eosin, methylene blue and riboflavin (Gordon, 1954). The riboflavin-sensitized process utilizes molecular oxygen and leads to the release of CO_2.

The present work reports on the chlorophyll-sensitized photooxidation of IAA.

MATERIALS AND METHODS

Chemicals

IAA, IAA-analogs, MES, TES and Triton X-100 were obtained from Sigma Chemical Co. (St. Louis, Mo.). [1-^{14}C]IAA, ammonium salt, 52 Ci/mole, was purchased from Amersham/Searle Corp. (Arlington Heights, Ill.). Other chemicals, all reagent grade, were obtained from Fisher Scientific Co. (Pittsburgh, PA).

Plant material

Pea (*Pisum sativum* L. cultivar Alaska W.R.) seeds were obtained from Ferry Morse Seed Co., sown in vermiculite and irrigated daily with half-strength Hoagland's solution (Arditti & Dunn, 1969). At planting the surface of the vermiculite was treated with Orthocide (Chevron Chemical Co.) to prevent fungal growth. Fourteen hours' daily illumination of 8×10^3 ergs $\times$ cm^{-2} $\times$ sec^{-1} was provided with fluorescent light (Cool White, Westinghouse) at 23 to 25° C.

Solubilized chloroplasts

Chloroplasts were isolated from leaves of 13–14-day-old plants at 0° C with a modified version of Nobel's (1967) technique. A sample of 10 g leaves was rinsed and briefly chopped in 10 ml of isolation medium containing 0.4 M sucrose and 20 mM TES buffer at pH 7.9 using six razor blades inserted 5 mm apart in a chopping device. The brei was placed in a nylon cloth bag having 40μm holes (GM Mfg. and Instrument Corp., New York, N.Y.), ground in a mortar with a pestle for 30 seconds and squeezed through the bag. (Microscopic examination showed the chloroplast suspension to be virtually free of cell debris.) The suspension was centrifuged at 1000X *g* for 90 seconds. Unless otherwise noted, the pellet was solubilized using a medium that contained the components of the isolation medium plus 1% (v/v) of the detergent Triton X-100 and 0.8 M KCl with one minute of grinding in a Ten Broeck ground-glass homogenizer. Chlorophyll content was determined according to Arnon (1949). The solubilized chloroplast suspensions were routinely adjusted to a concentration of 0.3 mg chlorophyll/ml and stored at −70° C if not used immediately.

Chlorophyll solution

Chlorophyll was extracted with a modified procedure of Perkins & Roberts (1962). All operations were performed under dim light. Fifteen g of pea leaves were frozen over dry ice, ground to a powder with mortar and pestle and extracted with 85% (v/v) aqueous acetone (3 × 25 ml). The combined extract was filtered through a Gelman glass fiber filter (0.3 μm pore size) under suction and shaken with an equal volume of petroleum ether. The top phase was washed with water (3 × 50 ml), extracted with 80% (v/v) aqueous methanol (3 × 50 ml) to remove carotenoids, then washed with 50-ml portions of distilled water until no emulsion was formed and dried with Na_2SO_4. Most of the petroleum ether was evaporated under reduced pressure and the remainder was blown to dryness with nitrogen gas. The dry residue was taken up in 10 ml of 80% (v/v) acetone. Chlorophyll determination and storage was done as described above.

Measurement of oxygen uptake

Oxygen uptake was measured with a polarographic probe (Model 53 Biological Oxygen Monitor, Yellow Springs Instrument Co., Inc., Yellow Springs, Ohio) connected to a strip chart recorder (Model 610, Instrumentation Specialties Co., Lincoln, Neb.). Unless otherwise noted, the reaction medium contained 0.4 M sucrose, 50 mM MES buffer at pH 5.5, 1% (v/v) Triton X-100, 0.8 M KCl, and 1.14 mM IAA (or IAA-analog) in a final volume of 3 ml. For measurements at higher pH values (7 to 8) 50 mM TES buffer was used instead of MES. IAA was added to the medium in a solid form (Stafford & Bravinder-Bree, 1972) followed by stirring for 5 minutes in the dark. Preliminary experiments showed that oxygen uptake reached a steady rate within 5 minutes of adding IAA. Finally, solubilized chloroplasts or chlorophyll solution was added at a final concentration of 0.01 mg chlorophyll/ml, except as otherwise noted. Oxygen concentration was at or near air saturation. The reaction cuvette was held in a water bath maintained at 20° C and illuminated with a slide projector (Kodak Cavalcade 520) using a 500 W General Electric DFR projection lamp at an intensity of 7.5×10^4 ergs × cm^{-2} × sec^{-1} for one minute except as otherwise noted. Light intensity was measured with a radiometer (Model 65, Yellow Springs Instrument Co., Inc.).

Measurement of IAA decarboxylation

To 1.6 ml of reaction medium, containing 0.4 M sucrose and 50 mM MES at pH 5.5, 0.4 ml of solubilized chloroplasts were added. The flask was connected to a trap containing 3 ml of 30% (v/v) ethanolamine in methanol and placed in a shaker bath held at 25° C. Air passing over the reaction medium was bubbled through the trap at a rate of 15 to 20 ml/min. Preliminary experiments showed that all of the $^{14}CO_2$ was retained by the trap without significant loss at this flow rate. The reaction was started by the injection of [1−^{14}C]IAA at a specific activity of 0.18 mCi/mmole. The final concentration of IAA was 0.11 mM. Illumination was provided by General Electric flood lamps through a tank of water. After 10 minutes the traps were disconnected and their radioactive content determined with a liquid scintillation counter (Nuclear Chicago Unilux III). The efficiency of counting for each sample was automatically determined using an external standard (^{133}Ba).

RESULTS

Effect of solubilization of chloroplasts on oxygen uptake

Chloroplasts were homogenized in the presence of various concentrations of the detergent, Triton X-100, and of KCl. The preparations were illuminated and their rate of oxygen uptake was measured with or without IAA. Table 1 shows that in the presence of IAA the reaction rate was greatly stimulated by using up to 1% Triton X-100 and 0.8 M KCl, the highest concentration tested. While high reaction rates could be obtained with 1% Triton X-100, addition of 0.8 M KCl increased the rate by about 50%. KCl alone had no effect. Unsolubilized chloroplasts did not show any oxygen uptake and homogenization without Triton X-100 had only a negligible effect. These results indicate that detergent-solubilization of chloroplast components is essential for high rates of IAA-dependent oxygen uptake, while partial physical disruption of the chloroplast structure is insufficient to produce active preparations (electron microscopic observations showed that homogenization in the absence of detergent and KCl converted a suspension of chloroplasts with intact double outer membranes to chloroplasts showing absence of the outer membrane and extensive loss of stromal content and granal structure [Schwartz, J. W., Tamas, I.

Table 1. Effect of Triton X–100 and KCL on oxygen uptake by chloroplasts[1]

Triton X–100 (%)	KCL (M)	IAA (mM)	Oxygen uptake[2]
			μmol O_2/mg chlorophyll/hr
0	0	0	0[3]
0	0	1.14	0[3]
0	0	0	0
0	0	1.14	6.1 ± 4.4
0	0.8	0	0
0	0.8	1.14	5.5 ± 5.5
0.05	0.04	0	9.8 ± 2.7
0.05	0.04	1.14	67.7 ± 4.3
0.1	0.08	0	15.2 ± 4.3
0.1	0.08	1.14	215 ± 3.6
1	0	0	27.5 ± 4.6
1	0	1.14	287 ± 5
1	0.8	0	34.2 ± 1.6
1	0.8	1.14	428 ± 10

[1] chloroplasts homogenized for 1 min. (see exceptions).
[2] mean ± S.E. of 3 measurements.
[3] not homogenized.

A., Axelrod, R.H. and Telford, J.N., unpublished]). Results in Table 1 also show that oxygen uptake strongly depends on IAA, since in its presence the rate of oxygen uptake was about ten-times greater than in its absence. It may also be noted that oxygen uptake by control samples, lacking IAA, was stimulated by increasing concentrations of Triton X-100 and KCl. This fact shows that oxygen uptake by control samples also depends on solubilization of chloroplast components.

Dependence of oxygen uptake on IAA concentration and light intensity

Aliquots of solubilized chloroplasts were incubated with various concentrations of IAA, and their rate of oxygen uptake was measured under illumination. According to Fig. 1, the rate of oxygen uptake depended greatly on IAA concentration. The reaction rate increased sharply up to about 5 mM IAA and approached saturation at about 10 mM IAA, showing a maximum of about 430 μmoles O_2 taken up per mg chlorophyll per hour. In comparison, the reaction rate was very low (about 16 μmoles/mg chlorophyll/hr) in control samples without added IAA.

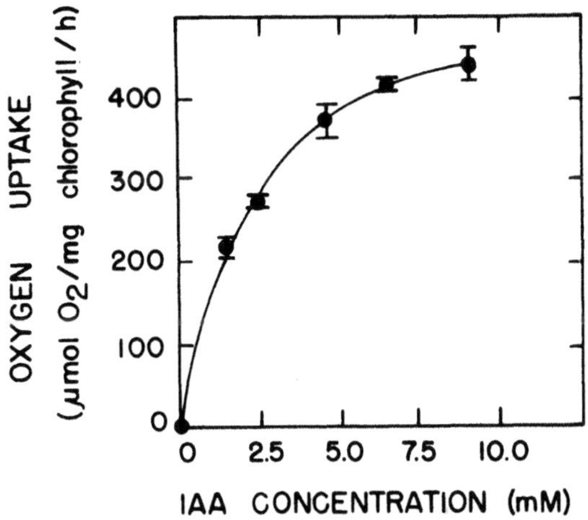

Fig. 1. Dependence of oxygen uptake by solubilized chloroplasts on IAA concentration. The control rate (no IAA present) was 16 μmoles O_2/mg chlorophyll/hr. This value was subtracted from all data shown. Chlorophyll concentration was 0.03 mg chlorophyll/ml. Data presented are from one of three similar experiments. Results represent the mean ± S.E. of three measurements.

Fig. 2 shows the dependence of oxygen uptake on light intensity with and without IAA. Samples with 1.14 mM IAA responded sharply to light intensity. Saturating light intensity appears to be around 1.2×10^6 ergs $\times$ $cm^{-2} \times sec^{-1}$. At this intensity about 1800 μmoles of oxygen were taken up per mg of chlorophyll per hour in the presence of 1.14 mM IAA. The maximum rate could possibly be much higher since 1.14 mM IAA is substantially below the saturating concentration (Fig. 1). At intensities above light saturation the reaction rate decreased with the length of illumination, probably because of light-induced damage. In darkness no measurable oxygen uptake took place. Control samples, lacking IAA, also responded to light, but their rate of oxygen uptake remained between 5 and 10% of that of the IAA-containing samples.

Results of Figs. 1 and 2 suggest that oxygen uptake is due to light-dependent oxidation of IAA by detergent-solubilized chloroplasts.

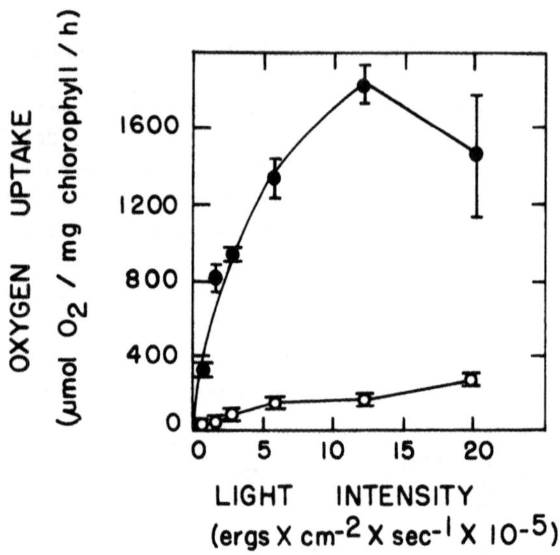

Fig. 2. Dependence of oxygen uptake by solubilized chloroplasts on light intensity. ●——●, 1.14 mM IAA; ○——○, control. Data presented are from one of three similar experiments. Results represent the mean ± S.E. of three measurements.

Role of chlorophyll in IAA-dependent oxygen uptake

To determine whether chlorophyll may be responsible for the light-requiring oxidation of IAA, a chlorophyll solution was incubated under illumination in reaction medium. A parallel experiment was performed using detergent-solubilized chloroplasts. Table 2 shows no significant difference between the oxygen uptake rates of the two preparations. Boiling the preparations for 10 minutes did not decrease their rate of oxygen uptake, indicating that all the activity was mediated by chlorophyll without the participation of enzymes.

The fact that IAA-dependent oxygen uptake can occur in chlorophyll solution shows that IAA itself is being oxidixed rather than some unknown chloroplast component. Chlorophyll itself may be oxidized but the amount present in the reaction medium appears to be insufficient to support the observed rates of oxygen uptake. In a typical experiment about 300 μmoles of oxygen are taken up per mg chlorophyll per hour. Given 0.03 mg chlorophyll per cuvette, this rate is equal to 0.15 μmole of oxygen per cuvette per minute. Since the molecular

Table 2. Effect of boiling on oxygen uptake by solubilized chloroplasts and chlorophyll solution.

Treatment	IAA concen-tration (mM)	Oxygen uptake Solubilized chloroplasts[3]	Chlorophyll solution[3]
		μmole O_2/mg chlorophyll/hr[4]	
Control[1]	0	18 ± 0.9	15 ± 0.6
Control[1]	1.14	292 ± 6	281 ± 11
Boiled[2]	0	21 ± 2	14 ± 0.5
Boiled[2]	1.14	324 ± 12	303 ± 10

[1] reaction medium at 0° C for 10 min.
[2] reaction medium at 100° C for 10 min.
[3] 0.03 mg chlorophyll/ml.
[4] mean ± S.E. of 6 measurements.

weight of chlorophyll a is 893.5, 0.03 mg chlorophyll per cuvette gives 0.034 μmole of chlorophyll a per cuvette. This is equivalent to about one-fifth of the amount of oxygen utilized per minute. Since these preparations can support undiminished oxygen uptake for at least 5 minutes, it is clear that chlorophyll is not the substrate of the reaction. In contrast, in a typical experiment the concentration of 1.14 mM IAA amounts to 3.42 μmoles of IAA per cuvette. This is equivalent to the amount of oxygen consumed in about 23 minutes at the stated average reaction rate. The foregoing calculations give strong support to the view that oxygen uptake in these experiments is due to chlorophyll-mediated oxidation of IAA.

Fig. 3 shows oxygen uptake by both solubilized chloroplasts and chlorophyll solution at pH values ranging from 4.5 to 8. While in both preparations the rate of oxygen uptake by IAA-containing samples rose with pH, the increase was greater with chlorophyll solution. A comparable but smaller response was shown by control preparations not containing IAA. The essential similarity in the pH response of the two preparations and the lack of a clear-cut pH optimum argues against the participation of enzymes in the oxidative activity of the solubilized chloroplasts.

The requirement for chlorophyll in the oxidation of IAA was established by incubating IAA with various concentrations of chlorophyll from 0 to 0.03 mg chlorophyll/ml and measuring oxygen uptake by the illuminated solution. No oxygen uptake was found in the absence of chlorophyll. In samples containing the pigment, oxygen uptake increased linearly with chlorophyll concentration with a slope of about 6.5 μmoles of O_2/hr per 0.01 mg chlorophyll/ml. These results suggest that chlorophyll directly participates in the oxidation of IAA.

The efficiency of light in different spectral regions in promoting the oxidation of IAA was determined. Table 3 shows that oxygen uptake by chlorophyll solution was greatest with red light, followed by blue and white, in that order. Oxygen uptake with green light was insignificant. Since the spectral regions of transmission of the red and blue filters coincide with the absorption regions of

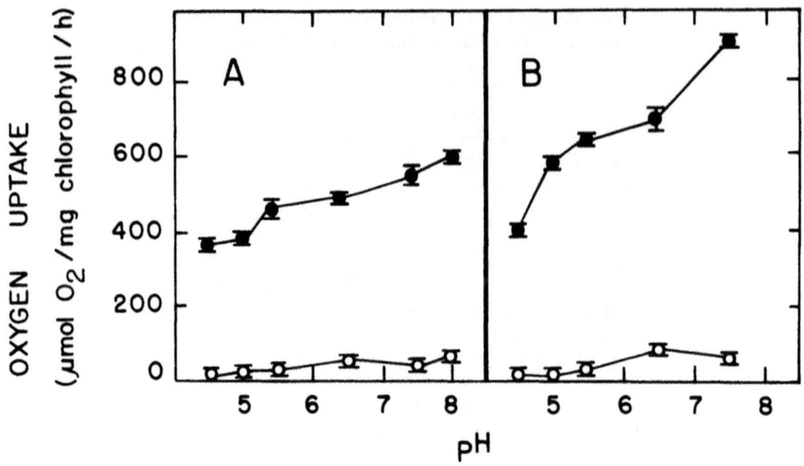

Fig. 3. Effect of pH on oxygen uptake by solubilized chloroplasts (A) and chlorophyll solution (B). Buffers used were succinate (pH 4.5 to 5), MES (pH 5.5 to 6.5) and TES (pH 7 to 8), each at a concentration of 50 mM. ●——●, 1.14 mM IAA; ○——○, control. Data presented are from one of two similar experiments. Results represent the mean ± S.E. of three measurements.

chlorophyll, while that of the green does not, these results give further support for the view that IAA oxidation depends on chlorophyll. It should be noted that control samples, not containing IAA, responded to the wavelength of illumination similarly to the IAA-containing ones. This fact suggests that oxygen uptake by control samples is also a chlorophyll-dependent process. Results of the experiment with solubilized chloroplasts were similar to those with chlorophyll solution except oxygen uptake under white light was higher than under red or blue (Table 3). The reason for this difference is not known.

IAA decarboxylation by solubilized chloroplasts

The oxidation of IAA by peroxidases or riboflavin involves decarboxylation as well as oxygen uptake (Hinman & Lang, 1965; Gordon, 1954). The ability of solubilized chloroplasts to decarboxylate IAA was tested by incubating them with [^{14}C]IAA labeled in the carboxyl position. According to Table 4, the rate of $^{14}CO_2$ release was dependent on light intensity. At an intensity of 10^5 ergs × cm^{-2} × sec^{-1} the reaction rate was more than ten-times greater than in darkness. IAA decarboxylation also depended on the presence of molecular oxygen. When samples were flushed with N_2 for 10 minutes before and during incubation with [1 −^{14}C]IAA (light intensity: 5 × 10^4 ergs × cm^{-2} × sec^{-1}), only 0.08 μmole of $^{14}CO_2$/mg chlorophyll/hr was released. This is less than one-fifth of the rate in air. The light-dependent, O_2-requiring decarboxylation of IAA by solubilized chloroplasts is in agreement with the view that IAA can undergo chlorophyll-dependent oxidation under illumination.

Table 3. Effect of illumination at various spectral regions on oxygen uptake by solubilized chloroplasts and chlorophyll solution.

Light color[1]	IAA concentration (mM)	Oxygen uptake Solubilized chloroplasts	Chlorophyll solution
		μmole O_2/mg chlorophyll/hr[6]	
White[2]	0	15 ± 5	14 ± 1
White[2]	1.14	331 ± 9	281 ± 3
Red[3]	0	19 ± 3	32 ± 1
Red[3]	1.14	260 ± 3	380 ± 17
Blue[4]	0	10 ± 4	28 ± 4
Blue[4]	1.14	166 ± 9	359 ± 11
Green[5]	0	0	0
Green[5]	1.14	16 ± 2	15 ± 3

[1] reaction medium illuminated at 4.9×10^4 ergs $\times$ cm^{-2} $\times$ sec^{-1} through Corning glass filters (Saxton, 1975).
[2] no filter.
[3] Corning no. 2404, 610 to 700 nm transmitted in the visible region.
[4] Corning no. 5030, 330 to 560 nm.
[5] Corning nos. 3384 and 5030, 360 to 460 nm (about 10% of transmitted light) and 470 to 560 nm (about 90% of transmitted light).
[6] mean ± S.E. of 3 measurements; data from one of two similar experiments.

Table 4. Effect of light on IAA-decarboxylation by solubilized chloroplasts[1]

Light intensity (ergs $\times$ cm^{-2} $\times$ sec^{-1} $\times$ 10^{-4})	$^{14}CO_2$ released
	μmole CO_2 /mg chlorophyll/hr[2]
0	0.07 ± 0.00
5	0.44 ± 0.02
7.5	0.56 ± 0.05
10	0.83 ± 0.02

[1] incubated with [1 −^{14}C] IAA for 10 min.
[2] mean ± S.E. of 3 determinations.

Oxygen uptake by solubilized chloroplasts in the presence of various IAA-analogs

We tested the ability of solubilized chloroplasts to oxidize various chemical and biological analogs of IAA, including indoles, carboxylated indoles and those lacking the indole ring. All compounds tested have either close structural relationship to IAA or possess some of its growth-promoting properties. The results show that the carboxylated indoles, such as IAA, indolepropionic acid, indolebutyric acid and tryptophan, supported high rates of oxygen uptake (Table 5). The rate declined slightly with increasing sidechain length. Indoles without a free carboxyl group, such as indole itself, IAN and ethyl indoleacetate, showed low rates of oxygen uptake compared to IAA. Compounds lacking the indole function, such as NAA and 2,4-D, showed no oxygen uptake. These results suggest that the indole and carboxyl functions are both required for maximal rates of oxidation. The indole function alone can support only modest rates of oxidation.

Table 5. Oxygen uptake by solubilized chloroplasts in the presence of various IAA-analogs[1].

Compound	Concentration of compound 1.14 mM	2.28 mM
	μmole O_2/mg chlorophyll/hr[2]	
Indoleacetic acid	132 ± 12	222 ± 3.5
Indolepropionic acid	114 ± 1.4	207 ± 4.1
Indolebutyric acid	99 ± 15	145 ± 1.0
Tryptophan	65.3 ± 2.3	–
Indole	22.6 ± 1.1	57 ± 1.9
Ethyl indoleacetate	15.2 ± 1.2	59 ± 3.7
Indoleacetonitrile	5.3 ± 0.4	16.3 ± 3.7
2, 4-dichlorophenoxy-acetic acid	0	0
Naphthaleneacetic acid	0	0

[1] 0.03 mg chlorophyll/ml.

[2] Mean ± S.E. of 3 measurements, obtained by subtracting a control (no substrate) rate of 15 μmole O_2/mg chlorophyll/hr from each value. Data from one of two similar experiments.

DISCUSSION

The present work demonstrated light-dependent oxidation of IAA and other carboxylated indoles by detergent-solubilized chloroplasts. It was shown that the reaction depended on chlorophyll. This was in part based on the high efficiency of blue and red light in promoting IAA oxidation. Since boiling did not destroy the activity of solubilized chloroplasts, and a solution of chlorophyll was able to mediate the reaction, it appears that enzymes were not involved.

Various detergents are known to solubilize chloroplasts and produce chlorophyll-containing fragments of varying size (Kupke & French, 1960). At the concentration range of Triton X-100 used in this study (1%), chlorophyll was predominantly in its dissolved, monomeric form (Shaposhnikova *et al.*, 1971), and also as a part of photochemically active subchloroplast particles (Kahn, 1964; Vernon *et al.*, 1966; Vernon & Klein, 1975). Chlorophyll in solutions, along with other pigments, was shown to sensitize light-dependent oxidation-reduction reactions, including oxidation of various substrates by molecular oxygen (Seely, 1966). Chlorophyll itself, dissolved in an aqueous solution of Triton X-100, may be oxidized by oxygen in light (Shaposhnikova *et al.*, 1971). The photooxidation of chlorophyll was stimulated by increased concentrations of Triton X-100 or NaCl (Shaposhnikova & Krasnovskii, 1973), in a manner similar to the effect of Triton X-100 and KCl on O_2 uptake in the present work. It is likely, therefore, that in samples without added substrate chlorophyll itself was oxidized by molecular oxygen.

In addition to oxygen uptake, light-dependent decarboxylation of IAA by detergent-solubilized chloroplasts was also demonstrated in the present study. The occurrence of both oxygen uptake and decarboxylation suggests oxidative decarboxylation of the side chain, in agreement with the reaction scheme proposed by Hinman & Lang (1965) for the enzymic oxidation of IAA by peroxidase. This scheme assumes one mole of oxygen utilized and one mole of CO_2 released per mole of IAA. In the present work, rates of CO_2 evolution were much lower than oxygen uptake rates, probably as a result of much lower substrate concentration and a lower concentration of detergent in the reaction medium for decarboxylation. However, it is possible that other oxidative reactions also occur that do not lead to decarboxylation of the substrate.

The present experiments show that IAA and other carboxylated indoles including tryptophan, were oxidized rapidly, whereas oxidation of indoles lacking a free carboxyl group was slow. Thus, the susceptibility of IAA to oxidation by chlorophyll or peroxidase enzymes may be attributed to the presence of both indole and carboxyl functions in its structure. Tryptophan is known to undergo photosensitized oxygenation, and a mechanism for the reaction has been proposed (Nakagawa *et al.*, 1977). In the present work, nonindolic biological analogs of IAA (those possessing some of its growth-regulating properties), such as 2,4-D and NAA, did not show chlorophyll-sensitized photooxidation. NAA and 2,4-D are known to undergo decomposition induced by sunlight or UV light (Crosby, 1973). Plants and bacteria are able to decompose 2,4-D at varying rates (Ray, 1958). It is interesting to note, however, that the rate of 2,4-D decomposition is generally much lower than that of IAA, and 2,4-D is able to persist in some plants for prolonged periods of time (Ray, 1958). Resistance of the molecule to decomposition by different systems may have a common structural ground.

The question whether intact chloroplasts are able to oxidize IAA at significant rates through a chlorophyll-dependent reaction cannot be answered with certainty. Only detergent-solubilized chloroplasts showed high rates of IAA oxidation. The apparent reaction rate for intact chloroplasts was zero. It is possible however that IAA oxidation in intact chloroplasts did occur at a low rate, not detected under the conditions used. Prevailing concentrations of endogenous IAA in various tissues (Schneider & Wightman, 1974) and optimum concentrations of applied IAA to elicit certain physiological responses are characteristically low (Thimann, 1972). Maximal stimulation of photophosphorylation in isolated chloroplasts occurs at 0.1 μM IAA (Tamas *et al.*, 1972). Therefore, even very low rates of IAA oxidation, perhaps several orders of magnitude lower than the rates attained in this study, could be significant in altering physiologically effective concentrations.

Very little is known about the presence, function and metabolism of plant hormones in chloroplasts. IAA was shown to stimulate CO_2 uptake in leaves (Bidwell *et al.*, 1968) and ATP synthesis and bicarbonate uptake in isolated chloroplasts (Tamas *et al.*, 1974). The presence of endogenous IAA in intact chloroplasts has not been demonstrated. Abscisic acid was synthesized in isolated chloroplasts from mevalonic acid (Milborrow, 1974). Various gibberellins were found to be specifically associated with chloroplasts by Browning & Saunders (1977). Thus, chloroplasts may be an important source of hormones in plants, and they may influence the activity of other parts of the plant through these substances. The question whether these and other hormones play a regulatory function within the chloroplasts themselves needs to be explored. Further work on the synthesis and metabolism of hormones in chloroplasts is necessary to evaluate the role of chloroplasts in the hormonal regulatory activity of plants.

ABBREVIATIONS

2,4-D = 2,4-dichlorophenoxyacetic acid
IAA = indoleacetic acid
IAN = indoleacetonitrile
MES = 2 (N-morpholino) ethane sulfonic acid
NAA = naphthaleneacetic acid
S.E. = standard error
TES = N-tris (hydroxymethyl) methyl-2-aminoethane sulfonic acid.

REFERENCES

Arditti, J. & A. Dunn. Experimental plant physiology. Experiments in cellular and plant physiology, p. 265. Holt, Rinehart & Winston, New York (1969).

Arnon, D.I. Copper enzymes in isolated chloroplasts. Polyphenol oxidase in *Beta vulgaris*. *Plant Physiol.* 24: *1–15* (1949).

Bidwell, R.G.S., W.B. Levin & I.A. Tamas. The effects of auxin on photosynthesis and respiration. pp. 361–376, in: Wightman, F. & G. Setterfield (eds.), Biochemistry and physiology of plant growth substances. The Runge Press, Ottawa (1968).

Browning, G. & P.F. Saunders. Membrane localized gibberellins A_9 and A_4 in wheat chloroplasts. *Nature* 265: *375–377* (1977).

Crosby, D.G. The fate of pesticides in the environment. *Ann Rev. Plant Physiol.* 24: *467–492* (1973).

Gordon, S.A. Occurrence, formation and inactivation of auxins *Ann. Rev. Plant Physiol.* 5: *341–378* (1954).

Hinman, R.L. & J. Lang. Peroxidase-catalyzed oxidation of indole-3-acetic acid. *Biochemistry* 4: *144–158* (1965).

Kahn, J.S. A soluble protein-chlorophyll complex from spinach chloroplasts. I. Isolation of a photochemically active complex. *Biochim. Biophys. Acta* 79: *234–240* (1964).

Kupke, D.W. & C.S. French. Relationship of chlorophyll to protein and lipoids; molecular and collodial solutions. Chlorophyll units. pp. 298–322, in: Ruhland, W. (ed.), Encyclopedia of plant physiology, vol. 5, part 1. Springer-Verlag, Berlin (1960).

Milborrow, B.V. The chemistry and physiology of abscisic acid. *Ann. Rev. Plant Physiol.* 25: *259–307* (1974).

Nakagawa, M., H. Okajima & T Hino. Photosensitized oxygenation of N_b-methoxycarbonyltryptamines. A new pathway to kynurenine derivatives. *J. Am. Chem. Soc.* 99: *4424–4429* (1977).

Nobel, P.S. A rapid technique for isolating chloroplasts with high rates of endogenous photophosphorylation. *Plant Physiol.* 42: *1389–1394* (1967).

Perkins, H.J. & D.W.A. Roberts. Purification of chlorophylls, pheophytins and

pheophorbides for specific activity determinations. *Biochim. Biophys. Acta* 58: *486–498* (1962).
Ray, P.M. Destruction of auxin. *Ann. Rev. Plant Physiol.* 9: *81–118* (1958).
Saxton, R.G. Transmission of Corning colored filters. pp. E231–E237, in: Weast, R.C. (ed.), Handbook of chemistry and physics. CRC Press, Cleveland (1975).
Schneider, E.A. & F. Wightman. Metabolism of auxin in higher plants. *Ann. Rev. Plant Physiol.* 25: 487–*513* (1974).
Seely, G.R. Photochemistry of chlorophylls *in vitro*. pp. 523–568, in: Vernon, L.P. & G.R. Seely (eds.), The chlorophylls. Academic Press, New York (1966).
Shaposhnikova, M.G., N.N. Drozdova & A.A. Krasnovskii. Chlorophyll oxidation in aqueous solutions of Triton X-100 detergent. *Biokhimiya* 36: *704–711* (1971).
Shaposhnikova, M.G. & A.A. Krasnovskii. Comparative study of photooxidation of chlorophyll analogs in aqueous detergent solutions. *Biokhimiya* 38: *193–201* (1973).
Stafford, H.A. & S. Bravinder-Bree. Peroxidase isozymes of first internodes of *Sorghum*. Tissue and intracellular localization and multiple peaks of activity isolated by gel filtration chromatography. *Plant Physiol.* 49: *950–956* (1972).
Tamas, I.A., B.D. Atkins, S.M. Ware & R.G.S. Bidwell. Indoleacetic acid stimulation of phosphorylation and bicarbonate fixation by chloroplast preparations in light. *Can. J. Bot.* 50: *1523–1527* (1972).
Tamas, I.A., J.W. Schwartz, J.M. Hagin & R. Simmonds. Hormonal control of photosynthesis in isolated chloroplasts. pp. 261–268, in: Bieleski, R.L., A.R. Ferguson & M.M.Cresswell (eds.), Mechanism of regulation of plant growth. Bull. 12, Royal Soc. New Zealand, Wellington (1974).
Thimann, K.V. The natural plant hormones. pp. 3–332, in: Steward F.C. (ed.), Plant physiology. A treatise, vol. VI B. Academic Press, New York (1972).
Vernon, L.P. & S.M. Klein. Nature of plant chlorophylls *in vivo* and their associated proteins. *Ann. N. Y. Acad. Sci.* 244: *281–296* (1975).
Vernon, L.P., E.R. Shaw & B. Ke. A photochemically active particle derived from chloroplasts by the action of the detergent Triton-X100. *J. Biol. Chem.* 241: *4101–4109* (1966).

Authors' address:
Biology Department
Ithaca College
Ithaca, New York 14850
U.S.A.

Present address J. D. Becker:
Philadelphia College of Osteopathic Medicine
Philadelphia, Pennsylvania 19131
U.S.A.

SESSION IV

PHOTOSYNTHESIS AND SOURCE-SINK RELATIONS

INTEGRATION OF ASSIMILATION AND CARBOHYDRATE UTILIZATION IN APPLE

D.J. AVERY, C.A. PRIESTLEY & K.J. TREHARNE
(Maidstone, U.K.)

ABSTRACT

Growth and yield ultimately depend on an adequate supply of carbohydrates and nutrients. When trees are grown under shade, total dry matter production per unit leaf area is directly proportional to irradiance. Contrasting stock/scion combinations in a given environment tend to have constant top/root ratios, but when shaded root increment is reduced more than shoot. Although internal competition for resources exists between fruit, shoots and other growth centres, differences in flowering, fruit set or cropping do not relate to limitations in carbohydrate supply. Failure of meristems to realize their full potential is likely to be related to plant hormones rather than limitations imposed by carbohydrate resources. Other than a depression at the time of spring bud-break, concentrations of soluble carbohydrates in perennial parts of the tree remain fairly constant even when fruit-load is heavy and shoot growth is reduced.

Various experiments involving deblossoming, partial defoliation and blocking translocation suggest that carbohydrate production is co-ordinated with utilization. Increases in crop demand are met by enhanced photosynthesis and prolonged leaf activity when the tree is carrying fruit. Leaf weight increases after deblossoming, fruit picking or the removal of rings of bark from shoots or branches. Ringing treatment does not result in accumulation of carbohydrates in the perennial parts above the ring, however, again suggesting integrated control of photosynthetic assimilation and resource demand. These findings are discussed in terms of modulation of leaf activity and the possible role of growth regulating substances.

INTRODUCTION

The form of the established composite apple tree is characteristic of the particular combination of scion variety grafted on a size-controlling rootstock in which growth and cropping are regulated by the interaction between the component partners. An understanding of the internal factors which govern tree efficiency with respect to size and regularity of cropping is required in order to optimize conversion of leaf assimilates into the economic product, the fruit, without unnecessary production of vegetative growth.

Heavy fruit loads are known to depress other growth centres leading in extreme cases to significantly reduced vegetative vigour (Barlow, 1963), poor root growth and failure to initiate fruit buds for the subsequent year's crop. It is important to determine the degree of competition between shoot, fruit and root, and the timing of such competition. This may be envisaged in terms of strength of sink demands for root factors, leaf assimilates and carbohydrate reserves or in terms of the different meristematic regions producing growth promoting or inhibitory substances which influence other growth centres through regulation of directional flow of materials or their utilization.

The proportion of annual vegetative increment converted into shoot growth is maintained within quite narrow limits over a wide range of total growth (Maggs, 1960) and it is principally between leaves and roots that growth adjustments occur. Sensitive regulatory mechanisms evidently exist which control the relative growth of component parts of the tree.

Towards achieving an understanding of tree growth regulation, changes in the light environment have been used to alter the balance between assimilate sources and sinks, and through manipulation of leaf surface and by deblossoming, assimilate supply and sink demand have been modified. Comparisons of distribution of growth, leaf assimilation and carbohydrate status resulting from such modifications are reported here.

MATERIALS AND METHODS

Various experiments were conducted with young trees grown outdoors either in the field or in pots of loam/sand mixtures in plunge-beds. A range of widely contrasting vigour and fruiting behaviour was obtained from materials as rooted shoots from stool-beds of apple rootstocks or young scion varieties such as Worcester Pearmain grafted onto rootstock varieties including 3426, M.2, M.7, M.9, M.25, M.26 and MM.111. At the end of their maiden year, potted trees were pruned to provide initial standardized size prior to subsequent growth and experimental treatments. Specific details of experimental manipulations are presented for greater convenience in the appropriate sections of the results and discussion.

Determination of growth parameters and estimations of soluble sugars, starch and hemicellulose contents have been described previously (Priestley, 1963; 1969).

Gas exchange measurements were made in a laboratory-based open system of infra red gas analysis on fully expanded, attached leaves. The assimilation chambers were each water cooled and fitted with a fan to provide both good distribution of air over the leaf surfaces and to create a rapid circulation imposed on the throughflow (Avery, unpublished). The irradiation sources were metal halide, 400 W MBIF, 'Kolorarc', Thorn Atlas, lamps. Irradiance level and leaf temperature were continuously monitored within the chambers during gas exchange determinations.

RESULTS AND DISCUSSION

Effects of shade on growth, photosynthesis and carbohydrate content

To investigate the pattern of growth distribution when light was a limiting factor, young trees of Worcester Pearmain on rootstocks giving a range of vigour were grown in walk-in cages made of timber frames covered with different layers of neutral plastic screening to give selected levels of shade from 20–65% natural irradiance (full sunlight). Monitored temperature throughout the growing season indicated those under shades were never more than 1.2° C below those in the open.

Dry weight increment per unit leaf area was linearly proportional to incident irradiance up to some ⅔ full sunlight with proportionately less increase in assimilation thereafter (Fig. 1). Shading resulted in an increased proportion of dry matter in leaves of Worcester Pearmain scions with a decrease in rootstock as shown by Maggs (1960) and this is reflected in changes in scion/stock ratio of weights of the perennial parts (Table 1). This effect was confirmed in

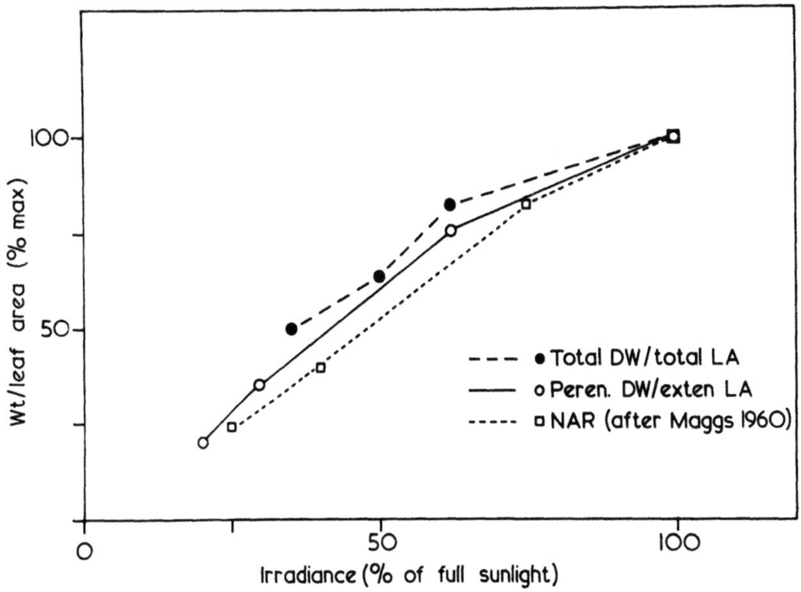

Fig. 1. The effect of shade on leaf efficiency in dry matter production.
●---● Total dry weight/total leaf area.
o—o Perennial dry weight/extension shoot leaf area.
▫---▫ Net Assimilation Rate (after Maggs, 1960).

comparisons between fruiting and deblossomed trees which also indicated that effects of shade and fruiting are similar in depressing rootstock growth (Table 2). With dwarfing rootstocks, fruiting and heavy shade combined resulted in a net loss in rootstock weight.

Table 1. Effect of shade on dry matter distribution in young apples trees.

	Scion/stock ratio (perennial parts)			
	% sunlight			
	100	62	28	20
Rootstock				
M.9	1.10	1.16	1.14	1.51
M.26	0.87	0.88	1.02	1.03
MM.111	1.15	1.10	1.31	1.21
M.25	0.94	1.01	1.16	1.19

Table 2. Interactions of shade and fruiting on tree growth.

	Rootstock					
	3426			M.25		
	Scion/stock*		fruit dry wt. (g)	Scion/stock		fruit dry wt. (g)
% sunlight	+	−		+	−	
100	0.99	0.69	56.7	1.17	1.01	67.9
60	0.90	0.70	51.3	1.18	1.10	59.5
23	1.00	0.96	10.5	1.24	1.14	11.6

* ratio of perennial parts + fruiting; − deblossomed.

Crude estimates of leaf photosynthesis of leaves grown in full sun and transferred to different levels of shade were obtained by leaf discing coupled with petiole steaming. Assimilation was decreased relatively little by shading to 60% of full sunlight but heavier shade had a proportionately greater effect (Table 3). Measurements of light-saturated CO_2 exchange by attached leaves

Table 3. Effect of shade on leaf weight gain.*

	% sunlight		
Stock	100	60	20
3426	0.92	0.86	0.43
M.9	1.11	0.54	0.21
M.26	0.62	0.74	0.34
MM.111	0.89	0.72	0.20
M.25	0.69	0.41	0.30

* dry weight increments ($mg/cm^2/$ 6hr) of Worcester Pearmain scion leaves after petiole steaming.

indicated that assimilation rates of leaves formed in 60% and 20–25% full sunlight were 80% and 60% respectively, of those of full sunlight grown leaves, when expressed on a leaf area basis. However, maximum net assimilation rates were attained at higher irradiances than experienced during growth under shades; light saturation ranged from 350 Wm^{-2} (visible) for full-sun leaves to 190 Wm^{-2} for leaves from 20% full light, as summarized in idealized curves in Fig. 2. Judging by the slopes of the light response curves, sun and shade leaves appear to have similar photochemical capacity at low irradiances. Low photosynthetic rates of shade leaves are associated with greater stomatal resistance to CO_2 diffusion but even more to greater 'residual' resistance (Table 4) normally associated with lower carboxylating capacity than sun leaves, as reported in other species (Björkman, 1968; Treharne & Eagles, 1969).

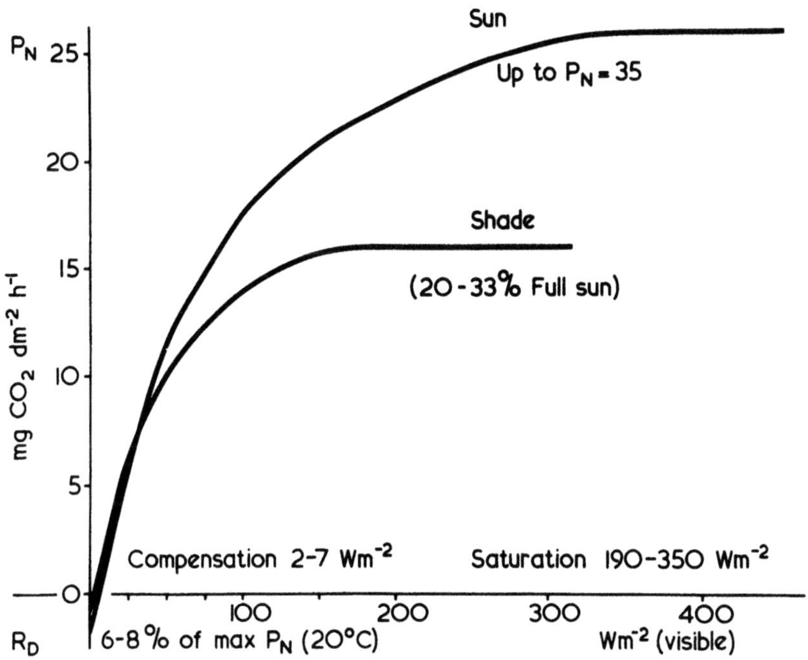

Fig. 2. Idealized curves for photosynthesis relative to irradiance in sun and shade leaves.

Table 4. Photosynthetic response of sun and shade leaves of Worcester Pearmain on rootstock 3426.

% sunlight	P_N*	E	r_s	r_M
100	30.4	1.37	2.1	4.7
58	19.5	1.16	3.8	8.2
24	12.6	0.70	4.8	12.2

* Details in methods. $P_N = mgCO_2dm^{-2}\ hr^{-1}$; $E = gH_2Odm^{-2}hr^{-1}$; $r_s\ r_M = scm^{-1}$.

Clearly, effects of shading on photosynthetic efficiency are confounded with effects on other processes such as leaf production and expansion, stem elongation, cambial activity and fruit retention.

As in other species, apple leaves grown in shade are modified in their gross morphology and anatomy (Jackson & Beakbane, 1970) with progressively lower specific leaf weight as shade increases. Chloroplasts of shade leaves have more deeply stacked grana (Skene, 1974) considered to be an adaptation to facilitate photon capture.

In previous experiments shading treatments had no effect on levels of extractable carbohydrates (Priestley, 1963). Although stem-reserve carbohydrates were depleted at budbreak in spring, they were subsequently recovered, shade level influencing only the rate of replenishment. There was no indication that level of available carbohydrate limited growth. The close similarity of sugar levels at growth rates which differed with irradiance indicated an internal mechanism of control.

Effects of partial defoliation on growth and carbon resources

To further examine regulation of assimilate utilization, photosynthetic capacity was reduced by removal of $\frac{2}{3}$ of each leaf from the period corresponding to the end of developing fruit cell division (mid-June) until fruit harvest (mid-September). Treatment was imposed on a biennial variety (Laxton's Superb) to test whether previously reported carbohydrate depletions (Priestley, 1970) could be accentuated by reducing photosynthesis and compared with a regular cropping variety (Worcester Pearmain), both on M.2 rootstocks. Measurements of growth of fruit, shoots and girth were made at the end of the cropping season and carry-over effects of defoliation determined on the subsequent year's production. Reduction in photosynthetic capacity was not directly proportional to the leaf area differences achieved (Fig. 3) due probably to enhanced photosynthesis of remaining foliage because of less competition for root factors, including cytokinins and possibly to better light distribution. However, a marked reduction in fruit yield was observed in both varieties but with less marked effects on shoot and girth growth (Fig. 3). In the following year, girth increment was again relatively unaffected but shoot growth was reduced by the previous year's partial defoliation. Also, the regular variety showed a significantly lower fruit yield, presumably through effects of defoliation on fruit-bud formation in the preceding season.

Irrespective of the reduction of photosynthetic capacity and marked effects on growth and development, the carbohydrate contents (soluble sugars, starch, hemicellulose) were essentially the same in defoliated trees and their respective untreated controls (Fig. 3). Thus it appears that even with changing assimilatory capacity, the availability of carbohydrate resources is not limiting and the question arises as to the mechanism whereby the tree maintains its soluble carbohydrate contents within narrow limits over a wide range of growth. The relative effectiveness of different meristems to 'compete' for assimilates implies sensitive regulatory control over translocation flow and possibly the processes following phloem unloading. It may be speculated that plant growth substances, particularly auxins, are involved in transport and unloading; these are known to be produced in shoot tips and certainly shoot growth was competitively better than fruit in the present investigation. No measurements were made on root growth or activity but since roots 'suffer' relatively more than shoots, particularly during cropping, as seen in the shading experiments, it is likely their activity was impaired by defoliation and may have contributed to the carry-over effect, particularly in the context of reduced source of cytokinin supply.

Relationships between assimilate supply and demand

The effect of removal of part of the leaf surface on the efficiency of dry matter production by remaining foliage was studied in vegetative single-shooted trees of M.7 rootstocks. Defoliation of 87% of the leaf surface, retaining only young expanding leaves at the shoot tip and 4 of the recently expanded leaves, had little

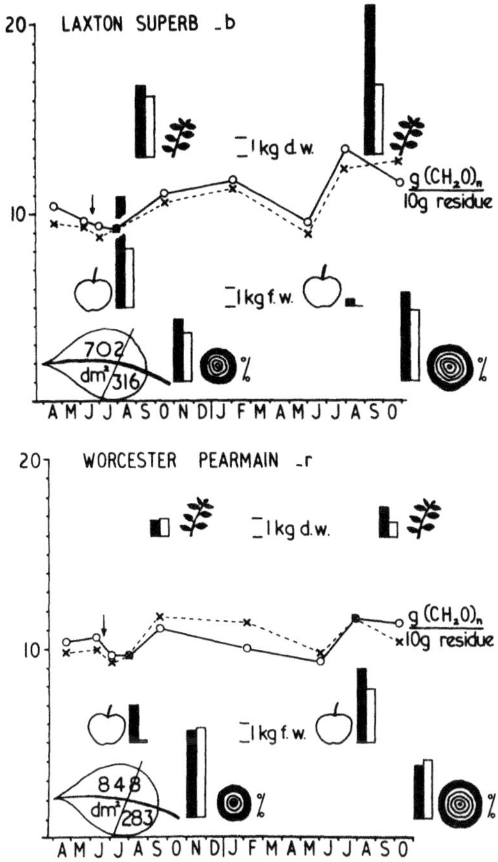

Fig. 3. Effects of leaf area reduction during fruiting on carbohydrate content and growth parameters in Laxton's Superb (biennial) and Worcester Pearmain (regular) apples on M.2 rootstock.

o——o Control
x - - - x Leaf area reduced

Carbohydrate content as g (Sugar + Starch + Hemicellulose)/10 g residue

Histograms: Kg fruit fresh weight; % increase in cross sectional area.

Kg shoot dry weight

effect on CO_2 exchange rate of the remaining expanded leaves in the first 3 days but subsequently a significant increase in photosynthesis was detected which persisted for 6 weeks (Table 5). Also, dark respiration rate was significantly lower than in comparable leaves of control plants. Measurements of ^{14}C translocation following $^{14}CO_2$ assimilation were made by taking discs of interveinal laminar tissue from leaves on a parallel series of plants, and showed a greater throughput of radioactivity in carbohydrates from leaves on the partially defoliated plants (Fig. 4)

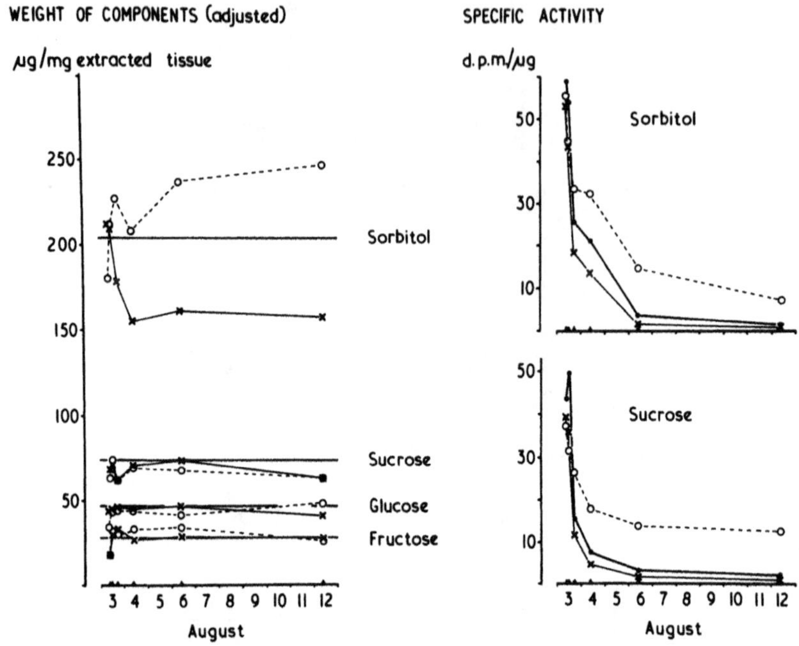

Carbohydrate contents of treated plants relative to mean control values.

Specific activities of sorbitol and sucrose.

Fig. 4. Soluble carbohydrate components in M.7 apple leaf discs after $^{14}CO_2$ assimilation.

----- Leaves with steamed petioles.
●——● Control leaves.
×——× Leaves on partially defoliated plants.

Table 5. Effects of translocation block and defoliation on CO_2-exchange of M.7 apple leaves.

Time	Net Photosynthesis*		Dark Respiration*	
(Days)	Steamed	Defoliated	Steamed	Defoliated
1	0.79	–	1.58	–
2–3	0.56	1.02	1.99	1.27
7–9	–	1.17	–	0.46
48–50	–	1.26	–	0.87

* Expressed as ratio of control values.

In an earlier investigation with Sunset/M.7 apple trees, assimilate flow from one branch was prevented by removal of a ring of bark at full bloom which resulted in improved fruit set and some reduction in growth above the ring. Production of dry matter per unit leaf area was higher when fruit was present than when branches were deblossomed (Priestley, 1976). In the absence of fruit demand above the ring, the rate of removal of ^{14}C-labelled assimilates from leaves on ringed shoots was slower than on unringed shoots (Priestley, 1973). In the present experiments with M.7, petioles of leaves were steamed to block translocation flow. A progressive decline in light-saturated photosynthesis occurred with a 2-fold increase in dark respiration (Table 5). Contrasting with the pattern observed in defoliated plants, a significant increase in leaf sugars, particularly sorbitol, and a slower decline in specific radioactivity after $^{14}CO_2$ exposure was observed in petiole-steamed leaves than in comparable controls (Fig 4).

It is evident that both photosynthesis and dark respiration are related to substrate levels, photosynthesis increasing and respiration decreasing when transport from the leaf is increased, the converse holding when translocation is prevented. Thus it appears that photosynthesis (source activity) is modulated by removal of carbohydrates from the leaf (sink demand), presumably involving some negative feedback via the phloem-transport system.

Fruiting is probably the most important natural phenomenon affecting the pattern of growth in trees, whose influence as a sink competing for assimilates has already been alluded to. Specific effects of fruiting have been derived from comparisons between cropping and deblossomed trees of Worcester Pearmain on contrasting rootstocks. Net photosynthesis by leaves of fruiting apple trees is higher than when not carrying fruit (Avery, 1975) and is related to smaller diffusion and residual resistances (Table 6). The effect of carrying fruit on enhanced CO_2 assimilation by leaves is further magnified by extending photosynthetic duration through retardation of leaf senescence. Dark respiration rates of leaves of fruiting trees were lower than on deblossomed trees (Table 6) and translocation of assimilates was faster, again demonstrating the integration between photosynthesis, respiration and carbohydrate utilization.

Table 6. Gas exchange properties of leaves of fruiting and deblossomed trees*.

	fruit status	P_N	E	r_s	r_M	R_D
July–August	+	29.7	1.43	1.99	4.95	1.25
	–	21.8	1.00	3.60	7.15	1.65
October	+	24.9	1.18	3.15	4.15	1.75
	–	5.2	0.36	33.20	82.35	2.60

* mean values for several scion/rootstocks.
+ fruiting; – deblossomed. Units as in Table 4.

Several explanations may be afforded as to how the effect of bearing fruit is manifested. Sink demand is lower in the absence of fruit and a lower photosynthetic supply is needed, and the converse is true. The question still remains as to how the message is received by the leaf. Moreover, when not bearing fruit, extra growth of shoot and other meristems provides alternative sinks, envisaged as competing for root factors and assimilates but possibly not to the same extent as fruit. However, cytokinin and gibberellin levels are reported as higher in leaves of grape when carrying fruit (Hoad, 1977) and it is possible that fruits do not compete with leaves for root-produced cytokinins because of the possible ability of fruit to *de novo* synthesize their own growth substances (Hoad, 1978) and even possibly to export them to leaves. Changes in abscisic acid metabolism of leaves in response to fruiting have been observed by Kriedemann & Loveys (1975) in grapevine.

Increased photosynthesis following applications of gibberellins was not attributable to enhanced enzyme synthesis (Oben & Marcelle, 1975) and the effect is likely to be related to changes in membrane permeability, perhaps facilitating mobilization of carbohydrates. The increases in photosynthesis and decreases in respiration described in the present investigations may be the result of similar phenomena. The roles of growth regulators in modulating leaf activity and controlling partition of assimilates to different meristematic regions are currently being studied in our laboratories.

REFERENCES

Avery, D.J. Effects of fruits on photosynthetic efficiency, in: 'Climate and the Orchard', (ed.) H.C. Pereira. *Res. Rev. Commonw. Bur. Hort. Plantn Crops*, No. 5, *110–112* (1975).

Barlow, H.W.B. An interim report on a long-term experiment to assess the effect of cropping on apple tree growth. Rep. E. Malling Res. Stn. for 1963, *84–93* (1964).

Björkman, O. Further studies on differentiation of photosynthetic properties in sun and shade ecotypes of *Solidago virgaurea*. *Physiol. Plantarum*, 21, *84–99* (1968).

Hoad, G.V., B.R. Loveys and K.G.M. Skene. The effect of fruit removal on cytokinins and gibberellin-like substances in grape leaves. *Planta* (Berl.), 136, *25–30* (1977).

Hoad, G.C. The role of seed-derived hormones in the control of flowering of apple, in: Proc. IIIrd. Internat. Symp. on Growth Regulators in Fruit Production, Posnan, 1977. *Acta Horticulturae* (in press).

Jackson, J.E. and A.B. Beakbane. Structure of leaves growing at different light intensities within mature apple trees. Rep. E. Malling Res. Stn. for 1969, *87–89* (1970).

Kriedemann, P.E. and B.R. Loveys. Hormonal influences on stomatal physiology and photosynthesis. pp 227–236, in: Environmental and Biological Control of Photosynthesis. R Marcelle (ed.), Dr. W. Junk b.v. Publishers, The Hague (1975).

Maggs, D.H. The stability of the growth pattern of young apple-trees under four levels of illumination. *Ann. Bot.*, N.S., 24, (96), *434–450* (1960).

Oben, G. and R. Marcelle. The effects of CCC and GA on some biochemical and photochemical activities of primary leaves of bean plants. pp 211–216, in: R. Marcelle (ed.), Environmental and Biological Control of Photosynthesis. Dr. W. Junk b.v. Publishers, The Hague (1975).

Priestley, C.A. The carbohydrate resources of young apple-trees under four levels of illumination. *Ann. Bot.*, N.S., 27, (107), *435–446* (1963).

Priestley, C.A. Some aspects of the physiology of apple rootstock varieties under reduced illumination. *Ann. Bot.*, 33, *967–980* (1969).

Priestley, C.A. Some observations on the effect of cropping on the carbohydrate content in trunks of apple trees over a long period. Rep. E. Malling Res. Stn. for 1969, *121–124* (1970).

Priestley, C.A. The use of apple leaf discs in studies of ^{14}Carbon translocation. Proc. Res. Inst. Pomology, Skierniewice, Poland. Series E, Conf. and Symposia No. 3, *121–128* (1973).

Priestley, C.A. Some effects of ringing branches on the distribution of dry matter in young apple trees. *J. Exp. Bot.*, 127, (101), *1313–1324* (1976).

Skene, D.S. Chloroplast structure in mature apple leaves grown under different levels of illumination and their response to changed illumination. *Proc. R. Soc. Lond.* B, (186), *75–78* (1974).

Treharne, K.J. and C.F. Eagles. Effect of growth at different light intensities on photosynthetic activity of two contrasting populations of *Dactylis glomerata* L. pp 372–382, in: H. Metzner (ed.), Progress in Photosynthesis Research, 1. Tübingen (1969).

Authors' address:
Physiology Department
East Malling Research Station
Maidstone, Kent ME19 6 BJ.
United Kingdom.

PHOTOSYNTHESIS IN TOMATO PLANTS WITH MODIFIED SOURCE-SINK RELATIONSHIP*

Z. STARCK, M. KOZIŃSKA & R. SZANIAWSKI
(Warsaw, Poland)

ABSTRACT

The mechanism of regulation of the rate and activity of photosynthesis and distribution of dry matter between various organs in tomato plants is the subject of the investigation. The effect of continuous removal of flower buds, or keeping the clusters in darkness on the photosynthesis and NAR as well as on the growth of particular organs was examined. Prevention of fruit production did not reduce the growth of the whole plant, but altered markedly the pattern of distribution of dry matter; more photosynthates accumulated in the leaves and stem, which under ordinary conditions act as much less active acceptors.

In the first 6–8 weeks, prevention of fruit development did not affect photosynthesis and NAR, but later the rate of photosynthesis decreased markedly. Owing to the higher mass of leaves, the photosynthetic activity of the whole plants was similar to that of control plants at the end of vegetation. Elimination of CO_2-absorption by clusters, by keeping them in darkness, did not increase the rate of CO_2-assimilation in the leaves. In the last few weeks the rate and activity of photosynthesis were even reduced. The maintenance of unchanged growth of the fruits, in spite of lower photosynthetic activity, was due to the growth retardation of vegetative organs and the depletion of reserve materials from leaves, stem and peduncles.

The results of reported study allow to discuss hypothesis concerning the mechanism controlling photosynthesis by its product inhibition as well as hormonal one, controlling both photosynthesis and the pattern of assimilates distribution.

INTRODUCTION

A relationship between an altered current demand for assimilates and the rate of photosynthesis has been documented in a large number of experiments.

The increase of current demand for assimilates in some experiments increases the rate of net photosynthesis and/or NAR. Elimination of some acceptors reduces the rate of assimilation. Another research group, discussed by Neales and Incoll (1968), Geiger (1976) and others did not find such a correlation.

Taking into account that each experimental modification of plants changes the natural balance in the whole organism, some experiments were done on intact radish plants with naturally variable size of the storage organ (Starck and Ubysz, 1976; Starck and Stradowska, 1977). From them it was concluded, that

* The authors are grateful to Misses B. Witek and M. Kołakowska for very good technical assistance. Experiments were coordinated and financially supported by the Inst. of Vegetable Crops in Skierniewice /No. 411/.

the high variability of the swollen hypocotyl size is not correlated with the rate and activity of photosynthesis. The above mentioned results as well as those with sugarbeet (Glasziau and Bull, 1971), did not support the hypothesis of end-product repression of photosynthesis.

Fruits, an important sink for assimilates, modulate the photosynthesis of adjacent leaves in many plant species (Flinn, 1974; Hansen, 1970; Lenz, 1974; Moss, 1962).

The relation between the physiology of leaves and fruit growth is reported in the papers of Moorby *et al*. (1974) and Moorby & Jarman (1975), where fruits removal in tomato plants had no effect on the rate of photosynthesis for the first 1–3 days, after which the rate of this process started to decline. In Tanaka's paper (1974) removal of tomato fruits decreased NAR. In *Capsicum* plants, after prevention of fruit development by continuous defloration, the photosynthetic capacity declined much more than in the leaves of control plants (Hall, 1977a, b). Nagarajah (1975) modified sink-source relationships by continuously debudding cotton plants; in that case the pattern of assimilate distribution changed, without reduction of the total plant growth. Similar results were presented (Lenz, 1974) in *citrus* plants with various number of fruits. Nevertheless the rate of photosynthesis in young cotton plants was not affected by debudding in contrast to *citrus*, plants where the rate of photosynthesis decreased with the reduced fruit number.

In the above mentioned papers, two main explanations are proposed: the simple feed-back control of photosynthesis by end-product inhibition, observed in biochemical pathways or the effect of acceptors, exporting hormonal or other signals directed to organs producing assimilates and affecting photosynthesis.

In the reported experiments, tomato plants were used as a model to study endogenous regulation of photosynthesis and growth pattern. Source-sink relationships were modified by continuous flower bud removal to prevent fruit production and in other series, by keeping the clusters in darkness.

MATERIAL AND METHODS

Tomato plants of Neederland variety DBR 90 were grown in the greenhouse, in polyvinyl rings, containing about 12 l of peat with a standard fertilizer supplement. Axillary shoots were successively removed; the plants formed 7–10 clusters.

At the beginning of flowering, the plants were divided into three experimental series: 1) control plants (intact), 2) plants with flower buds successively removed (defloration was repeated at 2–3 day intervals), 3) plants with all clusters kept in darkness by putting them into white and black paper backs.

The rate of apparent photosynthesis (APS) and dark respiration (R_D) was estimated by using an open circuit gas analysis system capable of handling five leaf chambers and fitted with Infralyt IV, Juncolor (DDR). Normal operating conditions were: temperature of air in leaf chamber: 20°C; light flux density on the leaf surface about 18 klux; illumination with 3 lamps, TELAM LRFR, 400 W each.

Gas exchange was measured six times at 2-week intervals, in 6 detached leaves, cut from 3 plants of each series. The leaves on the plants were situated near the first cluster. In older plants only the upper leaflets was measured. The leaf surface area was determined with an Automatic Photoelectric meter AAM-5 Hayashi Denko Co.

Beside leaves, gas exchange of the whole clusters or separated fruits was measured.

Parallel to each of the six measurements of gas exchange, the fresh and dry matter of particular organs from six plants was estimated and NAR (net assimilation rate) and RGR (relative growth rate) were calculated according to the formulae:

$$NAR = \frac{W_2 - W_1}{t_2 - t_1} \cdot \frac{\ln A_2 - \ln A_1}{A_2 - A_1}, \qquad RGR = \frac{\ln W_2 - \ln W_1}{t_2 - t_1}$$

where: A – successively changed dry matter of blades, W – successively increased dry matter of the whole aerial part, $t_2 - t_1$ – time interval between plant collection (usually 2 weeks).

In plant material frozen in dry ice and homogenized in 80% ethanol, soluble sugars were determined by the phenol method.

RESULTS

The dynamics of growth of aerial parts (expressed as dry matter accumulation) was similar in all examined series (Fig. 1). In the period between 6–10 weeks after the beginning of flowering, about 85% of dry matter accumulated in the aerial part was deposited in the fruits, causing a temporary retardation of growth of other parts. At that time, the dry matter of vegetative organs increased much more in deflorated plants than in intact ones. In the last 4 weeks (two successive harvesting intervals), growth of leaves and stem in plants with fruits was very small as compared with the deflorated plants (Figs. 1 and 4), where deposition of

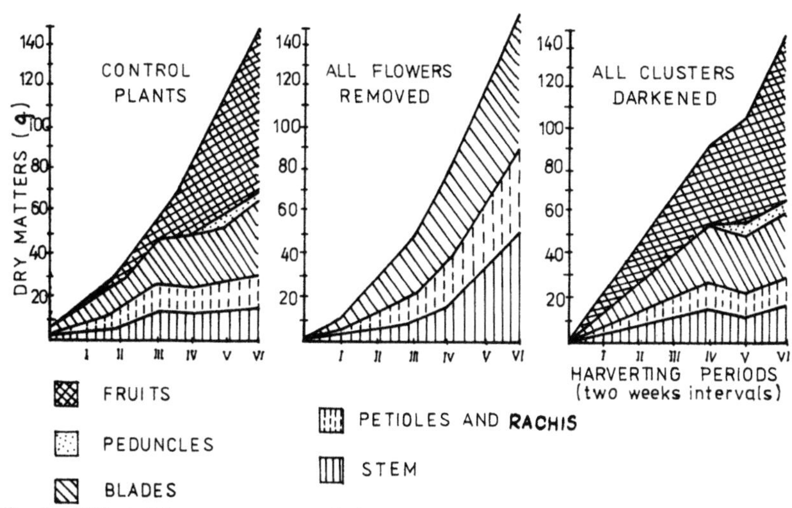

Fig. 1. Effect of donor⇄acceptor relationships on the dynamic of tomato plants growth (measurements done since beginning of the flowering).

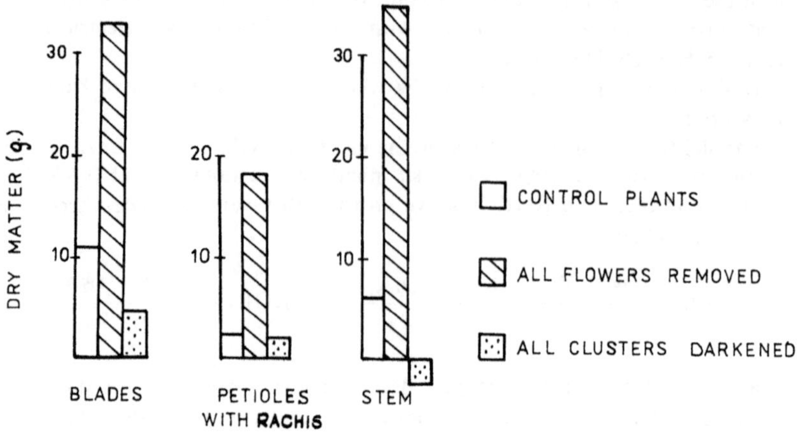

Fig. 2. Growth of vegetative organs in the last 4 weeks of vegetation (in g dry matter).

dry matter was still very high; the specific weight of leaves of that series increased to 465 mg d.m. × dm^{-2} as compared with 340 in intact, control plants. The domination of vegetative organ growth in plants without fruits is clearly seen in the RGR values, especially at the end of vegetation (Table 1).

In deflorated plant, organ tissue had a much higher content of dry matter, which in the last harvesting period increased in the petioles and stem to 11 and 17 per cent respectively as compared with 7 and 13 in control plants.

Table 1. Parameters of growth analysis in control and deflorated plants.

Growth parameters		Experimental series	harvesting periods /two week intervals/			
			I	II	III	VI
NAR	mg d.m. / g d.m. day	control	129	105	92	92
		deflorated	103	124	61	57
RGR mg / g. day	blades	control	85	43	25	2
		deflorated	47	63	27	22
	petioles + rachis	control	89	56	21	9
		deflorated	79	80	29	22
	stem	control	99	49	56	12
		deflorated	81	56	41	37
	total plant	control	72	50	38	21
		deflorated	59	65	31	26

In control and deflorated plants, the rate of photosynthesis and dark respiration significantly decreased in the first two weeks (Fig. 3); no effect of defloration on both processes was observed. After the next two weeks, the respiration rate did not change significantly in both experimental treatments, but the rate of APS decreased more in non-fruiting plants (Fig. 3) than in control ones. A serious retardation of the CO_2-assimilation rate in plants without fruits was observed as late as 6–8 weeks from the beginning of flowering, when the dry matter of fruits in intact plants contributed to the total aerial part for about 30 per cent. The differences increased up to the end of vegetation. Much lower differences in photosynthetic rate were noted when APS was calculated per leaf surface-area, owing to greater thickness of blades in the deflorated plants (data not presented).

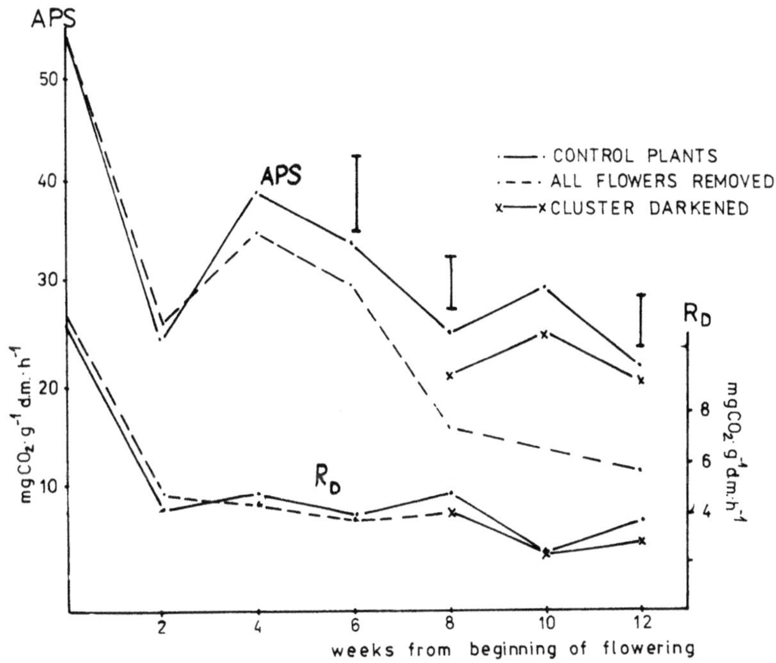

Fig. 3. Time course of the rate of apparent photosynthesis (APS) and dark respiration (R_D) in the blades; vertical bars represent LSD at $p = 0.05$.

Unexpected results were obtained in the series with clusters kept in darkness. In all examined periods, the rate of CO_2-assimilation slightly decreased as compared with the control plants in spite of elimination of cluster photosynthesis (Fig. 3). The photosynthetic activity of the whole plant was also reduced, especially in the last 4 weeks in contrast to higher NAR values (Fig. 2). NAR (calculated per unit of dry matter of blade) was similar in both intact and deflorated plants in the first 2 intervals examined, and then decreased in deflorated plants, (Table 1). In the last 2 weeks, NAR of control plants was almost at the same level as at the beginning of flowering but much lower in deflorated plants (Table 1). Higher NAR was observed in plants with darkened

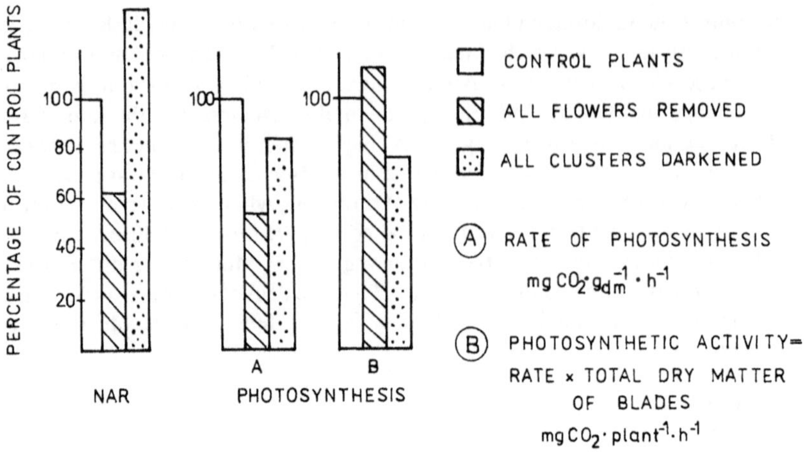

Fig. 4. Characteristic of NAR, rate and activity of photosynthesis in the last 4 weeks of vegetation, in relative values.

clusters (Fig. 2). Deflorated plants in spite of the much lower NAR value, produced dry matter in an amount similar as in control plants owing to twice higher blade dry matter. All the fruits on clusters kept in darkness accumulated dry matter in a similar proportion as that in control plants but they contained much less sugars soluble in 80 per cent of ethanol, 14 and 24 per cent of dry matter respectively). Fruits grow partly at the cost of leaf and stem growth (Figs. 1 and 2). The decrease of stem dry matter suggests remobilization of organic substances from those organs. This is in accordance with the significantly lower sugar content (Fig. 5) in vegetative organs, especially in peduncles.

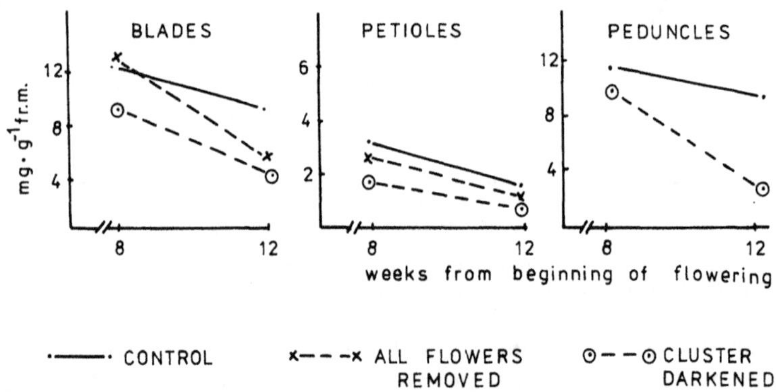

Fig. 5. Changes in the sugar content in the last 4 weeks of vegetation, in mg of fresh weight.

DISCUSSION

In the reported experiment, changes in the rate of photosynthesis in the blades were observed as late as a few weeks after preventing fruit development. The problem discussed is the fate of assimilates in nonfruiting plants. Assimilates accumulated in the leaves may affect the rate of photosynthesis in dependence, among other things, on their compartmentation. The reaction may be strong or none, if sugars or starch are deposited in the chloroplasts or in the storage pool (Moorby and Jarman, 1975) or in the free space of the leaves (Mokronosov, 1974). In tomato plants even a very high accumulation of starch in chloroplasts reduced slightly photosynthesis (Madsen, 1976). The relatively stable rate of tomato photosynthesis may be connected with continuous growth of vegetative organs, even during rapid growth of tomato fruits, as reported in the presented experiments as well as in the paper of Tanaka (1974b) who postulated that tomato leaf is itself one of the most important acceptors for its own assimilates. When plants do not form fruits, photosynthates are deposited in the leaves, causing their thickening, and also in the stem, which under ordinary conditions does not act as a so strong sink. A similar relation between growth of fruit and vegetative organs was observed in watermelon by Buttrose *et al.* (1978). Tanaka (1974a, c) concluded that in tomato plants the photosynthetic potential of the whole plant exceeds usually the current demand for assimilates for both growth of vegetative organs and development of fruits.

In the light of the above mentioned facts it seems possible that in the first few weeks after tomato defloration, the total current demand for assimilates did not change owing to their higher translocation to vegetative organs. Therefore a lack of reaction in the rate of photosynthesis did not eliminate the possibility of the mechanism of simple feed-back control of CO_2-assimilation.

The second possibility is that the total current demand for assimilates in nonfruiting plants decreased, but did not affect the rate of photosynthesis; therefore some other mechanism of regulation exists probably involving hormonal or other control.

An argument against the simple feedback control mechanism is obtained indirectly in the case of clusters kept in darkness creating probably a higher current demand for assimilates. Photosynthesis of the whole clusters in tomato plants is never high and in the reported experiments did not exceed 2–5% of that in the total plants. A similar conclusion is drawn by Tanaka (1974b). Nevertheless in plants with darkened clusters the rate of photosynthesis of the blades was never higher than in control plants but the same or even lower. The maintenance of fruit growth in that series at the same level as in intact plants, in spite of lower photosynthetic activity was obtained by retardation of growth and mobilization of reserves from the vegetative organs.

These results again suggest the hypothesis, that photosynthesis is not controlled by the reproductive sink *per se*, but may be influenced by growth substances produced by fruits or the other organs (Woodward & Rawson, 1976). A mechanism possibly involving hormonal control was suggested also by Bidwell (1974), Geiger (1976),Ješko *et al.* (1971, 1972, 1978), Lenz (1974) and others.Ješko observed that initiation of new roots in *Sorgum* and *Zea* increased the rate of APS. This experiment prompted the supposition, that information concerning the expected higher demand for assimilates is sent to the leaves; it precedes intensive growth of acceptors and affects photosynthesis.

A great number of data, indicate the effect of fruits on hormonal balance in the leaf. Luckwill (1975) reported that leaves in close proximity to developing fruits contain much less ABA and have a lower stomatal resistance than leaves distant from the fruits. Loveys & Kriedmann (1974) observed in many plants an increased level of ABA and PA as an effect of fruit removal. Hoad *et al.* (1977) reported increase of cytokinin and decrease of gibberellin content in vine leaves of plants deprived of fruits. In the present experiments, some hormonal changes in the leaves of deflorated plants are also suggested. Formation of flower buds on the petioles and rachis was observed and some morphological changes in the shape of leaves, as well as retardation of senescence of vegetative organs, as described by Wareing & Seth (1967) and Woodward & Rawson (1976) in the case of pod removal.

All the above described observations seem to suggest a causal relation governed probably by hormonal regulation, between the acceptors of assimilates (fruits) and their donors (leaves). Prevention of fruit development or even keeping them in darkness caused depression of photosynthesis probably indirectly by changes in the hormonal balance of the whole plant.

The reaction in the rate of photosynthesis, after removal of an important sink of assimilates, seems to depend on the ability of the plant to increase the mobilizing power of other organs.

ABBREVIATIONS

APS = apparent photosynthesis
R_D = dark respiration
NAR = net assimilation rate
RGR = relative growth rate
PA phaseic acid.

REFERENCES

Bidwell, R.G.S. A possible mechanism for the control of photoassimilate translocation, pp. 77–87, in: Proc. Res. Inst. Pomol. Transloc. 3rd Symp. on Accumulation and translocation, Proc. Res. Inst. of Pomol., Skierniewice (1973).

Buttrose, M.S. & M. Sedgley. Some effects of light intensity, day length and temperature on growth of fruiting and nonfruiting watermelon (*Citrullus lanatus*). *Ann. Bot.* 42: *599–608* (1978).

Flinn, A.M. Regulation of leaflet photosynthesis by developing fruit in the pea. *Physiol. Plant* 31: *275–278* (1974).

Geiger, D.R. Effects of translocation and assimilate demand on photosynthesis. *Can. J. Bot.* 54: *2337–2345* (1976).

Glasziou, K.T. & T.A. Bull. Feedback control of photosynthesis in sugarcane, pp. 82–88, in: Hatch, M.D., C.B. Osmond & R.O Slatyer (eds.), Photosynthesis and Photorespiration. Wiley Interscience, New York (1971).

Hall, A.J. Assimilate source-sink relationships in *Capsicum annum* L. I. The dynamics of growth in fruiting and deflorated plants. *Austr. J. Plant Physiol.* 4: *623–636* (1977).

Hall, A.J., C.J. Brady. II. Effects of fruiting and defloration on the photosynthetic capacity and senescence of the leaves. *Aust. J. Plant Physiol.* 4: *771–783* (1977).

Hansen, P. ^{14}C-studies on apple trees. VI. The influence of the fruit on the photosynthesis in the leaves and the relative photosynthetic yields of fruits and leaves. *Physiol. Plant.* 23: *805–810* (1970).

Hoad, G.V., B.R. Loveys & K.G.M. Skene. The effect of fruit removal on cytokinins and gibberellin-like substances in grape leaves. *Planta* (Berl.), 136: *25–30* (1977).

Ješko, T., K. Heinrichova & A. Lukacovic. Increase in photosynthetic activity during the

formation of the first node roots first tiller in *Sorgum saccharatum* (L) Moench. *Photosynthetica*. 5: *233–240* (1971).
Ješko, T. Removal of all nodal roots initiating the extention growth in *Sorgum saccharatum* (L.) Moench. I. Effect on photosynthetic rate and dark respiration. *Photosynthetica*. 6: *51–56* (1972).
Ješko, T. & A. Troughton. Root tip removal in relation to net photosynthesis and growth in *Lolium perenne* L., *Biologia* (Bratislava), 33: *65–71* (1978).
Lenz, F. Fruit effect on formation and distribution of photosynthetic assimilates. XIX-th Internat. Hort. Congress, 11–18 Sept. 1974, vol. II: *155*–166 (1974).
Loveys, B.R. & P.E. Kriedemann. Internal control of stomatal physiology and photosynthesis. I. Stomatal regulation and associated changes in endogenous levels of abscisic and phaseic acid. *Austr. J. Plant. Physiol.* 1: *407–415* (1974).
Luckwill, L.C. Long Ashton Research Station, Report 1975. Plant hormone studies, *43–46* (1976).
Madsen, E. Effect of Co_2-concentration on morphological, cytological and physiological processes in tomato plants. State Seed Testing Station, Skovbrynet 20, D.K.–2800, Lyngby, Denmark (1976).
Mokronosov, A.T. Transport of assimilates as factor in the endogenous regulation of photosynthesis in: Transport of assimilates and accumulation of organic substances. 71–84. Vladivostok, Inst. of Biol. and Pedology for East Sc. Centre Ac. of Science of USSR. (in Russian) (1973).
Moorby, J. & P.E. Jarman. The use of compartmental analysis in the study of the movement of carbon through leaves. *Planta* (Berl.), 22: *155–168* (1975).
Moorby, J., J.H. Troughton & B.G. Currie. 1974. Investigations of carbon transport in plants. *Jour. Expt. Bot.* 25: *937–944* (1974).
Moss, D.N. Photosynthesis and barrenness. *Crop Sci.* 2: *366–367* (1962).
Nagarajah, S. Effect of debudding on photosynthesis in leaves of cotton. *Physiol. Plant.* 33: *28–31* (1975).
Neales, T.F. & L.D. Incoll. The control of leaf photosynthesis rate by the level of assimilate concentration in the leaf: a review of the hypothesis. *Bot. Rev.* 34: *107–125* (1968).
Starck, Z. & M. Stradowska. Pattern of growth and ^{14}C-assimilates distribution in relation to photosynthesis in radish plants treated with growth substances. *Acta Soc. Bot. Pol.* 46: *617–627* (1977).
Starck, Z. & L. Ubysz. Source-sink relationships in radish plants. *Acta Soc. Bot. Pol.* 45: *477–493* (1976).
Tanaka, A., K. Fujita & K. Kikuchi. Nutrio-physiological studies on the tomato plant. I. Outline of growth and nutrient absorption. *Soil Sci. Plant Nutr.* 20: *57–68* (1974a).
Tanaka, A., K. Fujita & M. Shioya. II. Translocation of photosynthates. *Soil Sci. Plant Nutr.* 20: *163–171* (1974b).
Tanaka, A., K. Fujita & K. Kikuchi. III. Photosynthetic rate of individual leaves in relation to the dry matter production of plants. *Soil Sci. Plant Nutr.* 20: *173–183* (1974c).
Tanaka, A. & K. Fujita. IV. Source-sink relationship and structure of the source-sink unit. *Soil Sci. Plant Nutr.* 20: *305–315* (1974d).
Wareing, P.E. & A.K. Seth. 1967. Ageing and senescence in the whole plant in: Aspects of the biology of ageing: 543–599. Symp. of Soc. Experimental Biol. No. 21, Cambridge Univ. Press (1967).
Woodward, R.G. & H.M. Rawson. Photosynthesis and transpiration in dicotyledonous plants. II. Expanding and senescing leaves of soybean. *Aust. J. Plant Physiol.* 3: *257–267* (1967).

Authors' address:
Warsaw Agricultural University
Institute of Plant Biology
02–526 Warsaw
Rakowiecka 26/30
Poland.

Welsh Plant Breeding Station
[illegible]
[illegible]
[illegible]
[illegible]

PARTITIONING OF ^{14}C-ASSIMILATE WITHIN INDIVIDUAL TOMATO LEAVES IN RELATION TO THE RATE OF EXPORT

L.C. HO
(Littlehampton, U.K.)

ABSTRACT

Effects of leaf development and sink demands on the partitioning of ^{14}C-assimilates within the leaf, in relation to export, have been examined in the tomato.

A tomato leaf starts to export ^{14}C-leaf assimilates when it is 10% expanded. During the first 20% of leaf area expansion, the capacity of the leaf to export increases rapidly. Meanwhile, an increasing amount of primary ^{14}C-assimilates is used for sucrose synthesis at the expense of the synthesis of protein and structural carbohydrate.

In a fully expanded leaf, the translocate in the light period is mainly derived from the ^{14}C-assimilates initially fixed as sugars and other water- and ethanol-soluble assimilates. When the concurrent photosynthetic rate is lower than the sink demand, the supply of translocates is principally maintained by the early breakdown of ^{14}C-starch. On the other hand, when the export rate is low, due to low sink demand, the flow of ^{14}C for sucrose synthesis is reduced during the initial partitioning of assimilates. During the subsequent partitioning of assimilates, more ^{14}C-assimilates are stored as sugars, starch and structural carbohydrates.

The metabolism of sucrose and starch is a vital intermediary process linking photosynthesis and the export of assimilate in the leaf.

INTRODUCTION

The productivity of a fruiting crop can be improved by:

1. increasing the rate of net photosynthesis;
2. increasing the proportion of assimilates available for export;
3. increasing the proportion of translocates being imported by fruits.

To manipulate both photosynthesis and translocation a greater understanding is needed of interactions between (a) the metabolism of the assimilates and the loading of labile assimilates in leaves and (b) the unloading of translocates and the metabolism of imported assimilates in fruits. This paper analyses the relationship between leaf assimilation and the export of assimilates from the leaf.

In our laboratory, the rate of assimilate translocation from a single leaf is measured as the difference between the net amount of carbon being fixed and that being accumulated over the same period. The relationship between carbon metabolism and export in a leaf is established by monitoring the changes in its principal carbohydrates during translocation, while the involvement of the newly fixed carbon in the overall carbon metabolism is determined by observing the changes in the percentage distribution pattern of ^{14}C-assimilates after a pulse application of $^{14}CO_2$.

The partitioning of ^{14}C-assimilates was measured at different time intervals. The initial partitioning of ^{14}C-assimilates, 5 min. after $^{14}CO_2$ application is used as a measure of the initial assimilation pattern and as an index of the overall net carbon flow from $^{14}CO_2$ to the primary products of photosynthesis. The subsequent partitioning of ^{14}C-assimilates, at the end of the light period or dark period, is used as a measure of the effect of export and leaf growth on the metabolism of ^{14}C-assimilates . The partitioning of ^{14}C-assimilates has been investigated in both growing and mature leaves when the rate of concurrent photosynthesis was either higher or lower than the rate of export. It is in this context that the effects of leaf development and sink demand on leaf assimilation are demonstrated by the partitioning of ^{14}C-assimilates.

PARTITIONING OF ^{14}C-ASSIMILATES DURING THE TRANSITION OF A TOMATO LEAF FROM A NET CARBON IMPORTER TO NET CARBON EXPORTER

The carbon economy of a tomato leaf, *i.e.*, carbon fixation, accumulation, respiration and transport, was measured systematically on the seventh basal leaf (Ho & Shaw, 1979). The leaf was grown under a light flux density of 70 Wm^{-2} (PAR) for 7 hr daily and the air temperature (20° C) and CO_2 concentration (350 vpm) were maintained constant. The net photosynthetic rate of that leaf, 10 days after reaching 1 cm in length, was about 0.1 mg C mg $DW^{-1}d^{-1}$ or about 80% of the maximum rate. Since the amount of carbon being self-fixed was relatively high, the net amount of carbon being imported (which was less than 1 mg C $leaf^{-1}$ d^{-1}) accounted for less than 20% of that required for growth, *i.e.*, net accumulation plus night respiration. Three days later, the leaf fixed more carbon than was required for growth and so became a net carbon exporter.

In fact, import and export of concurrently formed leaf assimilates occurred simultaneously in the same leaf over a much longer time (Ho & Shaw, 1977). When the compound leaf was a net carbon importer, some leaflets, particularly the distal ones, simultaneously exported ^{14}C-assimilate to the rest of the compound leaf and to the plant as a whole. Conversely, long after becoming a net carbon exporter, the leaf continued to import ^{14}C leaf assimilates from older leaves, until it reached full expansion. Apparently, the capacity of a tomato leaf to export increases rapidly while its capacity to import starts to decline during the first 20% of leaf expansion. Similar time courses of translocation transitions in leaves have been reported for sugar beet (Fellows & Geiger, 1974; Giaquinta, 1978), bean (Swanson, Hoddinott & Su, 1976) and squash (Turgeon & Webb, 1975).

The pattern of initial partitioning of ^{14}C-assimilates in the tomato leaf changed most rapidly during the first 20% of leaf expansion (Ho & Shaw, 1977). Over this period, the ^{14}C initially incorporated into ethanol- and water-soluble compounds 5 min. after $^{14}CO_2$ application increased from 63% to 82% of the total $^{14}CO_2$ fixed, while that incorporated into residue, *i.e.*, protein and structural carbohydrates, decreased from 27% to 4%. Over the same period, the proportion of ^{14}C incorporated into starch increased from 10% to 14%.

The rapid increase in the synthesis of soluble ^{14}C-assimilates was mainly due to the increasing capacity of the leaf to synthesize ^{14}C-sucrose. The proportion of soluble ^{14}C as sugars (predominantly sucrose) increased from 10 to 40%. The initially high proportion of ^{14}C as residue in a very young leaf was due to both a high initial conversion of primary ^{14}C-assimilates to residues and also to the high subsequent conversion of other stored ^{14}C-assimilates to residues over 24 hr. The capacity of both conversions to residue declined rapidly during the first 20% of leaf expansion.

Towards the end of leaf expansion, the proportion of ^{14}C fixed initially as starch and residue was maintained at about 14 and 4%, respectively, meanwhile, the proportion of soluble ^{14}C as sugars increased from 40 to 65%.

The sharp increase in sucrose synthesis has been regarded as a preparatory step for sugar export (Fellows & Geiger, 1974). Recently, Giaquinta (1978) reported that during leaf expansion, at the time of the transition in the direction of translocation in sugar beet leaves, the activity of sucrose phosphate synthetase (for sucrose synthesis) increased while that of acid invertase and sucrose synthetase (for sucrose hydrolysis) were relatively unchanged. Therefore, the increasing proportion of $^{14}CO_2$ initially incorporated into sucrose during the development of export ability in tomato leaves may be due mainly to an increasing capacity for sucrose synthesis rather than a decreasing capacity for sucrose hydrolysis. Probably, therefore, the capacity for export in an expanding leaf is enhanced by increasing both the total leaf sucrose concentration and the labile sucrose at the site of phloem loading (Outlaw & Fisher, 1975; Outlaw, Fisher & Christy 1975).

PARTITIONING OF ^{14}C-ASSIMILATES IN MATURE TOMATO LEAVES WITH PHOTOSYNTHETIC RATES LOWER THAN THE RATE OF EXPORT

It has long been recognized that the sources of translocate are the newly fixed and the stored assimilates. Therefore, the rate of export during the light period is a function of the rate of concurrent net photosynthesis, modified by the supply of mobilized reserve in the leaf (Ho, 1976a).

The role of reserves, particularly starch, in maintaining the rate of export has been examined by monitoring the subsequent partitioning of ^{14}C-assimilates in a mature leaf under low light (Ho, 1976b). When such a leaf was fed with $^{14}CO_2$ for 1 min, at 90 Wm^{-2} and maintained at the same light flux density for 6 hr, about half of the ^{14}C assimilated was subsequently lost from the leaf through respiration and export. The ^{14}C lost was mainly derived from that initially fixed as sugars and other soluble assimilates, *i.e.*, sugar phosphates, amino and organic acids. On the other hand, the amount of ^{14}C incorporated into starch and residue did not fall throughout this period.

When the ^{14}C fed leaf was transferred to 4 Wm^{-2} for 6 hr, the rate of carbon export was reduced to below 1 mg C $dm^{-2}h^{-1}$ (39% less than that under 90 Wm^{-2}) while the ^{14}C loss increased to 62% of the initial uptake (32% higher than that under 90 Wm^{-2}). The increased loss of stored ^{14}C-assimilates in leaves, with a net photosynthetic rate less than the rate of carbon export, was achieved by an early breakdown of ^{14}C-starch and residue rather than by further loss of ^{14}C from soluble assimilates. The effect of the extra demand on the stored assimilates to sustain the export has also been examined in the single remaining leaf of a two-truss tomato plant 15 days after removing the other leaves (Ho, unpublished). In this situation, the leaf became the principal source of carbon for the further growth of the trusses and the sink demand was much higher than the amount of labile assimilates which could be supplied by concurrent photosynthesis. As expected, the loss of ^{14}C-assimilates from this leaf was about 10% more than the leaf in control plants. The higher loss of ^{14}C-assimilates , possibly due to higher export, was mainly due to a higher depletion of ^{14}C-starch and, to a lesser degree, of ^{14}C-sugars.

Therefore, when concurrent photosynthesis is lowered, or the sink demand on a leaf is enhanced, the subsequent partitioning of assimilates is altered by increased mobilization of starch.

PARTITIONING OF ^{14}C-ASSIMILATES IN MATURE LEAVES WITH PHOTOSYNTHETIC RATES HIGHER THAN THE RATE OF EXPORT

A tomato plant consists of several sources and sinks, so that regulation of the movement of assimilates between sources and sinks must be dynamic and well adjusted. When a single leaf was exposed temporarily to a higher light for 6 hr, more newly fixed assimilate was exported. Possibly, the sink organs as a whole could accommodate the extra translocates; the assimilates accumulated in the 6 hr light period represented only about 30% of the net carbon fixed (Ho, 1976a). However, accumulation of assimilate could be further enhanced when the rate of export was limited by sink demand despite the higher photosynthetic rate induced in the leaf (Ho, in press). In a one-leaf/one-fruit translocation system, increasing the net photosynthetic rate from 178 to 294 mg C d^{-1} did not increase the rate of export above 120 mg C d^{-1}, a rate similar to the maximum import rate of the fruit. When the net photosynthetic rate was well above the rate of export, there was an exceptionally high accumulation of assimilates (58% of the carbon net fixed) and the rate of net photosynthesis declined towards the end of the 24 hr light period. Accumulation of starch in chloroplasts may have reduced net photosynthesis in this situation, as suggested by Chatterton, Carlson, Hungerford & Lee (1972).

The effect of low sink demand on the partitioning of ^{14}C-assimilates in the source leaf was further examined over 15 days after trusses had been removed from a two-truss tomato plant (Ho, unpublished). Table 1 summarizes the carbon balance of the source leaf, which was the middle leaf between the two trusses, at 0, 7 and 15 days after truss removal and the net subsequent change over 15 days.

Over the first week after truss removal, the amount of carbon exported from the source leaf was only about 43% of that of the control, while both the carbon accumulation (mainly starch) and night respiration increased substantially. However, the net photosynthetic rate of the source leaf was similar to the control. During the second week, both the decrease in export and the increase in night respiration and carbon accumulation were lower than those of the first week, but the net photosynthetic rate decreased substantially. A much faster response of the leaf to truss removal by the reduction of net photosynthesis has been reported in a determinate variety (Moss, 1962). Clearly, the decrease in export and the increase of night respiration and carbon accumulation preceded the reduction in net photosynthesis. Possibly, the net photosynthesis was inhibited by the accumulation of assimilates.

Table 1. Effects of truss removal on net photosynthesis (Pn), night respiration (Rn), carbon content (C) and export of total or ^{14}C-leaf assimilate (T) in tomato leaves.

(A) At day of measurement:

Days after treatment	Pn ($mg\ C\ leaf^{-1}\ day^{-1}$)		Rn ($mg\ C\ leaf^{-1}\ day^{-1}$)		C ($mg\ C\ leaf^{-1}$)		T (% initial ^{14}C)	
	Control	Treated	Control	Treated	Control	Treated	Control	Treated
0	98	98	7	7	405	405	57	57
7	111	105	9	13	405	755	56	45
15	103	67	11	10	540	760	56	30

(B) Net changes:

Days after treatment	Pn (mg C)		Rn (mg C)		C (mg C)		T (mg C)	
	Control	Treated	Control	Treated	Control	Treated	Control	Treated
0–7	732	711	55	70	0	350	677	291
7–15	856	688	83	94	135	5	638	589
0–15	1,588	1,399	138	164	135	355	1,315	880

The patterns of ^{14}C-assimilate partitioning in the source leaves 7 and 15 days after truss removal further demonstrate the effect of assimilate accumulation on leaf photosynthesis (Table 2). At day 7, the source leaf without trusses retained 25% more ^{14}C-assimilate than the leaf with trusses in 24 hr. The patterns of initial partitioning in the source leaves were similar except that leaves without trusses had slightly more ^{14}C in phosphate esters (*i.e.*, 3-phosphoglyceric acid, phosphoenolpyruvic acid and hexose phosphates) and starch, thus less in sugars (*i.e.*, sucrose, glucose and fructose). Since intensified starch synthesis would require abundant ADP-glucose and 3-PGA (Preiss, Ghosh & Wittkop, 1972), the high retention of ^{14}C in phosphate esters is not unexpected. In the subsequent partitioning, more ^{14}C was retained as starch, sugars and structural carbohydrates (*e.g.*, cellulose, etc.) in leaves without trusses. At day 15, 61% more ^{14}C-assimilates was retained in such leaves. The flow of ^{14}C into sugars in the initial partitioning was further reduced, so that more ^{14}C was found in amino acids (*i.e.*, serine, glycine, aspartic acid, glutamic acid, glutamine and alanine) and malic acid as well as in the phosphate esters and starch. However, over 24 hr the increased retention of ^{14}C was still due to the accumulation of sugars, starch and structural carbohydrates rather than other assimilates.

It is clear that the effect of limited sink demand for assimilates on the accumulation of assimilates can effect not just the subsequent partitioning of the stored assimilates but also the initial partitioning of the newly formed assimilates, and furthermore, the effect on the initial and the subsequent partitioning of assimilates is different.

CONCLUSIONS

Leaf assimilation, in terms of the ^{14}C-assimilate partitioning, is controlled by endogenous factors (*i.e.*, developmental stage of the leaf and the export of assimilates) as well as by environmental factors (*e.g.*, light, CO_2 and temperature).

The change of ^{14}C-assimilation is markedly affected by the export of assimilates. By increasing the activity of sucrose phosphate synthetase, an expanding leaf increases the ^{14}C flow into sucrose and thus the capacity for export. In a mature leaf, the synthesis of labile sucrose can be reduced in response to low sink demand. As a result, more ^{14}C is initially retained as phosphate esters and more ^{14}C is incorporated into amino acids and starch. Subsequently, assimilates are mainly in the forms of sugars and starch. In contrast, when the sink demand is higher than photosynthesis more ^{14}C-starch is mobilized to sustain or increase the rate of export.

Therefore, through the regulation of the metabolism of sucrose and starch, leaf assimilation can be affected by the export of assimilate or the sink demand. This feedback between photosynthesis and assimilate utilization may be mediated by a change of enzymic activity (*cf.*, Wareing, Khalifa & Treharne, 1968).

Information on the subsequent intermediary ^{14}C metabolism is essential to understand the relation between photosynthesis and the other physiological processes of growth, respiration and transport with the ultimate aim of improving crop productivity.

Table 2. Effects of truss removal on percentage distribution of ^{14}C-assimilates in leaf 11 of two-truss tomato plants.

Days after truss removal	^{14}C-assimilates	Percentage distribution of ^{14}C-assimilates to initial uptake, minutes after $^{14}CO_2$ application					
		5		480		1440	
		Control	Treated	Control	Treated	Control	Treated
0	phosphate esters	10.7		0.5		0.6	
	sugars	54.7		37.3		19.3	
	starch	12.7		22.4		17.0	
	structural CHO	0.8		1.7		2.2	
	amino acids	18.1		0.7		0.8	
	organic acids	1.4		4.4		0.1	
	protein	1.6		3.4		3.2	
	Total	100.0		70.4		43.2	
7	phosphate esters	12.1	16.8	0.5	0.6	0.8	0.7
	sugars	45.2	35.2	40.7	30.2	14.2	17.5
	starch	14.8	18.6	29.4	40.9	17.0	24.8
	structural CHO	0.6	0.9	2.5	3.0	2.7	3.5
	amino acids	21.8	21.9	1.5	1.3	3.0	1.4
	organic acids	3.7	4.6	3.6	3.0	1.6	2.0
	protein	1.8	2.0	4.9	5.0	4.7	4.9
	Total	100.0	100.0	83.1	84.0	44.0	54.8
15	phosphate esters	11.9	17.9	0.4	0.5	0.5	0.5
	sugars	43.9	18.8	33.0	32.1	11.4	28.1
	starch	16.2	17.4	24.8	33.5	17.7	28.8
	structural CHO	0.8	0.6	2.2	2.4	2.3	3.0
	amino acids	21.7	37.1	1.0	0.9	3.6	2.4
	organic acids	3.4	5.7	5.2	3.7	2.5	2.7
	protein	2.1	2.5	5.7	5.4	5.7	5.0
	Total	100.0	100.0	72.3	78.5	43.7	70.5

REFERENCES

Chatterton, N.J., G.E. Carlson, W.E. Hungerford & D.R. Lee. Effect of tillering and cool nights on photosynthesis and chloroplast starch in Pangola. *Crop Sci.* 12: *206–208* (1972).

Fellows, R.J. & D.R. Geiger. Structural and physiological changes in sugar beet leaves during sink to source conversion. *Plant Physiol.* 54: *877–885* (1974).

Giaquinta, R. Source and sink leaf metabolism in relation to phloem translocation. Carbon partitioning and enzymology. *Plant Physiol.* 61: *380–385* (1978).

Ho, L.C. The relationship between the rates of carbon transport and of photosynthesis in tomato leaves. *J. exp. Bot.* 27: *87–97* (1976a).

Ho, L.C. The effect of current photosynthesis on the origin of translocates in tomato leaves. *Ann. Bot.* 40: *1153–1162* (1976b).

Ho, L.C. Regulation of assimilate translocation between leaves and fruits in the tomato. *Ann. Bot.* (in press).

Ho, L.C. & A.F. Shaw. Carbon economy and translocation of ^{14}C in leaflets of the seventh leaf of tomato during leaf expansion. *Ann. Bot.* 41: *833–848* (1977).

Ho, L.C. & A.F. Shaw. Net accumulation of minerals and water and the carbon budget in an expanding leaf of tomato. *Ann. Bot.* 43: *45–54* (1979).

Moss, D.N. Photosynthesis and barrenness. *Crop sci.* 2: *366–367* (1962).

Outlaw, W.H. & D.B. Fisher. Compartmentation in *Vicia faba* leaves. I. Kinetics of ^{14}C in the tissue following pulse labeling. *Plant Physiol.* 55: *699–703* (1975).

Outlaw, W.H., D.B. Fisher & A.L. Christy. Compartmentation in *Vicia faba* leaves. II. Kinetics of ^{14}C-sucrose redistribution among individual tissues following pulse labeling. *Plant Physiol.* 55: *704–711* (1975).

Preiss, J., H.H. Ghosh & J. Wittkop. Regulation of the biosynthesis of starch in spinach leaf chloroplasts. pp. 131–152 in: Goodwin, T.W. (ed.), Biochemistry of Chloroplasts, vol. II. Academic Press, London & New York (1967).

Swanson, C.A., J. Hoddinott & J.W. Su. The effect of selected sink parameters on translocation rates. pp. 347–356 in: Wardlaw, I.F. & J.B. Passioura (eds.), Transport and transfer processes in plants. Academic Press, New York (1976).

Turgeon R. & J.A. Webb. Leaf development and phloem transport in *Cucurbita pepo*: carbon economy. *Planta* (Berl.), 123: *55–62* (1975).

Wareing, P.F., M.M. Khalifa & K. Treharne. Rate-limiting processes in photosynthesis at saturating light intensities. *Nature* 220: *453-457* (1968).

Author's address:
Glasshouse Crops Research Institute
Littlehampton
Sussex
United Kingdom.

PHOTOSYNTHESIS AND TRANSLOCATION IN RELATION TO GROWTH IN LAMINARIALES*

RICHARD G. BUGGELN
(St. John's Nfld., Canada)

ABSTRACT

Earliest photosynthesis studies involving Laminariales (1930s) were mostly comparative, ecological investigations aimed at clarifying the role of light intensity and complimentary chromatic adaption in marine algae. Phloem-like cells had been described in Laminariales much earlier (1880s), accompanied by speculation on their role in translocation. However, it is from the classic study by Sargent & Lantrip in 1952 on photosynthesis and growth in the giant kelp, *Macrocystis*, that major tenets of the relationship between photosynthesis, translocation and development in all kelp can be traced: photosynthetic capacity of meristems is low compared with mature portions of kelp thalli and growth utilizes carbon translocated from mature regions to meristems. Subsequent physiological and biochemical studies have characterized gradients in photosynthetic capacity along kelp blades, determined products of both light and dark CO_2 fixation, identified solutes in the translocation stream and demonstrated the integration of source-sink relationships within intact algae. A speculative mechanism of translocation involving symplastic flow via vacuoles and vesicles is proposed.

INTRODUCTION

The main objective of the paper is to describe the roles of photosynthesis and translocation in relation to growth in economically important brown algae, the Laminariales, more commonly known as kelp. However, photosynthesis (PS) in kelp will be broadly reviewed before discussing the process in context of growth of the macroscopic kelp sporophyte. Though translocation of organic matter has been given considerable attention in Laminariales, its mechanism is still obscure. A more critical analysis of this phenomena is therefore included and, given author's license, a final speculation on the nature of the mechanism is presented. Aspects of translocation of inorganic solutes have been reviewed recently (Penot *et al*., 1976).

PHOTOSYNTHESIS

Earliest data on photosynthesis in Laminariales were obtained by Kniep (1914) with *Laminaria saccharina* (L.) Lamour. Many subsequent investigators

* MSRL Contribution Number 320; manuscript preparation supported by grant from National Research Council of Canada A-6866.

addressed the question of the PS performance of the alga in its niche in the sea, attempting to obtain evidence in support of either of the two propositions explaining verticle distribution of marine algae, *i.e.*, complimentary chromatic adaption *vs*. light intensity (*cf.* Seybold, 1934; Levring, 1947). With knowledge of light penetration and changing spectral composition of visible radiation with water depth, PS of a number of kelp species was studied *in situ* at different depths (Gail, 1922; Montfort, 1933, 1934, & 1936), PS *vs*. irradiance curves were obtained in field (Moore, *et al.*, 1923) and laboratory experiments (Montfort, 1935; Stocker & Holdheide, 1937), compensation depths were determined (Prinz, 1939), and PS in different regions of the visible spectrum was compared (Montfort, 1934).

Montfort (1936) showed that *L. hyperborea* (Gunn.) Fosl. growing in deep water suffered an irreversible inhibition of PS when exposed to surface irradiances. He also recorded a mid-day depression in diurnal PS which was coincident with peak solar radiation. More recently, Drew (1974) compared light shock susceptibility in *L. hyperborea* and *L. digitata* (Huds.) Lamour. The former (even from shallow water) showed greater sensitivity than the latter with PS reduction of 50% after 5 hr in full sunlight; PS in deep water specimens was completely inhibited after 1.5 hr. Similar stress on PS was also reported for *Macrocystis pyrifera* (L.) C. Ag. blades from deep water (Clendenning, 1971).

Using combinations of light filters, Montfort (1938) derived assimilation curves for *L. digitata* for comparison with light absorbance of chlorophyll and carotenoid extracts; a light absorption curve for *L. saccharina* was obtained by Seybold & Weissweiler (1943). However, most precise PS action spectra for algae, including a *Laminaria* species, are found in the work of Haxo and Blinks (1950).

Interaction of temperature and light on PS was first studied in a *Laminaria* species by Erman (1928). Lampe (1935) compared PS of surface and deep water *L. hyperborea* at both summer and winter water temperatures and at several light intensities. Net assimilation in summer exceeded winter rates at all irradiances with considerable overlap in performance of shallow water and deep water specimens. Lüning's (1971) more recent seasonal study with *L. hyperborea* showed a shift in PS *vs*. irradiance with higher net PS at saturating irradiances (ca. 400 μg-cal/cm^2/s) during May and August than during December and March (see also *L. longicruris* (Hatcher *et al.* 1977) and *L. dititata* (King & Schramm 1976b)), yet a maximum winter rate could be obtained by reducing irradiance to 100 μg-cal/cm^2/s. PS capacity in *M. pyrifera* was higher in March than in June (Clendenning, 1971).

Bidwell (1958) was the first to study photosynthesis in kelp (3 species) using radioactive tracer ($^{14}CO_2$) (though kelp were later shown to efficiently assimilate carbon as bicarbonate (Joliffe & Tregunna (1970)). After 4–6 hr incubations, 60–90% of incorporated radioactivity was in the soluble fraction, principally mannitol (85–93%), with significant radioactivity in alanine followed by aspartic and glutamic acids. However with shorter incubations (10s), radioactivity in *L. saccharina* photosynthate appeared first in sugar mono- and diphosphates as well as phosphoglyceric acid (Kremer & Willenbrink, 1972). Activity in these compounds decreased rapidly with label subsequently appearing in mannitol and amino acids. Available evidence suggests that the pentose phosphate pathway is probably the major carbon fixation route in the PS cycle in Laminariales (*cf.* Kremer & Küppers, 1977). Although some uncertainty is apparent in the chromatographic identity of ^{14}C organic acids, Willenbrink *et al.*, (1975) later reported activity in both citrate and malate. Following a ^{14}C pulse specific activity of mannitol in *Eisenia bicyclus* (Kjellm.) Setchell decreased during the dark period (chase) suggesting conversion of mannitol to other products (Yamaguchi *et al.*, 1966). In *Alaria esculenta* (L.) Grev. seasonal differences were apparent in the ratio of ^{14}C incorporated in mannitol:amino acids; summer ratio gave mannitol > amino acids with a reversed situation in winter. Inhibition of PS in *Alaria* by growth inhibiting concentrations of auxins and their chemically related, non-auxin analogs caused a change in ^{14}C incorporation with a higher % of label in amino acids in both summer and winter (Buggeln & Bal, 1977).

Plastids are present throughout the thallus of Laminariales not only in the highly pigmented surface layers of cells; those in the cortex and medulla of *L. digitata* have PS capacity (Hellebust & Haug, 1972). Although PS and chlorophyll content were highest in the epidermal layers of *L. digitata* stipes (= stems), exercised cortex and medullary sections fixed carbon with greatest tracer accumulation in alginate in contrast with the epidermal layer where activity was highest in soluble fraction. The authors concluded that in the intact alga, overlapping plastids in the epidermal layers shade interior tissues and probably prevent PS in cortex and medulla. In pulse-chase experiments with whole stipe sections, alcohol- and acid-soluble fractions decreased with time while ^{14}C in alginates increased suggesting a precursor pool through which carbon entered en route to alginate synthesis.

Akazawa *et al.* (1972a, b) followed incorporation of ^{14}C into soluble assimilate of *E. bicyclus* maintained in darkness. 72–90% of total incorporated ^{14}C was in amino acids: aspartic acid > glutamic acid >> alanine, after 3 hr with about 15% in organic acids and only very low levels in the neutral fraction. Citrate was most strongly labelled after 0.5 hr but its activity decreased with time. A dark pulse of ^{14}C given to *L. hyperborea* followed by a chase in light showed initially high levels of tracer in glutamate which then decreased over time with concomitant increase in activity in mannitol and serine/glycine (Kremer & Willenbrink, 1972). The enzyme responsible for dark fixation is PEP-carboxykinase (PEP-CK) (*cf.* Akazawa *et al.*, 1972c; Weidner & Küppers, 1973; Kremer & Küppers, 1977). Calculations of dark fixation rates based solely on exogenous uptake of ^{14}C bicarbonate may underestimate true dark fixation capacity of these algae. Plastids deep in the cortex and medulla which have the potential for PS (Hellebust & Haug, 1972) probably refix respiratory CO_2 in darkness. In short term experiments diffusion limits the availability of carbon in sea water to cells other than those in the epidermal layers (Buggeln, *et al.*, 1978). Though little studied, photorespiration may not be significant in kelp (Hatcher *et al.*, 1977).

GROWTH AND PHYSIOLOGICAL ORGANIZATION WITHIN THE BLADE

Two principles of blade growth in Laminariales can be credited to the original work of A.L. Fallis: (A) blade growth (elongation and expansion) occurs at the blade base (Fallis, 1916; Sundene, 1964; Nicholson, 1970; Parker, 1971; Buggeln, 1974): (B) blade growth rate increases with increasing blade length (Fallis, 1916; Nicholson, 1970; Lüning *et al.*, 1973; Buggeln, 1976b). In kelp with more complex thallus structure, *e.g.*, *Macrocystis*, individual blades are sequentially produced by division of an expanded, scimitar-shaped blade at the apex of the frond. While meristematic activity is located at the base of each blade, frond elongation is internodal, occurring between blades. The following data relate to the division of kelp blades into meristematic, mature (non-growing) and senescing portions as one progresses distally from the blade-stipe junction. Chlorophyll content is generally higher in the mature portions than in meristems (Weidner & Küppers, 1973; Johnston *et al.*, 1977; Buggeln, *et al.*, 1978) and PS likewise, is highest in the former (Hellebust & Haug, 1972; Schmitz *et al.*, 1972; Towle & Pearce, 1973; Weidner & Küppers, 1975; Willenbrink *et al.*, 1975; Johnston *et al.*, 1976; King & Schramm, 1976a; Buggeln *et al.*, 1978). An analogous situation occurs in *Macrocystis* where PS capacity increases with increasing blade age (Sargent & Lantrip, 1952; Clendenning, 1964, 1971) however PS capacity of older blades (probably senescing), at the base of the frond in deep water and dim light, were *ca.* 50% lower than young blades. When PS of a *Macrocystis* blade is calculated on a fresh weight basis, the rate increases with distance from the blade base; on an area basis, PS is higher near the blade base decreasing slightly toward the blade end (Clendenning, 1971). In *L. saccharina* and *L. hyperborea* PS is saturated at higher irradiances in older than younger portions of the blades (Lüning, 1971).

Comparison of relative distribution of ^{14}C assimilated by old and young portions of *L. digitata* blades showed the former with greatest activity in alcohol-soluble and acid (laminarin) fractions while the latter had more activity in alginate and protein (Hellebust & Haug, 1972). Enzyme activity of both RuDPase and PEP-CK (on a chlorophyll basis) was higher in the meristem than in mature parts of *L. hyperborea* blades (Weidner & Küppers, 1975) which contrasts with PS capacity of this and other species. In a comparison of absolute carboxylation rates of these enzymes in 3 species of *Laminaria*, 2 showed RuDPase > PEP-CK; in *L. hyperborea* PEP-CK > RuDPase. Dark fixation of ^{14}C bicarbonate was usually lower in the mature portion than in the young part of kelp blades (Towle & Pearse, 1973; Willenbrink *et al.*, 1975; Johnson *et al.*, 1977; Buggeln *et al.*, 1978), ranging between 3–10% of PS with occasional higher values *e.g.*, 40% (Johnson, *et al.*, 1977).

INTEGRATION OF PHYSIOLOGICAL PROCESSES

Movement of matter from senescing *A. esculenta* blades into the stipe was proposed by Wille (1897) and he is credited with the first suggestion of translocation in Laminariales. However, it was not until Sargent & Lantrip's study with *M. pyrifera* in 1952 that the roles of photosynthesis and translocation in relation to growth were studied experimentally, although evidence for translocation was still indirect. Nevertheless, their analysis of probable source-sink relationships within *Macrocystis*, and consideration of alternative fates of the large carbon surplus not required for growth, provided generalizations which have been largely substantiated in subsequent studies using morphologically simple Laminariales species and employing radioactive tracers. Sargent & Lantrip found that young, growing blades in the apical region of the frond did not photosynthesize sufficient carbon for their own growth and respiration. This deficit was assumed to be offset by importing carbon from mature, non-growing blades, further down the frond, whose photosynthetic rates exceeded their endogenous requirement for carbon. Such blades had a calculated export rate of *ca.* 200 mg C/blade/day. Concluding that only about 1/6 of frond assimilate was needed for growth Sargent & Lantrip proposed that excess carbon was probably stored in the holdfast (= root-like anchor) and/or excreted from the alga. Thus, mature blades of *Macrocystis* were 'sources' of carbon while younger blades and potential storage organs, *i.e.*, haptera (root-like projections of holdfast), stipe, and growing sporophylls (specialized reproductive blades), were 'sinks'. In simple Laminariales where there is only a single blade terminating a stipe, a 'source-to-sink' relationship exists within each alga. That is, blade, stipe and haptera meristems are principal sinks (Titlyanov, 1973; Schmitz & Lobban, 1977; Lobban, 1978) while the mature, non-growing distal portion of blades is the principal source.

Important contributions to the study of translocation in Laminariales have been made by Lüning and co-workers in investigations with *L. hyperborea* and *L. saccharina*. From pulse-chasing experiments (96 hr duration) they found *ca.* 50% of ^{14}C assimilate was exported from distal blade in *L. hyperborea*; 4–7% in *L. saccharina*. In similar experiments with other species Hellebust & Haug (1972) found 1.2% of source photosynthate was exported (after 5 hr) from the old blade of *L. digitata*; 4.8% (after 4 hr) in *A. maringata* (Schmitz & Srivastava, 1975); and 75% (8 days) in *Nereocystis leutkeana* (Mertens) Postels & Ruprecht (Nicholson & Briggs, 1972). Translocation exhibited definite polarity with less than 1% of radioactivity moving in direction opposite to the blade meristem. ^{14}C translocation rates supported growth data for *L. saccharina* which indicated that 70% of carbon for meristem growth was imported from the distal blade. While similar calculations were also made for *L. hyperborea*, Lüning *et al.* (1973) proposed that stored carbon in the distal blade was remobilized, *i.e.*, during winter when radiation for PS was below compensation point, and exported to the blade meristem. Previous growth experiments in darkness (Lüning, 1970) showed that either blade or stipe could provide organic matter for growth. It is interesting to note however that ^{14}C assimilated by the stipe of *L. hyperborea* could not be detected in the main blade.

A hierarchy of 3 potential sources was identified in *A. esculenta*: 1°, distal main blade; 2°, sporophylls; 3°, stipe. Translocation to the blade meristem from the 2° source did not occur until the blade beyond the meristematic portion was removed; likewise the 3° source only exported when the 1° & 2° sources were excised (Buggeln, 1977). As the morphology of kelp becomes more complex, source-sink interactions also increase in complexity. Youngest *Macrocystis* blades import ^{14}C assimilate into their meristematic lower half; during blade maturation there is a period of transition where they both import and export; finally the non-growing blade exclusively exports. The direction of export within the frond *i.e.*, apically or basally, depends on position of the source blade on the frond. Those toward the frond apex export apically; next is a section of the frond where translocation is bi-directional; below this zone, transport is exclusively basipetal toward several sinks: new fronds arising from the base of the alga, and sporophylls (Lobban, 1978). Although haptera did not accumulate ^{14}C, they do act as sinks in other genera (Nicholson & Briggs, 1972; Schmitz *et al.*, 1972; Schmitz & Srivastava, 1975; Schmitz & Lobban, 1977).

A more detailed investigation of the general relationship of growth *vs.* blade length showed that in *L. saccharina* and *A. esculenta*, enhancement of growth plateaus with blades 40–50 cm long (Lüning *et al.*, 1973; Buggeln, 1977). This contrasts with translocation of ^{14}C assimilate into the blade meristem of *Alaria* from the distal portion up to 200 cm away (Buggeln, unpublished). Although a slow rate of translocation might account for apparent non-utilization of organic matter from distal blade beyond 50 cm, a more likely explanation is the limited growth capacity of the blade meristem itself. As proposed by Sargent & Lantrip (1952) a substantial amount of assimilated carbon must either be stored or excreted. Recent carbon budgets for *L. saccharina* (Johnson *et al.*, 1977) and *L. longicrurus* (Hatcher *et al.*, 1977) respectively allocated 13–30% and 35% of assimilated carbon for excretion.

In growth experiments to determine the rate of transport of carbon in the *Alaria* blade, Buggeln (1976b) tacitly assumed that an isolated blade meristem would perform the same as a meristem in an intact blade except that the major source of carbon for growth would have been removed. However, by 'amplifying' meristem elongation so that daily increments could be detected, it was shown that growth after excision of meristems was slow for several days, but then increased to a constant rate (Buggeln, 1976a). In addition, *ca.* 50% increase in PS capacity occurred within 24 hr after meristem excision and the elevated rate was maintained for a similar number of days (Buggeln & Bal, 1977). Thus in the case of *Alaria*, the isolated meristem appears to be physiologically different from the meristem in the intact blade.

TRANSLOCATION — PATHWAY AND MECHANISM

'There is possibly no other aspect of plant physiology which compares with translocation in the way in which the problem is tied to structure as well as organization of the plant as a whole' (Zimmermann, 1969).

The controversy regarding the ontogeny and terminology of Laminariales conducting cells, discovered in 1876 by Reinke, has been reviewed (see Sykes, 1908; Esau, 1969; Schmitz & Srivastava, 1974b and references cited therein). These specialized cells are located in the medulla of all parts of a kelp thallus, except the root-like holdfast, and will be referred to as sieve cells or sieve filament members. Wille (1897) made important observations on the cytology of sieve cells (*A. esculenta*) which are relevant today and applicable to other Laminariales. He described the cells as having 'foam-like' contents — a clear allusion to cytoplasm containing many small vacuoles (Fig. 10 in Wille, 1897) — and porous sieve plates presumably pierced by cytoplasmic threads. In addition he mentions a nucleus and chromatophores in young sieve cells. Numerous EM studies have confirmed and extended Wille's observations (*e.g.*, Parker & Philpott, 1961; Ziegler & Ruck, 1967; Schmitz & Srivastava, 1974b, 1975, 1976; Schmitz in Behnke, 1975) of multivacuole cytoplasm containing mitochondria, golgi, ER; vesicles (some of which are coated), *etc.*, with plasmodesmata connecting adjacent sieve filament members. Diameters of sieve plate pores range from 0.04 μm in *Laminaria* spp. to 6.0 μm in *Macrocystis* (*cf.* Sideman & Scheirer, 1977) and vary inversely with pore number and density. It is not surprising that chemical tests of medullary tissue (Rigg, 1925) and exudate (Parker, 1966; Nicholson & Briggs, 1972) showed high protein content, with substantial amounts of amino acids and mannitol. Exudate from *M. pyrifera* also incorporated ^{32}P into ATP and other phosphorylated metabolites (Schmitz & Srivastava, 1974a) while that from *L. saccharina* reportedly contained c-AMP (Becker & Ziegler, 1973).

The exudation process, first reported by Crafts (1939), is slowest near the apex of the *Macrocystis* frond (Schmitz & Srivastava, 1974a) and blade meristem in *Alaria* (Buggeln, 1978); rate of flow can be affected by light or removal of lateral portions of the blade in *Alaria* (Buggeln, 1978).

Although sieve cells actively involved in translocation are perimedullar in location (Steinbiss & Schmitz, 1973), the number of actively translocating sieve cells/mm^2 of conducting tissue is not known. Callose eventually occludes sieve plate pores in older cells thus preventing translocation but perhaps acting to shunt assimilate to more efficient sieve cells (Nicholson, 1976). Branching sieve cells, lateral connections between sieve cells and multi-junction sieve plate areas have been reported but their frequency and location in the medulla has not been indicated (Wille, 1897; Parker & Huber, 1965; Steinbiss & Schmitz, 1973; Schmitz & Srivastava, 1974b; Nicholson, 1976). In the blade of *Alaria* (perhaps in other genera as well), there are two distinct parts to the translocation system: the centrally located, longitudinal (long distance) pathway; and a peripheral system for transport from lateral portions of the blade to the longitudinal pathway. In *Alaria* location of the former is easily identified by the morphologically prominent midrib; similar anatomical specialization appears to be visualized in autoradiographs of species with flat blades, *e.g.*, *L. hyperborea* (Lüning, *et al.*, 1972).

While it is recognized that regions of high metabolic activity (sinks) accumulate photoassimilate, little is known of the factor(s) which control the translocation process. If the *Alaria* meristem is removed from the blade, little ^{14}C assimilate leaves the source 80 cm away (Buggeln, unpublished). When the *Macrocystis* apex is removed, blades normally exporting to the apex begin exporting basipetally (Parker, 1965; Lobban, 1978). If the longitudinal conducting pathway is severed in *Alaria*, ^{14}C does not bipass the cut via the lateral conducting system (Buggeln, 1977). An artificial sink in the distal end of blades of *Alaria* spp. can be created by darkening this normally non-sink area (Schmitz & Srivastava, 1975; Buggeln, unpublished). It appears then that cytoplasmic continuity between sieve filament members in the longitudinal pathway connecting source with sink is an essential requirement for translocation.

As ^{14}C labelled assimilate (mannitol plus *ca.* 12 amino acids) moves from source to sink, recovery of alcohol soluble radioactivity from tissues (not exclusively medullary) along the pathway shows a decrease in the initial source ratio of mannitol > amino acids, implying that metabolism of mannitol is occurring in transit (Schmitz *et al.*, 1972; Schmitz & Srivastava, 1975).

What is the mechanism of translocation in Laminariales? Autoradiographic evidence suggests a symplastic route through perimedullary sieve filaments (Steinbiss & Schmitz, 1973). Polarity of translocation and rapid velocities (*e.g.*, Schmitz & Lobban, 1977) argue against diffusion as a mechanism, while the cytoplasmic nature of sieve cells, plasmodesmata-filled sieve plate pores and lack of osmotic gradient in sieve cell exudate between source and sink exclude a Münch-type pressure flow system (Buggeln, 1977). Other higher plant mechanisms (*cf.* Zimmermann, 1969) are as difficult to evaluate in kelp as in higher plants. In view of the highly vacuolate and vesiculate nature of sieve cells, together with the large numbers of mitochondria at sieve plates, I cannot resist speculating that translocation in Laminariales occurs *via* intervacuolar and vesicular transfer with polarity of transport within sieve tubes maintained by an active transport pump through plasmodesmata of sieve plates. Difficulty in finding "non-metabolized" organic substances which will be both absorbed by kelp and translocated (except see Parker, 1965) has hampered progress in investigations of translocation mechanism.

REFERENCES

Akagawa, H., T. Ikawa & K. Nisizawa. Initial pathway of dark $^{14}CO_2$-fixation in brown algae. *Bot. Mar.* 15: *119–125* (1972a).

Akagawa, H., T. Ikawa & K. Nisizawa. $^{14}CO_2$-fixation in marine algae with special reference to the dark-fixation in brown algae. *Bot. Mar.* 15: *126–132* (1972b).

Akagawa, H., T. Ikawa & K. Nisizawa. The enzyme system for the entrance of $^{14}CO_2$ in the dark CO_2-fixation of brown algae. *Plant & Cell Physiol.* 13: *999–1016* (1972c).

Becker, D. & H. Ziegler. Cyclisches Adenosin-3′:5′-monophosphat in pflanzlichen Leitbahnen? *Planta* (Berl.), 110: *85–89* (1973).

Behnke, H.-D. Phloem tissue and sieve elements in algae, mosses and ferns. pp. 187–224 in: S. Aronoff, J. Dainty, P.R. Gorham, L.M. Srivastava & C.A.Swanson (eds.), Phloem Transport. Plenum Press, N.Y. (1975).

Bidwell, R.G.S. Photosynthesis and metabolism of marine algae. II. A survey of rates and products of photosynthesis in $^{14}CO_2$. *Can. J. Bot.* 36: *337–349* (1958).

Buggeln, R.G. Physiological investigations on *Alaria esculenta* (L.) Grev. (Laminariales). I. Elongation of the blade. *J. Phycol.* 10: *283–288* (1974).

Buggeln, R.G. Auxin, an endogenous regulator of growth in algae? *J. Phycol.* 12: *335–358* (1967a).

Buggeln, R.G. The rate of translocation in *Alaria esculenta* (Laminariales, Phaephyceae). *J. Phycol.* 12: *439–442* (1976b).

Buggeln, R.G. Physiological investigations on *Alaria esculenta* (Laminariales, Phaeophyceae). II. Role of translocation in blade growth. *J. Phycol.* 13: *212–218* (1977).

Buggeln, R.G. Physiological investigations on *Alaria esculenta* (Laninariales, Phaeophyceae). III. Exudation by the blade. *J. Phycol.* 14: *54–56* (1978).

Buggeln, R.G. & A.K. Bal. Effects of auxins and chemically related non-auxins on photosynthesis and chloroplast ultrastructure in *Alaria esculenta* (Laminariales). *Can. J. Bot.* 55: *2098–2105* (1977).

Buggeln, R.G., S. Lucken & R. Hooper. Heterotrophic CO_2 fixation by Arctic kelp species. (Manuscript in preparation.)

Clendenning, K.A. Photosynthesis and growth in *Macrocystis pyrifera*. pp. 55–65 in: A. Davy de Virville & J. Feldman (eds.) IVth Intern. Seaweed Symposium. Pergamon Press, New York (1964).

Clendenning, K.A. Photosynthesis and general development in *Macrocystis*. pp. 170–190 in: W.J. North (ed.). The biology of giant kelp beds (*Macrosystis*) in California. Nova Hedwigia *32* (1971).

Crafts, A.S. The relations between structure and function of the phloem. *Amer. J. Bot.* 26: *172–177* (1939).
Drew, E.A. Light inhibition of photosynthesis in macroalgae. *British Phycol. Jour.* 9: *217–218* (1974).
Erman, C. Förberedande undersökningar över kolsyreassimilationen–hos *Laminaria. Botaniska Notiser*: *331–342* (1928).
Esau, K. The phloem. Handbuck der Pflanzenanatomie. Band 5 Vol. 2: 263–68. Gebruder Borntraeger, Berlin-Stuttgart (1969).
Fallis, A.L. Growth in some Laminariaceae. *Puget Sound Biol. Sta. Publ.* 1: *137–155* (1916).
Gail, F.W. Photosynthesis in some of the red and brown algae as related to depth and light. *Puget Sound Biol. Sta. Publ.* 3: *177–193* (1972).
Hatcher, B.G., A.R.O. Chapman & K.H. Mann. An annual carbon budget for the kelp, *Laminaria longicruris. Mar. Biol.* 44: *85–96* (1977).
Haxo, F.T. & L.R. Blinks. Photosynthesis action spectra of marine algae. *J. Gen. Physiol.* 33: *389–422* (1950).
Hellebust, J.H. & A. Haug. Photosynthesis, translocation and alginic acid synthesis in *Laminaria digitata* and *Laminaria hyperborea. Can. J. Bot.* 50: *169–176* (1972).
Johnston, C.S., R.G. Jones, & R.D. Hunt. A seasonal carbon budget for a laminarian population in a Scottish sea-loch. *Helgölander wiss. Meeresunters.* 30: *527–545* (1977).
Jolliffe, E.A. & E.B. Tregunna. Studies on HCO_3^- ion uptake during photosynthesis in benthic marine algae. *Phycologia* 9: *293–303* (1970).
King, R.J. & W. Schramm. Determination of photosynthetic rates for the marine algae *Fucus vesiculosus* and *Laminaria digitata. Mar. Biol.* 37: *209–213* (1976a).
King, R.J., & W. Schramm. Photosynthetic rates of benthic marine algae in relation to light intensity and seasonal variation. *Mar. Biol.* 37: *215–222* (1976b).
Kniep, H. Über die Assimilation und Atmung der Meeresalgen. *Internat. Rev. der Gesamten Hydrobiol. und Hydrograph.* 7: *1–38* (1914).
Kremer, B.P. & J. Willenbrink. CO_2-Fixierung und Stofftransport in benthischen marinen Algen. I. Zur Kinetik der $^{14}CO_2$-Assimilation bei *Laminaria saccharina. Planta* (Berl.) 103: *55–64* (1972).
Kremer, B.P. & U. Küppers. Carboxylating enzymes and pathway of photosynthetic carbon assimilation in different marine algae – evidence for the C_4-pathway? *Planta* (Berl.) 133: *191–197* (1977).
Lampe, H. Die Temperatureinstellung des Stoffgewins bei Meeresalgen als plasmatische Anpassung. *Protoplasma* 23: *534–578* (1935).
Levring, T. Submarine daylight and the photosynthesis of marine algae. *Göteborgs K. Vetensk.-O. Vittersamh. Handl. Ser. B.* 5: *1–90* (1947).
Lobban, C.S. Translocation of ^{14}C in *Macrocystis pyrifera* (Giant kelp). *Plant Physiol.* 61: *585–589* (1978).
Lüning, K. Cultivation of *Laminaria hyperborea in situ* and in continuous darkness under laboratory conditions. *Helgoländer wiss. Meeresunters.* 20: *79–88* (1970).
Lüning, K. Seasonal growth of *Laminaria hyperborea* under recorded underwater light conditions near Helgoland. pp. 347–361 in: D.J. Crisp (ed.), Fourth European Marine Biology Symposium. Cambridge Univ. Press, London (1971).
Lüning, K., K. Schmitz & J. Willenbrink. Translocation of ^{14}C-labelled assimilates in two *Laminaria* species. pp. 420–425 in: K. Nisizawa. (ed.), 7th Intern. Seaweed Symp. Univ. of Tokyo Press, Tokyo (1972).
Lüning, K. CO_2 fixation and translocation in benthic marine algae. III. Rates and ecological significance of translocation in *Laminaria hyperborea* and *L. saccharina. Mar. Biol.* 23: *275–281* (1973).
Montfort, C. Über Beziehungen zwischen Farbton, Lichtausnutzung und Stoffgewinn bei roten und grünen Florideen sowie bei anderen Meeresalgen. *Biochem. Zeit.* 261: *179–201* (1933).
Montfort, C. Farbe und Stoffgewinn im Meer. Untersuchungen zur Theorie der komplementären Farbenanpassung nordischer Meeresalgen. *Jahrb. für wiss. Bot.* *493–592* (1934).

Montfort, C. Zeitphasen der Temperatur-Einstellung und jahreszeitliche Umstellungen bei Meeresalgen. *Ber. deutsch Bot. Ges.* 53: *651–674* (1935).
Montfort, C. Umwelt, Ergbut und physiologische Gestalt. I. Lichttod und Starklichtresistenz bei Assimilationsgeweben. *Jahrb. für wiss. Bot.* 84: *1–57* (1936).
Montfort, C. Funktionstypen des Assimilationsapparates und das Problem der gelben Blattfarbstoffe. I. Aufdeckung und vergleichende Analyse der Carotinoidwirkungen. *Kieler Meeresforschungen* 2: *301–344* (1938).
Moore, B., E. Whitely & T.A. Webster. Studies of photosynthesis in marine algae. I. Relative photosynthetic activity of green, brown and red seaweeds in light of varying intensity and from different regions of the spectrum. *Proc. and Trans. Liverpool Biol. Soc.* 37: *38–47* (1923).
Nicholson, N.L. Field studies on the giant kelp *Nereocystis*. *J. Phycol.* 6: *177–182* (1970).
Nicholson, N.L. Anatomy of the medulla of *Nereocystis*. *Bot. Mar.* 19: *23–31* (1976).
Nicholson, N.L. & W.R. Briggs. Translocation of photosynthate in the brown algae *Nereocystis*. *Amer. J. Bot.* 59: *97–106* (1972).
Parker, B.C. Translocation in the giant kelp *Macrocystis*. I. Rates, direction, quantity of ^{14}C-labelled products and fluorescein. *J. Phycol.* 1: *41–46* (1965).
Parker, B.C. Translocation in *Macrocystis*. III. Composition of sieve tube exudate and identification of the major ^{14}C-labelled products. *J. Phycol.* 2: *38–41* (1966).
Parker, B.C. The internal structure of *Macrocystis* in: W.J. North (ed.)., The biology of giant kelp beds (*Macrocystis*) in California. *Nova Hedwigia* 32: *99–121* (1971).
Parker, B.C., & J. Huber. Translocation in *Macrocystis*. II. Fine structure of the sieve tubes. *J. Phycol.* 1: *172–179* (1965).
Parker, J. & D.E. Philpott. The ultrastructure of sieve plates of *Macrocystis pyrifera*. *Bull. Torrey Bot. Club* 88: *85–90* (1961).
Penot, M., J.-Y. Floc'h et M. Penot. Physiologie comparée des transports à longue distance chez les végétaux supérieurs et chez les algues marines. *Phycologia* 15: *299–308* (1976).
Printz, H. Über die Kohlensäureassimilation der Meeresalgen in verschieden Tiefen. *Norske Vidensk. Akad. Skr.*, I. *Mat.-Naturv. Klasse* 1: *1–101* (1939).
Rigg, G.B. Some physiology of the sieve tubes of *Macrocystis*. *Puget Sound Biol. Sta. Publ.* 3: *311–329* (1925).
Sargent, M.C. & L.W. Lantrip. Photosynthesis, growth and translocation in giant kelp. *Amer. J. Bot.* 39: *99–107* (1952).
Schmitz, K. & C.S. Lobban. A survey of translocation in Laminariales (Phaeophyceae). *Mar. Biol.* 36: *207–216* (1977).
Schmitz, K., K. Lüning & J. Willenbrink. CO_2-Fixierung und Stofftransport in benthischen marinen Algen. II. Zum Ferntransport ^{14}C-markierter Assimilate bei *Laminaria hyperborea* und *Laminaria saccharina*. *Zeitschr. Pflanzenphysiol.* 67: *418–429* (1972).
Schmitz, K. & L.M. Srivastava. The enzymatic incorporation of ^{32}P into ATP and other organic compounds by sieve-tube sap of *Macrocystis integrifolia* Bory. *Planta* (Berl.), 116: *85–89* (1974a).
Schmitz, K. & L.M. Srivastava. Fine structure and development of sieve tubes in *Laminaria groenlandica* Rosenv. *Cytobiologie* 10: *66–87* (1974b).
Schmitz, K. & L.M. Srivastava. On the fine structure of sieve tubes and the physiology of assimilate transport in *Alaria marginata*. *Can. J. Bot.* 53: *861–876* (1975).
Schmitz, & L.M. Srivastava. The fine structure of sieve elements of *Nereocystis leutkeana*. *Amer. J. Bot.* 63: *679–693* (1976).
Seybold, A. Über die Lichtenergiebilanz submerser Wasserpflanzen, vornehmlich der Meeresalgen. *Jahrb. für wiss. Bot.* *593–654* (1934).
Seybold, A. & A. Weissweiler. Weitere spektrophotometrische Messungen an Laubblätern und Chlorophyll-Lösungen sowie an Meeresalgen. *Bot. Archiv.* 44: *102–153* (1943).
Sideman, E.J. & D.C. Scheirer. Some fine structural observations on developing and mature sieve elements in the brown alga *Laminaria saccharina*. *Amer. J. Bot.* 64: *648–657* (1977).
Steinbiss, H.-H. & K. Schmitz. CO_2-Fixierung und Stofftransport in benthischen marinen

Algen. V. Zur Autoradiographischen Lokalisation der Assimilattransportbahnen im Thallus von *Laminaria hyperborea*. *Planta* (Berl.), 112: *253–264* (1973).
Stocker, O. und W. Holdheide Die Assimilation Helgoländer Gezeitenalgen während der Ebbezeit. *Zeit. für Bot*. 32: *1–59* (1937).
Sykes, M.G. Anatomy and histology of *Macrocystis pyrifera* and *Laminaria saccharina*. *Ann. Bot*. 22: *291–325* (1908).
Sundene, O. The ecology of *Laminaria digitata* in Norway in view of transport experiments. *Nytt Mag. Bot.* 11: *83–107* (1964).
Titlyanov, E.A. & B.M. Peshekhodko. On the transport of assimilates in thalli of fringing seaweeds. Dal'nevostochnyi nauch'nogo tsentra. *Biologo-pochvenogo instituta Trudy* 20: *137–141* (1973).
Towle, D.W. & J.S. Pearse. Production of the giant kelp *Macrocystis*, estimated by *in situ* incorporation of ^{14}C in polyethylene bags. *Limmol. & Oceanogr*. 18: *155–159* (1973).
Weidner, M. & U. Küppers. Phosphoenolpyruvat-Carboxykinase und Ribulose-1,5-Diphosphat-Carboxylase von *Laminaria hyperborea* (Gunn.) Fosl.: Das Verteilungsmuster der Enzymaktivitäten im Thallus. *Planta* (Berl.) 114: *365–372* (1973).
Wille, N. Beiträge zur physiologischen Anatomie der Laminariaceen. Univers. Festskr. til H. Maj. Kong Oscar II, Det. Kgl. Norske. pp. 1–70 (1897).
Willenbrink, J., M. Rangoni-Kübbeler, and B. Tersky. Frond development and CO_2 fixation in *Laminaria hyperborea*. *Planta* (Berl.), 125: *161–170* (1975).
Yamaguchi, T., T. Ikawa & K. Nisizawa. Incorporation of radioactive carbon with $H^{14}CO_3$ into sugar constituents by a brown alga, *Eisenia bicyclis*, during photosynthesis and its fate in the dark. *Plant & Cell Physiol*. 7: *217–229* (1966).
Ziegler, H. & I. Ruck. Untersuchungen über die Feinstruktur des Phloems. III. Mitt. Die 'Trumpetzellen' von *Laminaria*-Arten. *Planta* (Berl.) 73: *62–73* (1967).
Zimmerman, M. Translocation of nutrients. pp. 381–417 in: M.B. Wilkins (ed.), The physiology of plant growth and development. McGraw-Hill, London (1969).

Author's address:
Marine Sciences Research Laboratory
Memorial University of Newfoundland
St. John's Nfld.
Canada A1C 3X9.

MODIFICATION OF THE STRAWBERRY RECEPTACLE ACCUMULATION ABILITY BY GROWTH REGULATORS*

E.K. LIS & R. ANTOSZEWSKI
(Skierniewice, Poland)

ABSTRACT

Influence of growth substances on the ability of the strawberry receptacle to accumulate ^{14}C-assimilates and $_{32}PO_4^{--}$ was studied.

Fourteen to fifteen days after pollination, the achenes were removed from the receptacles and the surface of each receptacle was divided into two parts by a thin barrier of lanolin. One half of the surface was treated with growth regulator (IAA or GA_3 or kinetin or IAA+GA_3+kinetin, at a concentration of 3×10^{-4}M), and the second half was treated with water and served as control. After three and ten to twelve days of hormone treatment, $^{14}CO_2$ was applied to the leaves and $^{32}PO_4^{--}$ to the roots. Twenty-four hours later, the receptacles were analyzed for accumulation of both markers.

GA_3 as well as kinetin did not influence the accumulation of both labeled substances, ^{14}C-assimilates and ^{32}P, whereas a marked effect was obtained with IAA or IAA+GA_3+kinetin. The accumulation of ^{14}C-assimilates was several times higher in halves treated with these regulators as compared to the control halves. The accumulation of phosphate was only slightly affected by IAA.

After a three day period of hormone application, IAA was the most effective. Longer treatment (10–12 days) showed the prevailing effect of the mixture of IAA+GA_3+kinetin over IAA alone.

These results show that growth regulators are able to evoke a sink for nutrients. However, the sink power for assimilates is much stronger than for phosphate.

INTRODUCTION

Hormonal regulation of sink power of fruits was the subject of many reviews: Bollard (1970), Crane (1964, 1969), Nitsch (1970, 1971). The question of whether the influence of hormones on sink power is connected with stimulation of metabolic activity of the accumulating organ or with activation of long-distance transport (Patrick & Wareing, 1976; Patrick, 1976) remains still unsolved. Bidwell (1976) emphasizes the possibility that phytohormones generated by the developing organs could influence the rate of formation and supply of assimilates to them. One may also postulate that growth regulators trigger some mechanism in the transport system.

Different hormones are capable of controlling sink power of fruits for assimilates in various species: Breen (1968), Chalmers *et al.* (1976),

* This study was supported in part by USDA Grant FG-Po-363. Our thanks are due to Dr. R. Marcelle for his stimulatory discussion of the objectives of this paper.

Kriedemann (1968), Weaver *et al.* (1969). However, the question raised by Bollard in 1970: 'is a sink for carbohydrates also a sink for inorganic ions?' remains unanswered.

Independence of sugar and cation transport was evidenced by Leep & Peel (1970) on a strip of bark. Stimulation by auxins of the transport of ^{14}C-sucrose but not ^{32}P in decapitated stems was established by Bowen & Wareing (1971) and was explained by the authors as a reflection of the operation of hormonally-controlled metabolite-selective sink.

The present paper reports on the effect of growth regulators on the accumulation of ^{14}C-labelled assimilates and mineral ions such as: $^{32}PO_4^{--}$ and $^{45}Ca^{++}$ by strawberry receptacle. Double- and triple-labelling techniques were used in investigations, since this technique provides the only reliable tool to study differential accumulation of nutrients.

Strawberry fruit and in particular the achenes, contain endogenous growth regulating substances such as auxins, cytokinins and gibberellins. Consequently, exogenous application of hormones does not induce a big increase in growth rate or accumulation of assimilates in fruits.

According to Nitsch (1950), in strawberry *cv.* Marshal, auxin was not detected in the receptacle, and the achenes are the source of auxin for the growth of this organ. In our determination (Lis *et al.*, 1978), on the cultivar Talisman, auxins were present in the achenes and also in the receptacle; in this last organ, auxin was only detected in the very young stage, until 8 days after pollination. The maximum content of auxin in achenes has been found 7–14 days after pollination. For this reason, strawberry fruits at the age of 14 days after pollination were used in our investigations. Removal of achenes at this time deprived the receptacle of any source of auxin.

MATERIAL AND METHODS

Strawberry plants (*Fragaria grandiflora cv.* Talisman) were potted and cultivated in greenhouse in long day photoperiod (14 hr). The study was restricted to the first fruit formed on the trusses.

Fourteen to eighteen days after pollination the achenes were removed from the receptacle by hand according to the scheme shown in Fig. 1. The surface of the receptacles was divided into two equal parts by putting a lanolin paste in a line perpendicularly to the base of the fruit. A solution of growth regulator(s) was supplied to one half of the receptacle and a 'control solution' to the other. The control solutions were the solvents used for dissolving growth regulators: 0.12% ethanol for IAA and GA_3 and 0.024N HCl for kinetin. The solutions (20 μl) were uniformly spread on one part of the receptacle surface and allowed to dry. Treatments were made twice a day during three or ten to twelve days.

In the first experiment the achenes were removed from one half of the receptacle, as in Fig. 1a, and these bare halves were treated with water or growth substances: IAA or IAA+GA_3 at concentration of 3×10^{-4}M for each ingredient. The same growth substances were applied in the second experiment but after removal of achenes from the whole surface of the receptacle, as it is shown in Fig. 1b. In both experiments, the application of growth substances

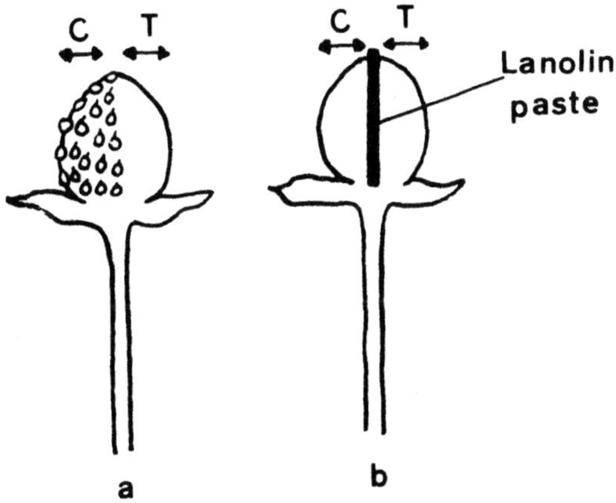

Fig. 1. Ways of preparing fruit for treatment: a) Achenes removed from one half of the receptacle. This surface (T) was treated with growth regulators. The other half (C) was treated with water.
b) Achenes removed from the whole surface: half (C) treated with water, half (T) treated with growth regulator(s).

lasted three days. After this time the plants were exposed to $^{14}CO_2$ for 20 minutes (25 μCi of $^{14}CO_2$ was generated from Na_2 $^{14}CO_3$ in a 5 liter chamber per plant). Fruit trusses were enclosed in black paper bags to prevent straight incorporation of $^{14}CO_2$. Six hours later the fruits were gathered, divided into proper halves as marked by the lanolin in order to analyze separately the treated half and compare it with the control half. The material was dried at 80° C and the dry weight was determined. Samples of 30 mg were prepared for measurement of radioactivity with a thin-end window of G-M counter.

In the experiment with ^{32}P, the fruits were prepared as on Fig. 1b, and treated with a solution of IAA or GA_3 or IAA+GA_3+kinetin. On the third day, the plants were removed from the pots and transferred to glass containers with 50 ml of Hoagland's nutrient solution containing 20 μCi of H_3 $^{32}PO_4$ per plant. After 20, 30 and 40 hours the fruits were harvested, and prepared for measurement of radioactivity with the G-M counter.

In the experiment with double-labelling, after the pretreatment of the fruits with growth regulators, the plants incorporated $^{14}CO_2$ through the leaves and $^{32}PO_4$ through the roots. Twenty-four hours later the fruits were analyzed for both radioisotopes with the G-M counter. Using an appropriate screen, ^{14}C radiation was distinguished from ^{32}P and both emitters quantitatively determined in one sample. In these experiments, the fruits were treated with growth regulators in a period of three days and ten to twelve days.

In the last experiment, strawberry plants were labelled with three isotopes: ^{14}C, ^{45}Ca and ^{32}P. For this experiment, plants were grown in Hoagland's nutrient solution. After feeding the plants with $^{14}CO_2$, two additional markers, 50 μCi of $^{45}CaCl_2$ and 20 μCi of H_3 $^{32}PO_4$ per plant, were added to 450 ml of the nutrient solution. Samples of 10 mg of dry matter was assayed for radioactivity. Using a set-up using a semiconductor-detector (Si/Li, produced by the Nuclear Research Institute, Poland), pre-amplifier, proper amplitude analyzer and counter, simultaneous determination of three emitters was done in one sample.

RESULTS

The results are presented as ratios of radioactivity per mg of dry matter of the receptacle half treated with growth regulator to the radioactivity of the control half, treated with water. Calculation was made for each fruit separately and the mean ratio drawn out.

Table 1 presents results on the influence of achenes on the accumulation of ^{14}C-assimilates in the receptacle in a six hour period. It is seen from experiment 1

Table 1. Effect of removal of achenes and treatment of the receptacle with growth regulators on accumulation of ^{14}C-assimilates in a six hour period. Values are ratios of the 'specific radioactivity' of treated (T) to control (C) half of receptacle, ± standard error of the mean.

Exp. No.*	Experimental systems	Hormone treatment		
		H_2O	IAA	IAA+GA_3
1.	as in fig 1a	0.20 ±0.04	0.44 ±0.09	0.56 ±0.10
2.	as in fig 1b	1.02 ±0.11	2.55 ±0.22	2.93 ±0.32

* Exp. 1 – nine replicates per treatment.
Exp. 2. – six replicates per treatment.
Growth substances were applied during three days before the exposure to $^{14}CO_2$; duration of $^{14}CO_2$ incorporation: 20 minutes; duration of transport: 6 hours.

that the part of the receptacle deprived of achenes accumulated five times less radioactive assimilates than the control one. The treatment with IAA or IAA+GA_3 could partly substitute for the presence of achenes: the treated half accumulated only two times less radioactivity than the control half. Removal of achenes from the whole surface and treatment of one half of the receptacle with IAA or IAA+GA_3 (exp. 2), increased also the accumulation of ^{14}C-assimilates in the treated part as compared to the bare half treated with water.

Data for the accumulation of $^{32}PO_4^{--}$ delivered through the roots are presented in Table 2. Similar results were obtained after various times of $^{32}PO_4^{--}$ accumulation. IAA significantly increased the accumulation of ^{32}P in the treated half of the receptacle. GA_3 alone was without effect. Application of the mixture of IAA, GA_3 and kinetin did not increase the accumulation of ^{32}P above the results obtained for IAA alone.

Results on the effect of growth regulators on accumulation of both markers delivered simultaneously to the plants are presented in Table 3. It is seen that only some of the tested hormones were able to influence the attraction power of the receptacle. GA_3 and kinetin did not effect the accumulation of ^{14}C-assimilates or mineral ions in the strawberry receptacle. In contrast to them

Table 2. Effect of treatment of the receptacle with growth regulators after removal of achenes on accumulation of $^{32}PO_4^{--}$ at various times. Values are ratios of the 'specific radioactivity' of treated to control half of receptacle ± standard error of the mean.

Duration of $^{32}PO_4^{--}$ accumulation (hr)	Hormone treatment* H_2O	GA_3	IAA	IAA+GA_3+ kinetin
20	1.16	1.04	1.93	1.97
	±0.38	±0.36	±0.10	±0.52
30	1.22	0.99	2.27	3.85
	±0.16	±0.27	±0.28	±0.80
40	0.99	1.03	1.82	1.51
	±0.11	±0.11	±0.41	±0.11
Average	1.13	1.02	2.01	2.44
	±0.10	±0.12	±0.16	±0.45

* Three replicates per treatment.
Application of growth regulators: see table 1; ^{32}P applied during 20, 30 or 40 hours.

Table 3. Influence of growth regulators on accumulation of ^{14}C-assimilates and $^{32}PO_4^{--}$ in strawberry receptacles in a 24-hours period. Values are ratios of the 'specific radioactivity' of treated to control half of receptacle ± standard error of the mean.

Radioisotopes detected	Hormone treatment GA_3	kinetin	IAA	IAA+GA_3+kin.
	3 days pretreatment*			
^{14}C-assimil.	[a]1.88	[a]0	[b]8.52	[b]11.08
	±0.62		±0.71	±2.51
$^{32}PO_4^{--}$	[a]2.12	[a]0.99	[b]2.20	[b]2.93
	±1.09	±0.21	±0.50	±0.85
	[c]10–12 days pretreatment			
^{14}C-assimil.	1.58	–	9.19	23.39
	±0.29	–	±1.64	±3.11
$^{32}PO_4^{--}$	1.66	–	1.75	2.63
	±0.52	–	±0.19	±0.47

* a – three replicates per treatment. b – six replicates per treatment.
c – nine replicates per treatment.
Growth regulator treatments during 3 or 10 to 12 days.

IAA was very effective in evoking the sink power of the receptacle. A three-day pretreatment with IAA causes a nine-fold increase of ^{14}C radioactivity in comparison to the part of the receptacle treated with water. The effect of IAA on ^{32}P accumulation was rather small, only about two-fold increase.

Three, and ten to twelve days of treatment with IAA stimulated the accumulation of ^{14}C-assimilates and ^{32}P to the same extent. Thus, the sink power of the strawberry receptacle was independent from the period of IAA treatment.

The mixture of growth regulators applied during three days gave the same effect as IAA alone, which means that gibberellin and kinetin did not manifest an additive or synergistic effect to IAA.

A different result was obtained after prolonged application of IAA+GA_3+kinetin. A twenty-fold increase in accumulation of ^{14}C-assimilates in comparison to control halves was noted. The mixture of IAA+GA_3+kinetin was most effective in the stimulation of normal enlargement and maturation of the treated half of the receptacle, which can be seen from the increase in fresh and dry matter in Table 4. Thus, the mixture of regulators which leads to normal fruit development acts on sink power towards assimilates, too.

Table 4. Fresh and dry weights of the receptacle halves as an effect of growth regulator treatment during a 10–12 day period. Fruit treatment started 14 days after pollination.

Hormone treatment	Receptacle half	Weight (mg) Fresh	Dry
GA_3	C	541 ± 78*	64 ± 6
	T	713 ± 104	67 ± 8
IAA	C	523 ± 63	70 ± 9
	T	954 ± 105	115 ± 13
IAA+GA_3+kin.	C	565 ± 151	71 ± 9
	T	1737 ± 286	191 ± 24

* Standard error of the mean. Nine replicates per treatment.

The results obtained in Table 3, show a different sink power of the receptacle in the attraction of ^{14}C-assimilates and $^{32}PO_4^{---}$ under stimulating action of IAA or IAA+GA_3+kinetin. Accumulation of ^{14}C-assimilates is strongly connected with the presence of IAA in the receptacle, however $^{32}PO_4^{---}$ to the contrary, is in a weak relation to hormonal control.

The low content of $^{45}Ca^{++}$ in the receptacle does not allow us to discuss the growth regulator effect on the receptacle sink power for this cation. Nevertheless, ^{45}Ca was detected in larger amounts in the petiole of the leaf and in the peduncle; it was also present in the fruit calyx. These results suggest that the fruit possesses distinctly lower sink power to calcium than to phosphate and assimilates.

DISCUSSION

As shown in Table 1 the achenes possess factor(s) essential for maintaining the receptacle sink power for assimilates. One factor is IAA which in strawberry, fourteen days after pollination, is exclusively present in the achenes (Lis *et al.*,1978). GA_3 and kinetin are without effect. Mixture of IAA with GA_3 or IAA with GA_3 and kinetin are not more effective than IAA alone in a three-day hormone treatment. Lack of synergistic or additive effect of GA_3 or kinetin with IAA could be explained by self-sufficiency of the receptacle of this age for gibberellins and cytokinins as these hormones are also present in the achenes and receptacle. A longer pretreatment of the receptacle with IAA+GA_3+kinetin is very effective in accumulating ^{14}C-assimilates, probably by supplementing the low content of endogenous cytokinins and gibberellins in the receptacle at this stage (Lis *et al.*, 1978). The effect of a longer pretreatment could be also a consequence of metabolical changes in maturing fruits, leading to a rapid increase in the concentration of carbohydrates as reported by Bollard (1970).

The sink power of the receptacle to $^{32}PO_4^{---}$ is stimulated by IAA in a very limited range is contrary to the effect on ^{14}C-assimilates accumulation; this finding corresponds with the results of Bowen & Wareing (1971), showing an IAA stimulation of the transport of ^{14}C-sucrose but not of ^{32}P in decapitated stems of *Coleus spp.* and *Helianthus annuus*. It would be interesting to know to what extent achenes influence phosphate accumulation in the receptacle. Taking into consideration the low accumulation of ^{32}P in the receptacle treated with hormone (IAA) one can deduce a rather low capacity of the receptacle for phosphate. Thus growth regulators can not stimulate it very much.

The results can be discussed also in terms of the influence of hormones on transporting channels. In our experiments, $^{32}PO_4^{---}$ was delivered through the roots, and it was transported probably in the xylem. The xylem water stream to strawberry fruit is limited by lack of stomata on the fruit epidermis and very low transpiration. Results of the marked effect of IAA and IAA+GA+benzyladenine on the accumulation of phosphate obtained by Seth & Wareing (1967) and Hatch & Powell (1971) deal rather with phloem transport.

Calcium transport to the receptacles was very small, which could be a result of the too low concentration of ^{45}Ca in the nutrient solution. Nevertheless, it could be connected with the specificity of calcium transport. Ferguson & Bollard (1976) described that movement of calcium in the xylem appears to be normally by exchange rather than by mass flow. It is also known from the literature that movement of Ca to the fruit is small in comparison with the leaves (Stösser, 1970; Bünemann & Grassia, 1973).

From the theoretical point of view the question of whether the sink-donor relation depends on the mass flow or is regulated by specific phloem ability remains unclear.

It is obvious that the chemical constitution of different plant parts differs essentially. To what extent this depends on the synthetic ability of the organ or to the differential transport ability is not known. In the case of fruits one can postulate that the fruit as a whole unit attracts all nutrients but that, before final accumulation takes place, some ingredients are 'filtered off' from the phloem contents. Thus, seeds or achenes accumulate proteins and lipids predominantly, whereas the receptacle accepts the excess of carbohydrates. On the other hand, different mechanisms can function here. Due to specific sink power different nutrients could be mobilized, thus differential accumulation should be observed. Our results support this view.

REFERENCES

Bidwell, R.G.S. A possible mechanism for the control of photoassimilate translocation. Proceedings of the Research Institute of Pomology, Skierniewice, Poland, No. 3: 77–89 (1973).

Bollard, E.G. The physiology and nutrition of developing fruits. pp 387–425, in: Hulme, A.C. (ed.) The biochemistry of fruits and their products. Acad. Press, London. v. I (1970).

Bowen, M.R., P.F. Wareing. Further investigations into hormone-directed transport in stems. *Planta* (Berl.) 99: *120–132* (1971).

Breen, P.J. Effects of hormones on translocation of nutrients to fruits. Diss. Abstr. Sect. B, 28: 4888. from Hort. Abstr. 1969, 39: 6813 (1968).

Bünemann, G & A. Grassia. Growth and mineral distribution in grafted tomato-potato plants according to sink number. *Scientia Horticulturae* 1: *13–24* (1973).

Chalmers, D.J. & P.H. Jerie. The effect of 2-chloroethyl-phosphonic acid on sink strength of developing peach (*Prunus persica* L.) fruit. *Planta* (Berl.) 131: *203–205* (1976).

Crane, J.C. Growth substances in fruit setting and development. *Ann. Rev. Plant Physiol.* 15: *303–326* (1964).

Crane, J.C. The role of hormones in fruit set and development. *Hort Sci.* 4: *108–111* (1969).

Ferguson I.B. & E.G. Bollard. The movement of calcium in woody stems. *Ann. Bot.* 40: *1057–1065* (1976).

Hatch, A. & L.E. Powell. Hormone-directed transport of ^{32}P in *Malus sylvestris* seedlings. *J. Amer. Soc. Hort. Sci.* 96: *230–234* (1971).

Kriedemann, P.E. An effect of kinetin on the translocation of ^{14}C-labelled photosynthate in citrus. *Aust. J. Biol. Sci.* 21: *399–400* (1968).

Leep, N.W. & A.J. Peel. Some effects of IAA and kinetin on the movements of sugars in the phloem of willow. *Planta* (Berl.) 90: *230–235* (1970).

Lis, E.K., B. Borkowska & A. Antoszewski. Growth regulators in the strawberry fruit. *Fruit Sci. Rept.* 5: (3), *17–29* (1978).

Nitsch, J.P. Growth and morphogenesis of the strawberry as related to auxin. *Am. J. Bot.* 37: *211–215* (1950).

Nitsch, J.P. Hormonal factors in growth and development. pp. 427–472 in: Hulme, A.C. (ed.), The biochemistry of fruits and their products. Acad. Press, London, v. I (1970).

Nitsch, J.P. Perennetion through seeds and other structures: Fruit development. pp. 413–501, in: Steward, F.C. (ed.), Plant Physiology. A treatise. v. VI A, Acad. Press, New York, London (1971).

Patrick, J.W. Hormone-directed transport of metabolites. pp. 433–446, in: Wardlaw, I.F. & J.B. Passioura (eds.), Transport and transfer processes in plants. Acad. Press, London (1976).

Patrick, J.W. & P.E. Wareing. Auxin-promoted transport of metabolites in stems of *Phaseolus vulgaris* L. Effects at the site of hormone application. *J. Exp. Bot.* 27: *969–982* (1976).

Seth, A.K. & P.E. Wareing. Hormone directed transport of metabolites and its possible role in plant senescence. *J. Exp. Bot.* 18: *65–77* (1967).

Von Stösser, R. Translokation und Verteilung von ^{45}Calcium und ^{32}Phospher bei Früchten von *Prunus avium* L. in Beziehung zum Fruchtfall. *Angew. Botanik* XLIV: *153–165* (1970).

Weaver, R.J., W. Shindy & W.M. Kliewer. Growth regulators induced movement of photosynthetic products into fruits of 'Black Corinth' grapes. *Plant Physiol.* 44: *183–188* (1969).

Authors' address:
Isotope Laboratory of Fruit Biochemistry
Research Institute of Pomology
Skierniewice
Poland.

FRUIT EFFECTS ON PHOTOSYNTHESIS, LIGHT- AND DARK-RESPIRATION

F. LENZ
(Bonn, Fed. Rep. of Germany)

ABSTRACT

Investigations of fruit effects on photosynthesis and respiration mainly of *Citrus madurensis* (Lour.) and *Solanum melongena* (L.) are reviewed. Net photosynthesis, gauged either by growth analysis or by direct measurement of gas exchange is always higher on plants with actively growing fruits. The higher photosynthetic efficiency of fruiting – as compared to non-fruiting plants – may be associated with differences in assimilate concentration, photorespiration rates and gaseous diffusive resistance of leaves in controlling several mechanisms of fruit dependent internal regulation of photosynthesis.

INTRODUCTION

Developing fruit have considerable effects on the growth of other plant organs. With increasing fruit load growth of roots, shoots and leaves is reduced (Leonard, 1962). Moreover, flower formation can be inhibited. This leads to irregular yields in perennial fruit crops, a phenomenon, which in horticulture is defined as 'alternate bearing'.

To gain a better understanding of the physiological reasons for 'alternate bearing' we have investigated the effects of fruit load on such growth reactions as water consumption, nutrient uptake, photosynthesis and respiration of *Citrus madurensis* (Lour.) and *Solanum melongena* L. (eggplants).

Dry matter distribution

Rooted citrus cuttings and eggplants were grown in glasshouses. The plants were cultivated in sand and supplied with modified Hoagland solution. The crop load was regulated by flower removal at an early stage of development (Lenz, 1967). At the stage of fruit maturity all plants were harvested and separated into different organs for dry matter determination.

In most such experiments it became obvious that with increasing fruit load, dry matter of roots, shoots and leaves were strongly reduced. Total dry matter of the plants was similar however. An example of this is shown in Fig. 1 in *C. madurensis*.

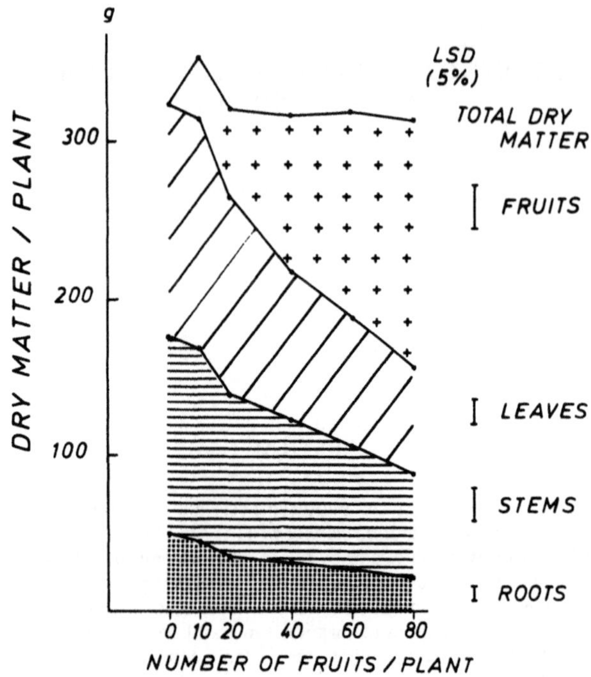

Fig. 1. Effect of fruit numbers on the dry matter distribution in plants of *Citrus madurensis* (Lour.).

At high fruit loads (80 fruits/plant) more than 50% of plant dry matter was in the fruits. This indicates that these organs obviously compete successfully with vegetative parts of the plant for photosynthetic products. This capacity seems to be of general occurrence (Wardlaw, 1969; Bünemann & Grassia, 1973).

As shown for eggplants (Claussen, 1976) dry matter formation in roots, leaves and shoots is strongly inhibited when fruits have the highest growth rates. At this stage 92–95% of assimilates produced are accumulated in the fruits. Despite reduced leaf area, fruiting plants produced under favourable growing conditions (light saturation, optimal water- and nutrient-supply and root temperatures) have more total dry matter than non-fruiting ones (Cary, 1970; Hansen, 1971). This indicates higher photosynthetic efficiencies of leaves in fruiting plants, if not the fruit themselves contribute substantially to dry matter production.

With *Citrus* (Lenz, 1967) and *S. melongena* (Claussen, 1975) it was demonstrated, that fruits at all developmental stages evolve more CO_2 than they could gain by light driven processes. They are therefore dependent on assimilates from the leaves for growth. Fruit photosynthesis might serve to limit CO_2 losses however, and some of the CO_2 evolved from the internal fruit tissue can be refixed (Kriedemann, 1968; Laval-Martin *et al.*, 1977).

In the experiments with fruiting and non-fruiting plants it became apparent that the dry matter production per unit leaf area, *i.e.*, the photosynthetic efficiency of leaves increased with higher fruit load. Similar results were found for other fruit crops, *e.g.*, apples (Maggs, 1963; Avery, 1969; Hansen, 1971) and strawberries (Lenz & Bünemann, 1967). There is however an upper limit to this photosynthetic response and in some species such as *Coffea arabica* and *Citrus* 'over-cropping' has been described (Leonard, 1962; Smith, 1976). This condition results in total inhibition of vegetative growth and premature senescence of leaves with an attendant reduction in rates of CO_2-fixation. Impaired root growth or nutritional stress under heavy crop loads can be important factors in such a decline.

Photosynthetic and dark respiration rates

In citrus plants (Lenz & Daunicht, 1971) as well as in eggplants (Hoffmann, 1973; Claussen, 1975) it can be shown that the photosynthetic rates of leaves in fruiting plants are higher than in non-fruiting ones. This is particularly pronounced at low root temperatures and fruit effects on photosynthesis can be intensified by raising the temperature of the developing organ.

Regardless of uncertainties about the mechanism of this physiological coupling between leaf and fruit, the photosynthetic response in adjacent leaves is unmistakable. When fruits on *C. madurensis* were held at 32° C (other aerial organs at 20° C) leaf photosynthesis was increased by 40% over the rate previously observed at 20° C.

The fruit effect on photosynthetic rates was most pronounced during time of intensive fruit growth. Flowers, small or mature fruits had no effect on photosynthetic rates (Claussen, 1977). In eggplants, differences in photosynthetic rates of plants with and without fruits were not always determined at the beginning, but at a later stage of a light period.

So far all the available data on *citrus* indicates that dark respiration in fruiting and non-fruiting plants is similar (Lenz & Daunicht, 1971).

Self-shading and photosynthesis

Plants without fruits often produce more shoots and leaves, so that self-shading within the leaf canopy becomes more accentuated in the absence of a crop. This alteration in microclimate could then limit photosynthesis due to reduced light intensity, especially in plants like strawberry, with leaves. The different degree of self-shading of leaves in fruiting and non-fruiting plants could be an explanation for different peaks of photosynthetic rates. In the experiments with citrus and eggplants however, self-shading was avoided. Despite this, leaves of fruiting plants produced more dry matter, *i.e.*, they had higher photosynthetic rates. Also measurements at light saturation in single leaves of apples showed higher photosynthetic rates as the result of fruit load (Kazaryan *et al.*, 1965; Hansen, 1970).

REGULATORY MECHANISMS

When considering how leaf photosynthesis might be coupled to fruit growth and development the question is whether the photosynthetic rate depends on current 'demand' for assimilate or whether a more direct form of regulation is involved. So far it appears that photosynthetic efficiency of leaves in fruiting and non-fruiting plants is probably regulated by several mechanisms.

Hormonal influences

As fruits can have relatively high concentrations of phytohormones (Crane, 1969; Luckwill, 1970; Nitsch, 1970) it has been assumed that hormones deriving from the fruits perhaps regulate photosynthetic rates by directly activating RudP-carboxylase (Wareing, 1968).

Some experiments have shown that auxins, cytokinins and gibberellins can stimulate photosynthetic rates of leaves (Kull, 1970). The amount and activity of RudP-carboxylase in leaves could be enhanced by gibberellins (Treharne *et al.*, 1970; Broughton *et al.*, 1970; Huber & Sankha, 1973). Tamas *et al.* (1972) showed that indole-3yl-acetic acid (IAA) can increase photosynthesis of chloroplasts by enhancement of photophosphorylation. Furthermore Hoad *et al.* (1977) using grape vines found that changes in gibberellin and cytokinin levels do occur when sink strength varies and that the rate of photosynthesis alters as a result of the observed changes in hormone levels. Inhibitors of plant growth such as abscisic acid and phaseic acid can also be involved in such source – sink feedback effects as shown by Kriedemann *et al.* (1976). Investigations with *Citrus madurensis* revealed that leaves of fruiting plants had a 70% increase in specific activity of RudP-carboxylase (Tung *et al.*, 1973).

It remains however to be seen whether phytohormones do in fact migrate from 'sink' to 'source' and so fulfil regulating functions on the photosynthetic enzymes. On the other hand assimilate turnover and transport could be affected by phytohormones (Bidwell, 1973; Bidwell & Chiu-Kwong Quong, 1975) thus regulating photosynthesis. It has been indicated that hormones can stimulate transport from 'source' to 'sink' by increasing metabolic activity of fruits (Booth *et al.*, 1962) or by affecting the transport mechanisms (Davis & Wareing, 1965; Mullins, 1970). This last possibility is supported by hormonal effects on the formation of transport tissues, sieve-tube loading and unloading as well as longitudinal transfer within the sieve-tubes (Patrick, 1976). For example indole-3yl-acetic acid (IAA) not only alters permeability of membranes (Polevoi & Salamatova, 1974) but also can promote differentiation of phloem tissue (Sachs, 1975). Finally photosynthesis can be influenced to some degree by both stomatal and internal restrictions on CO_2 movement from the air around the leaf to photosynthetic sites within the leaf. Hormonal effects can be expressed at both locations (Loveys & Kriedemann, 1974).

Assimilate concentration

Fruits not only effect photosynthetic rates but also assimilate concentrations in the leaves. Plants bearing fruits have lower assimilate concentrations in the leaves than non-fruiting ones (Hoffmann, 1973; Lenz & Küntzel, 1974: Quast, 1975). Causal connections between photosynthetic rates and leaf assimilate concentrations have therefore been suggested (Neales & Incoll, 1978). This assumes that an increase in concentration, particularly of sucrose, fructose and glucose, causes the decline of photosynthetic rates by end-product inhibition of enzyme activity. Perhaps an accelerated transfer of these carbohydrates from leaves to fruits could be a main reason for higher photosynthetic rates in fruiting as compared to non-fruiting plants.

Claussen (1975) in extensive investigations on photosynthetic rates and leaf contents of alcohol soluble sugars of *S. melongena* showed that sucrose, fructose, and glucose in concentrations occurring in leaves do not affect photosynthetic enzymes or photosynthetic rates. Similar results for other plants were reported by Bucke *et al.* (1966); Bowes & Ogren (1972), Buchmann & Schürmann (1973).

It cannot be excluded however that other metabolites may have a negative feedback effect on photosynthetic enzymes. For example accumulation of 6-P-gluconate or fructose-1,6-diphosphate could strongly inhibit RudP-carboxylase activity (Kelly *et al.*, 1976; Kluge, 1977) and ribulose-5-phosphate can regulate activity of RudP-carboxylase within a very small concentration range. This way fine regulation of photosynthesis would be possible.

Furthermore, an accumulation of starch in the chloroplasts is associated with reduced rates of photosynthesis (Ito, 1971). Hoffman's (1973) examination of chloroplasts from leaves on non-fruiting eggplants does reveal severe disruption of chloroplast fine structure compared to the situation for fruiting plants, and it seems quite remarkable that such plastids are still capable of sustaining further photosynthesis and starch accretion. Deterioration in both photochemical and biochemical properties of such plastids (Wildman 1967) would be expected.

Although starch does not seem to be involved directly in cell metabolism it appears possible that coupling of chloroplast enzymes to starch could cause reduction of their activity as well as changes of their substrate affinity and specifity as assumed for starch-synthetase (Meisel, 1974). Moreover starch binds Mg-ions and by this changes in Mg-concentrations and pH values in the stroma occur. As Mg-ions have importance for the regulation of enzyme activities, their binding onto starch could have a regulating effect on photosynthesis (Wishnick & Lane, 1970; Klemme & Jacobi, 1974). Despite this Claussen (1975) after numerous investigations of leaves in fruiting and non-fruiting eggplants found that starch effects on photosynthesis if they occurred at all, would be very small.

Photorespiration

An accumulation of intermediates of the Calvin cycle in particular of ribulose-5-phosphate, and ribulose-1,5-diphosphate could increase photorespiration and by this affect photosynthetic efficiency (Zelitsch, 1975).

In fact several experiments with *C. madurensis* have shown that non-fruiting plants have higher photorespiration rates than fruiting ones. An example is demonstrated in Tables 1 and 2.

Table 1. Photosynthesis and photorespiration in *Citrus madurensis* dependent upon fruit load*.

	without fruit	with fruit
Photosynthetic rate ($ml\ CO_2 \cdot dm^{-2} \cdot h^{-1}$) at 21% O_2	2.9	4.8
Photorespiration ($ml\ CO_2 \cdot dm^{-2} \cdot h^{-1}$)	2.6	1.6
Photorespiration in % of photosynthetic rate at 2% O_2	47.9	24.8

* Each value is the mean of 12 measurements on 3 plants.

Table 2. Photosynthetic rates and photorespiration as dependent upon fruit load and root temperature in *Citrus madurensis**

		12°C		28°C	
	Experiment	− Fruit	+ Fruit	− Fruit	+ Fruit
Photosynthetic rate	II	1.9	6.3	5.8	6.9
($ml\ CO_2 \cdot dm^{-2} \cdot h^{-1}$)	III	2.1	6.4	5.8	8.1
Photorespiration rate	II	2.2	2.7	3.5	2.9
($ml\ CO_2 \cdot dm^{-2} \cdot h^{-1}$)	III	1.8	2.1	4.2	2.4
Photorespiration in %	II	53.6	30.0	37.6	29.6
of photosynthetic rate at 2% O_2	III	46.2	24.7	42.0	22.8

* Each value is the mean of 12 measurements on 8 plants.

The measurements were carried out when fruits showed highest growth rates. During the measurements whole plants were enclosed in a hermetically sealed chamber (1 m^3) with close control over CO_2 and O_2 concentration, air temperature (20° C) and humidity (RH 70%). Light intensity at plant level was 40 klx (Mercury halide lamps). CO_2 uptake under illumination was monitored by automatic compensation at 400 vpm CO_2 (see Daunicht, 1970, for details on method). Photorespiration rates were determined by two hour measurements of photosynthetic rates at 21% O_2 and then at 2% O_2. The differences in CO_2 uptake at the two O_2-levels are caused largely by photorespiration and therefore designated as such.

Data from these experiments are shown in Table 2. Again it is obvious that leaves of fruiting plants had higher photosynthetic rates than non-fruiting ones. This was particularly pronounced at low root temperatures. At the higher root temperatures non-fruiting plants had considerably higher photorespiration rates. At low root temperatures photorespiration in non-fruiting plants was somewhat lower than in fruiting ones, probably due to the very low photosynthetic rates.

However in non-fruiting plants a considerably higher portion of CO_2 was evolved by photorespiration as compared with fruiting ones.

As shown in apples (Hansen, 1971), vine (Loveys & Kriedemann, 1974) and soybeans (Lenz & Williams, 1973; Woodward & Rawson 1976) leaves of fruiting plants often have lower gaseous diffusive resistances than non-fruiting ones. In soybeans stomates of leaves in fruiting plants remain wider and longer open during the light period (Lenz & Williams, 1973). That such stomatal reactions could be hormonally controlled has been suggested by Loveys & Kriedemann, 1974. They found close relationships between stomatal resistance of leaves, and the abscisic acid and phaseic acid contents. Also of course differences in assimilate concentration in the leaves of fruiting and non-fruiting plants could be reasons for the different stomatal behaviour.

Possibly accumulation of assimilates found mainly in leaves of non-fruiting plants, other resistances along the CO_2 diffusion pathways (cell walls, membranes and cytoplasma *etc.*) could affect CO_2 concentrations of the intercellular spaces and by this stomata movement.

If by accumulation of metabolites, photorespiration is increased, rising CO_2 concentrations in the intercellular spaces can be expected. As stomata react sensitively towards intercellular CO_2 levels (Raschke, 1976) closure of stomata must be assumed. In fact in recent investigations in fruiting and non-fruiting citrus plants using a continuously recording diffusive porometer (Block & Sterzelmeyer, 1978) it was found that leaves of non-fruiting plants stomata open after the beginning of the light period. Possibly due to accumulation of intercellular CO_2 however, they partly close and after oscillatory movement remain nearly closed for the rest of the light period (Fig. 2).

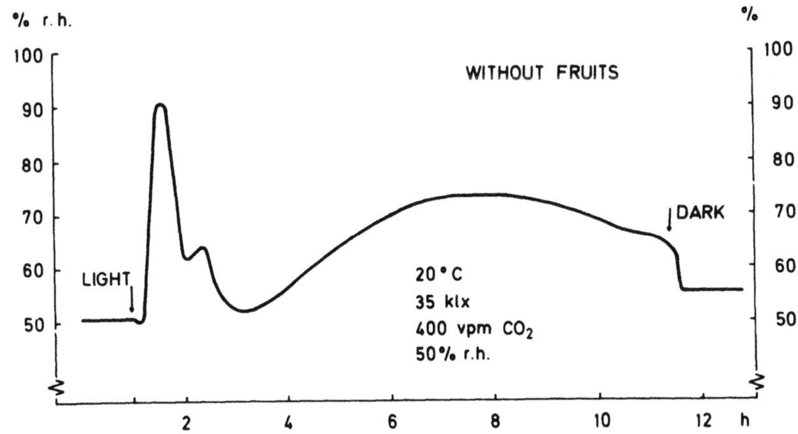

Fig. 2. Fine measurements on leaf transpiration of a non-fruiting plant in *Citrus madurensis* as an indicator for stomata movement.

This also reflects the course of photosynthetic rates observed in the leaves of non-fruiting plants (Claussen, 1975). In contrast leaves of fruiting plants with lower photorespiration rates and probably intensified CO_2-uptake by chloroplasts have lower intercellular CO_2 concentrations. Therefore, as shown in Fig. 3 stomata open at the beginning of the light period and remain open until darkness even during prolonged light periods.

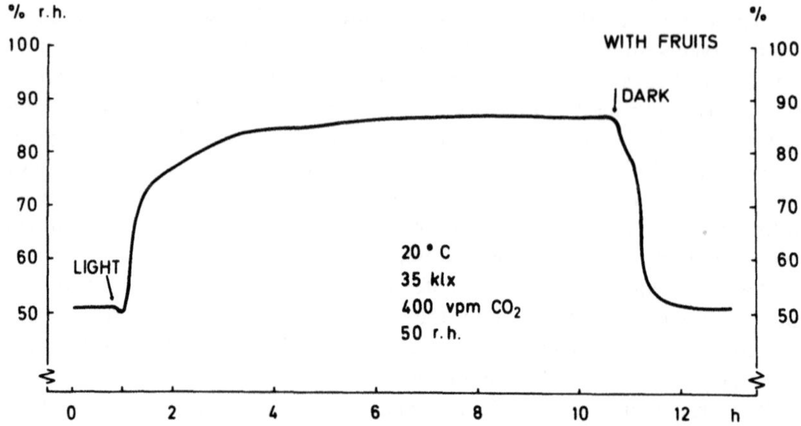

Fig. 3. Effect of fruit load on leaf transpiration, *i.e.*, stomata movement in *Citrus madurensis*.

CONCLUSION

Numerous investigations have shown that the photosynthetic efficiency of leaves in a plant can be increased by fruits. Possibly the rate of photosynthesis can be regulated by several mechanisms some of which were discussed in the paper. If such mechanisms were damaged others could replace their function. Therefore the risk of leaf damage by uncontrolled CO_2 uptake or assimilate accumulation is small. This unfortunately aggravates any attempt to improve photosynthetic efficiency by manipulating the regulatory mechanisms.

On the other hand it is possible by inhibition of photorespiration to increase the photosynthetic rates for several hours. Our investigations on *S. melongena* have shown however that the photosynthetic rate then declined steeply, probably as the result of metabolite accumulation in the leaf. For continuous increase of photosynthetic efficiency the efficiency of assimilate transport from leaves to fruits *i.e.* 'sinks' would also have to be increased.

REFERENCES

Avery, D.J. Comparisons of fruiting and deblossomed maiden apple trees and of non-fruiting trees on a dwarfing and an invigorating rootstock. *New Phytol.* 68: *323–336* (1969).

Bidwell, R.G.S. & E. Chiu Kwong Quong. Indoleacetic acid effect on the distribution of photosynthetically fixed carbon in the bean plant. *Biochem. Physiol. Pflanzen* 168: *361–370* (1975).

Bidwell, R.G.S. A possible mechanism for the control of photoassimilate translocation. p. 17–30 in: Trans. 3rd Symp. on Accumulation and Translocation of Regulators in Plant Organisms. Warszawa, 3 (1973).

Buchanan, B.B. & P. Schürmann. Regulation of ribulose-1,5-diphosphate carboxylase in the photosynthetic assimilation of carbon dioxide. *J. Biol. Chem.* 248: *4956–4964* (1973).

Bucke, C.D., D.A. Walker & S.W. Baldry. Some effects of sugars and sugar phosphates on carbon dioxide fixations by isolated chloroplasts. *Biochem. J.* 101: *636–641* (1966).
Bünemann, G. & A. Grassia. Growth and Mineral distribution in grafted Tomato/Potato Plants according to sink number. *Scientia Horticulturae* 1, *13–24* (1973).
Booth, A., J. Moorby, C.R. Davies, H. Jones & P.F. Wareing. Effects of indolyl-3-acetic acid on the movement of nutrients within plants. *Nature* (London) 194: *204–205* (1962).
Bowes, G. & W.L. Ogren. Oxygen inhibition and other properties of soybean ribulose-1,5-diphosphate carboxylase. *J. Biol. Chem.* 217: *2171–2176* (1972).
Broughton, W.J., E.O. Hellmuth & D. Yeung. Role of glucose in development of the gibberellin response in peas. *Biochim. Biophys. Acta* 222, *491* (1970).
Cary, P.R. Growth, yield and fruit composition of 'Washington Navel' orange cuttings as affected by root temperature, nutrient supply and crop load. *Hort. Res.* 10: *20–33* (1970).
Claussen, W. Untersuchungen über den Einfluß der Frucht auf die Netto-Photosyntheseraten und den Saccharose-und Stärkestoffwechsel der Blätter und Wurzeln von Auberginen (*Solanum melongena* L.). Diss. Fachbereich Internat. Agrarentwicklung der T.U. Berlin (1975).
Claussen, W. Einfluß der Frucht auf Netto-Assimilationsleistung und Netto-Photosyntheseraten der Aubergine (*Solanum melongena* L.). *Gartenbauwiss.* 42: *61–65* (1977).
Crane, J.C. The role of hormones in fruit set and development. *Hort. Science* 4: *108–111* (1969).
Daunicht, H.J. Ein Verfahren zur exakten automatischen Photosynthesekompensation. *Ber. Dt. Bot. Ges.* 83: *499* (1970).
Davies, C.R. & P.F. Wareing. Auxin directed transport of radiophosphorus in stems. *Planta* (Berl.), 65: *139–156* (1965).
Hansen, P. ^{14}C-studies on apple trees. VI. The influence of the fruit on the photosynthesis of the leaves, and the relative photosynthetic yields of fruits and leaves. *Physiol. Plant.* 20: *382–391* (1970).
Hansen, P. The effect of fruiting upon transpiration rate and stomatal opening in apple leaves. *Physiol. Plant.* 24: *181–183* (1971).
Hansen, P. The effect of cropping on the distribution of growth in apple trees. *Tidssk. Planteavl.* 75: *119–127* (1971).
Hoad, G.V., B.R. Loveys & K.G.M. Skene. The Effect of Fruit-removal on Cytokinins and Gibberellin-like Substances in Grape Leaves. *Planta* (Berl.) 136: *25–30* (1977).
Hoffmann, E. Der Einfluß der Früchte auf die Photosyntheseraten und die Assimilatverteilung bei fruchttragenden und nichtfruchttragenden Auberginen und Erdbeerpflanzen. Diss. T.U. Berlin (1973).
Huber, W. & N. Sankhla. Effect of gibberellic acid on the activities of photosynthetic enzymes and $^{14}CO_2$-fixation products in leaves of *Pennisetum typhoides* seedlings. *Z. Pflanzenphysiol.* 71: *275–280* (1973).
Ito, T. *Engei Gakkai Zasshi* 40: *41–47* (1971).
Kazaryan, M.O., N.K. Balagezjan & K.A. Karapetjan. Effect of fruits on the physiological activity of apple leaves. (Russ.). *Fiziol. Rast.* 12: *313–319* (1965). Ref. Landw. Zbb. 10, 3116 (1966).
Kelly, G.J. & E. Latzko. Regulatory Aspects of Photosynthetic Carbon Metabolism. *Ann. Rev. Plant Physiol.* 27: *181–205* (1976).
Klemme, B. & G. Jacobi. Der Einfluß von Stärke auf die Aktivität der Pyrophosphatase aus isolierten Chloroplasten. *Planta* (Berl.), 120: *155–162* (1974).
Kluge, M. Regulation of Carbon Dioxide Fixation in Plants, pp. 155–175, in: Jennings, D.H. (ed.), Integration of Activity in the Higher Plant. Cambridge University Press (1977).
Kriedemann, P.E. Observations on gas exchange in the developing Sultana berry. *Aust. J. Biol. Sci.* 21: *907–916* (1968).
Kriedemann, P.E., B.R. Loveys, J.V. Possingham & M. Satoh. Sink Effects on Stomatal Physiology and Photosynthesis. pp. 401–414, in: Wardlaw, I.F. & J.B. Passioura (eds.), Transport and Transfer Processes in Plants. Academic Press, New York (1976).

Kull. U. Wirkungen von Wuchsstoffen auf Speicherung und Stoffwechsel in vegetativen Pflanzenteilen (unter besonderer Berucksichtigung des Kohlenhydrathaushaltes). Ser. Bot. Studien Bd. 19. VEB. Gustav Fischer Verlag, Jena (1972).
Laval-Martin, D., J. Farineau & J. Diamond. Light versus Dark Carbon Metabolism in Cherry Tomato Fruits. I. Occurrence of Photosynthesis. Study of the Intermediates. *Plant Physiol.* 60: *872–876* (1977).
Lenz, F. Relationship between the vegetative and reproductive growth of Washington Navel orange cuttings (*Citrus sinensis* L. Osbeck). *J. Hort. Sci.* 42: *31–39* (1967).
Lenz, F. & G. Bünemann. Beziehungen zwischen dem vegetativen und reproduktiven Wachstum in Erdbeeren (Var. *Senga Sengana*). *Gartenbauwiss.* 32: *227–236* (1967).
Lenz, F. & H.J. Daunicht. Einfluß von Wurzel und Frucht auf die Photosynthese bei Citrus. *Angew. Bot.* 35: *11–20* (1971).
Lenz, F. & C.N. Williams. Effect of fruit removal on net assimilation and gaseous diffusive resistance of soybean leaves. *Angew. Bot.* 47: *57–63* (1973).
Lenz, F. & U. Küntzel. Carbohydrate content of citrus leaves as affected by fruit load. *Gartenbauwiss.* 39: *99–101* (1974).
Leonard, E.R. Interrelation of vegetative and reproductive growth with special reference to indeterminate plant. *Bot. Rev.* 28: *253–410* (1962).
Loveys, B.R. & P.E. Kriedemann. Internal control of stomatal physiology and photosynthesis. I. Stomatal regulation and associated changes in endogenous levels of abscisic and phaseic acids. *Aust. J. Plant Physiol.* 1: *407–415* (1974).
Luckwill, L.C. The control of growth and fruitfulness of apple trees. pp. 237–253, in: Luckwill, L.C. & C.V. Cutting (eds.), The Physiology of Tree Crops. Academic Press Inc., London, New York (1970).
Maggs, D.H. The reduction in growth of apple trees brought about by fruiting. *J. Hort. Sci.* 38: *85–94* (1963).
Meisel, P. Die Biosynthese der Stärke. Handbuch der Stärke VI – 4, Paul Parey, Berlin und Hamburg (1974).
Mullins, M.G. Hormone-directed transport of assimilates in decapitated internodes of *Phaseolus vulgaris* L. *Ann. Bot.* 34: *897–909* (1970).
Neales, T.F. & L.D. Incoll. The control of leaf photosynthesis rate by the level of assimilate concentration in the leaf: A review of the hypothesis. *Bot. Rev.* 34: *107–125* (1968).
Nitsch, J.P. Hormonal factors in growth and development, pp. 428–464, in: Hulme, A.C. (ed.), The Biochemistry of Fruits and their Products. Vol. 1. Academic Press, London, New York (1970).
Patrick, J.W. Hormone-directed Transport of Metabolites. pp. 433-446, in: Wardlaw, I.F. J.B. Passioura (eds.), Transport and Transfer Processes in Plants. Academic Press, New York (1976).
Polevoi, V.V. & T.S. Salamatova. The mechanism of auxin action on membrane transport of hydrogen ions. Translated from *Fiziol. Rast.* 22: *519–526* (1974).
Quast, P. Gibberellinbestimmung in Verbindung mit Kohlenhydratgehalten und Trockensubstanzverteilung bei Solanaceen. Diss. T.U. Berlin (1975).
Rascke, K. Transfer of Ions and Products of Photosynthesis to Guard Cells, pp. 203–215, in: Wardlaw, I.F. & J.B. Passioura (eds.), Academic Press, New York (1976).
Sachs, T. The induction of transport channels by auxin. *Planta* (Berl.), 127: *201–206* (1975).
Smith, P.F. Collapse of 'Murott' tangerine trees. *J. Amer. Hort. Sci.* 101: *23–25* (1976).
Tamas, I.A., B.D. Atkins, S.M. Ware & R.G.S. Bidwell. Indoleacetic acid stimulation of phosphorylation and bicarbonate fixation by chloroplast preparation in light. *Can. J. Botany* 50: *1523–1527* (1972).
Treharne, K.J. & J.L. Stoddart. Effects of gibberellin and cytokinins on the activity of photosynthetic enzyme and plastid ribosomal RNA synthesis in *Phaseolus vulgaris* L. *Nature* 228: *129–131* (1970).
Tung, H.F., W.J. Broughton & F. Lenz. Effects of fruit on ribulose diphosphate carboxylase activity in citrus madurensis leaves. *Experientia* 29: *271* (1973).
Wardlaw, I.F. The Control and Pattern of Movement of Carbohydrates in Plants. *Bot. Rev.* 34: *70–105* (1968).

Wareing, P.F. The physiology of the whole tree. Rep. E. Malling Res. Stn. for 1967, *55–68* (1968).
Wildman, S.G. The organization of grana containing chloroplasts in relation to location of some enzymatic systems concerned with photosynthesis, protein synthesis and ribonucleic acid synthesis. pp. 295–319, in: Goodwin, T.W. (ed.), Biochemistry of Chloroplasts, Vol 2. Academic Press, London and New York (1967).
Wishnick, M. & M.D. Lane. The interaction of metal ions with ribulose-1,5-diphosphate carboxylase from spinach. *J. Biol. Chem.* 245, *4939–4947* (1970).
Woodward, R.G. & H.W. Rawson. Photosynthesis and transpiration in dicotyledonous plants. II. Expanding and senescing leaves of soybean. *Aust. J. Plant Physiol.* 3, *257–267* (1976).
Zelitsch, J. Improving the efficiency of photosynthesis. *Science* 188, *626–633* (1975).

Author's address:
Institut für Obstbau und Gemüsebau
der Universität Bonn
Auf dem Hügel 6
5300 Bonn
Federal Republic of Germany.

SESSION V

PHOTOSYNTHESIS AND NITROGEN METABOLISM

EFFECT OF LIGHT INTENSITY ON NET ASSIMILATION RATES OF NITRATE-SUPPLIED OR NITROGEN-FIXING LEGUMES*

W.J. BROUGHTON

(Kuala Lumpur, Malaysia and Bonn, Fed. Rep. of Germany)

ABSTRACT

Two legumes, *Glycine max* L. Merr., and *Vigna unguiculata* L. Walp. were grown in pots under conditions in which the nitrogen supply was solely mineral (NH_4NO_3 with *G. max* and KNO_3 with *V. unguiculata*) or atmospheric via *Rhizobium* induced nodules. The normally encountered difficulties of comparing the rates of growth of the differently treated plants were eliminated by measuring their response to varying light intensity. To do this, the plants were raised under different levels of overhead sunlight, harvested periodically, and subjected to classical plant growth analysis. Both dinitrogen-fixing and nitrate-assimilating plants responded in identical ways to varying light intensity. These data show that within the limits of experimental error (minimum 8.8%), the energy costs of a legume assimilating either gaseous or mineral nitrogen are not different, or that if they are, either one of the treatments produces plants which are inherently more photosynthetically efficient. Partitioning of the relative growth rate within the plants during the first weeks after imposition of the different treatments suggests that most of the nitrate-reduction occurs in the leaves.

INTRODUCTION

Given satisfactory agronomic practices, seed yields of *Cajanus cajan*, *Glycine max*, *Vicia faba* and *Vigna unguiculata* are but little improved by application of nitrogenous fertilizers (Weber, 1966a; Hanway & Weber, 1971; Dart,1973; Hardy, Burns & Holsten, 1973; Rachie and Roberts, 1974; Olsen, Hamilton, and Elkins, 1975; Ham, Lawn & Brun, 1976; Havelka & Hardy, 1976; Summerfield *et al.*, 1977). This is true for any level or type of fertilizer normally applied, as well for plants grown under both glass-house and field conditions. Perhaps the most convincing observations of this type are those of Weber (1966a) who was unable to demonstrate yield increases of non-nodulating versus nodulating *G. max* plants, despite application of up to two tonnes of ammonium nitrate per hectare.

This equality between seed yield of nitrogen-supplied or nodulated legumes also extends to other parameters of growth. Weber (1966b) showed that the nitrogen contents of *G. max* plants supplied with nitrogen in the two different

* I wish to thank T. Puskaran for his excellent technical assistance, Prof. F. Lenz for his encouragement, and the Alexander von Humbolt Stiftung for a fellowship. Financial assistance was provided by the International Foundation for Science (Stockholm) and the University of Malaya.

ways were not very different. Gibson (1966), and Broughton *et al.* (1978) observed comparable rates of vegetative growth in nitrate-supplied or nodulated plants of *Trifolium subterraneum* and *Centrosema pubescens* and *V. unguiculata* respectively.

Minchin & Pate's (1973) detailed carbon budgets in *Pisum sativum* confirmed these observations showing that 5.9 mg of carbon was respired for each mg of nitrogen fixed in nodulated plants. The corresponding figure for nitrate supplied plants was 6.2 mg of carbon respired for each mg of nitrate-nitrogen reduced. Thus, it has been concluded that the energy costs to a plant assimilating dinitrogen or nitrate are not significantly different (see Gibson, 1966; 1974; 1979).

Despite the above, numerous reports showing poorer growth of nodulated- as opposed to ammonia- or nitrate-assimilating plants exist (*e.g.* Bouma, 1970; Kunelius & Clarke, 1970). Whilst it is easier to produce healthy plants with mineral fertilizers than it is to effectively husband the *Rhizobium*-legume symbiosis, these contradictory reports are worrying. For this reason I decided to investigate growth in relation to photosynthesis in nitrate-supplied or nitrogen-fixing *G. max* and *V. unguiculata* plants. The result of these investigations are reported below.

MATERIALS AND METHODS

Seeds of *Glycine max* L. Merr. *cv* Palmetto were obtained from the Rubber Research Institute of Malaysia, Kuala Lumpur, while seeds of *Vigna unguiculata* L. Walp cv China were purchased from the Soon Huat Seeds Co., 330 Beach Street, Pulau Pinang. The rhizobial strains used (31/b 143 with *G. max* and UMKL 76 with *V. unguiculata*) were obtained from the University of Malaya/Rubber Research Institute of Malaysia Rhizobial Culture Collection (see Broughton & John, 1979). For innoculation purposes each strain was raised to 10^9 cells/g in the coir dust-soil compost of Faizah, Broughton, & John (1979).

Growth of the plants

Seeds were sterilized by immersion in dry concentrated sulphuric acid for one min., then washed with copious quantities of sterile water. In preliminary experiments to ascertain the optimal levels and form of mineral nitrogen for growth of these two plants, five sterilized seeds were planted directly into each bag. Imposition of the various treatments (listed in the legends to Figs. 1 and 2) was initiated immediately.

In the main experiments, the surface-sterilized seeds were pre-germinated on water agar (15 g agar$\cdot l^{-1}$ tap water) for one or two days, and transplanted into plastic bags containing 700 ml of washed, sterile river sand (five seeds per bag thinned out to three uniform plants when treatments were initiated). Until the nitrogen treatments were imposed, all plants were watered daily with 100 ml of the nutrient solution of Broughton & Dilworth (1971) (see Broughton & John, 1979). Twelve days after transplanting the *G. Max* seedlings, they were split into two different groups. One group was then watered daily with 100 ml of 2.5 mM NH_4NO_3 dissolved in the nutrient solution (and treated with 0.5 to 1.0 g of sterile compost per bag), while the other was inoculated with 0.5 to 1.0 g per bag of compost containing 31/b 143 (inoculation was repeated seven days later). Inoculated plants continued to receive 100 ml of the nutrient solution daily.

Seven days after *V. unguiculata* seeds germinated, the following treatments were imposed: half the plants were watered daily with 100 ml of 10 mM KNO_3 (and treated with 0.5 to 1.0 g of sterile compost per bag), while the other half was inoculated with 0.5 to 1.0 g compost containing UMKL 76, and watered daily with 100 ml of the nutrient solution. Plants in the preliminary experiments were raised on benches in the glasshouse, while those in the other experiments were also subjected to different light regimes. These were affected by placing the plants in large wooden boxes (lined inside with reflective material) covered with zero, one, two, or four layers of green mosquito-proof mesh, providing (at the soil level in the bags) 41, 25, 16, and 6% of daylight respectively. Temperature within the glasshouse varied from $24 \pm 1°$ C (night) to $32 \pm 1°$ C (day), light intensity (on the open benches) was restricted to a maximum of 1,400 μ Einsteins m^{-2} sec^{-1} with a twelve hour day, and the relative humidity varied from $83 \pm 3\%$ (night) to $62 \pm 6\%$ (day).

Five replicate bags (five seeds per bag) were used for each treatment in the preliminary experiments, while three replicate bags (three seedlings per bag) were used in the main experiments. Each of the main experiments was repeated several times, and results are reported from each of these.

Harvest

In the preliminary experiments, germination was assessed two weeks after planting, while the plants were harvested two weeks later. In the main experiments, three bags (nine plants) were harvested at weekly intervals beginning 12 days after the imposition of treatment in *G. max* and 14 days after treatment was initiated in *V. unguiculata*. Parameters measured included fresh and dry weights (after oven drying at 100° C for 16 hr) of the various plant parts: roots, stems, and leaves), the number of nodules, and the leaf area (determined by means of an 'Automatic Area Meter' – Hayashi Denko Co., Ltd., Tokyo, Japan). All data were generated from determinations on individual plants.

RESULTS

Levels and sources of applied nitrogen

In preliminary experiments the effect of NH_4NO_3, $(NH_4)_2SO_4$, KNO_3, and urea on germination and growth of the plant were investigated.

Potassium nitrate was the best utilized of all the nitrogen sources by both plants (Fig. 1 & 2). *V. unguiculata* had a much greater tolerance for all forms of nitrogen applied than did *G. max*, even surviving concentrations as high as 160 mM. Functional, acetylene reducing nodules were observed in plants watered with KNO_3 concentrations as high as 40 mM (Fig. 2). Low concentrations of all the compounds supported growth of both plants, but high concentrations of urea and ammonium sulphate were especially toxic to *G. max* and *V. unguiculata* respectively. Total dry weight production was maximal in *G. max* plants watered

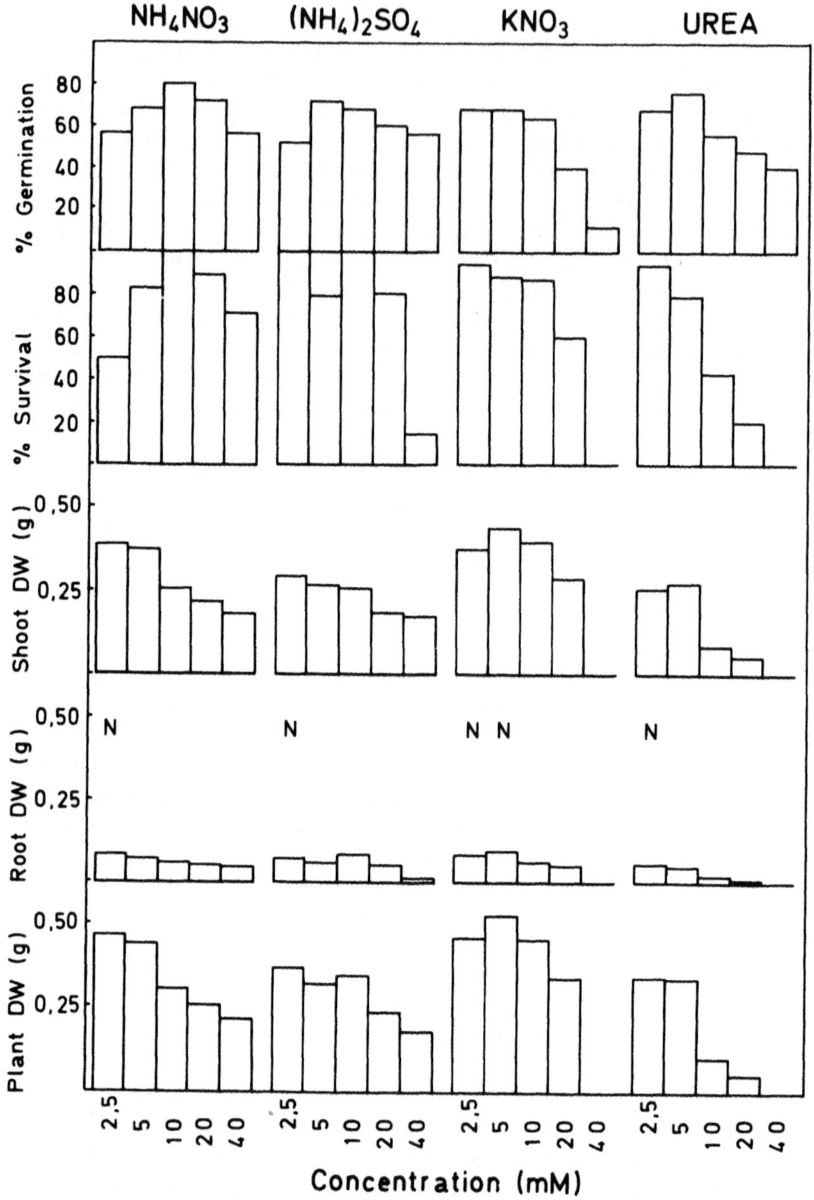

Fig. 1. Effect of NH_4NO_3, $(NH_4)_2SO_4$, KNO_3, and urea on germination and growth of *G. max*. Plants were raised in plastic bags containing sterile sand, and the various treatments imposed immediately after planting. All bags were inoculated with 0.5 to 1.0 g of compost containing 31/b 143, and the different nitrogen treatments were superimposed on top of this. These were applied daily in 100 ml portions made up in the nutrient solution of Broughton & Dilworth (1971). N – plants possessing nodules.

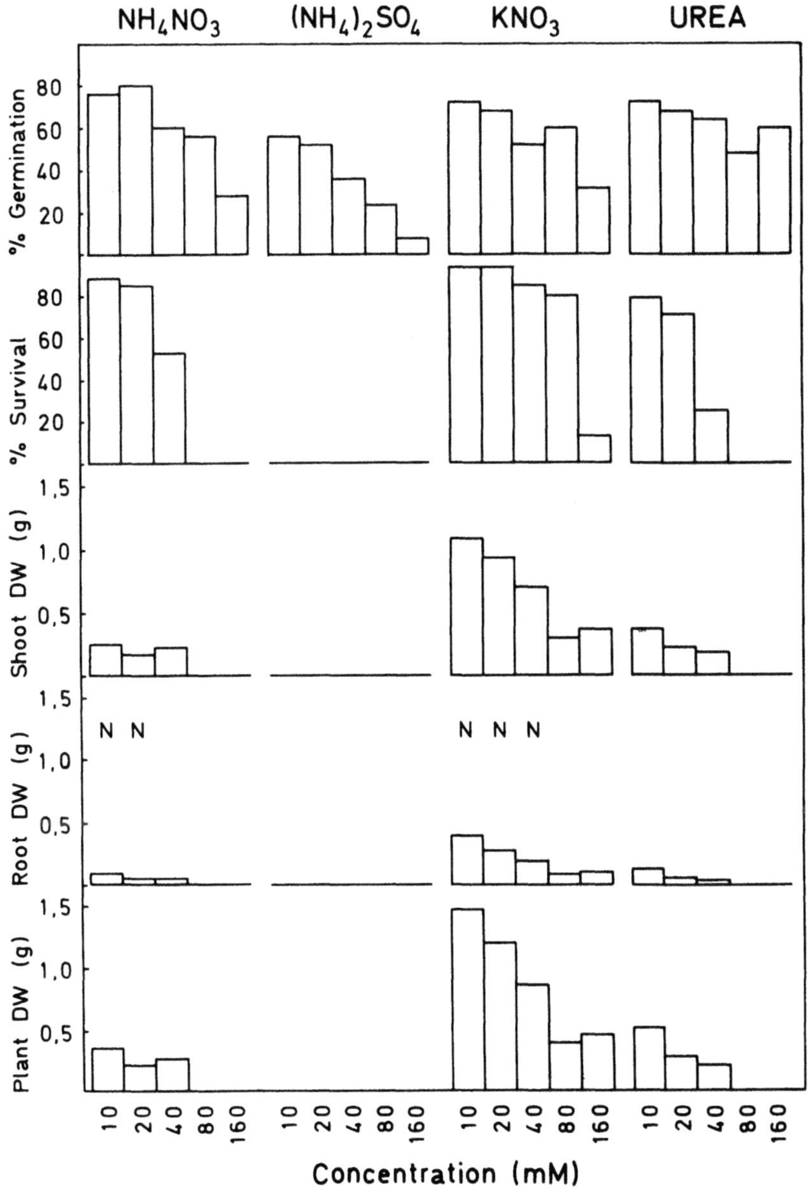

Fig. 2. Effect of NH_4NO_3, $(NH_4)_2SO_4$, KNO_3, and urea on germination of *V. unguiculata*. The experiment was similar to the one described in Fig. 1 except that the inoculum was UMKL 76.

with 5 mM KNO_3 while in *V. unguiculata* the optimum concentration was 10 mM. The next best *G. max* treatment was 2.5 mM NH_4NO_3 (Fig. 1). As there was no significant difference between these top ranked treatments, it was decided to investigate the effect of this compound in *G. max*, and KNO_3 in *V. unguiculata*.

Plant growth

Prior to comparing the different treatments on growth, it was first necessary to establish that the plants used were both normal and healthy. *G. max* plants had higher shoot to root ratios, a greater percentage dry matter, and a slower initial rate of growth than *V. unguiculata* plants, but otherwise they were not very different. The mean maximum relative growth rate (RGR) obtained for *G. max* (107 $mg \cdot g^{-1} \cdot d^{-1}$) was 80 to 82% of that obtained for *Trifolium subterraneum* in the experiments of Blackman & Wilson (1951) and Gibson (1966), while the comparable figure for *V. unguiculata* (97 $mg \cdot g^{-1} \cdot d^{-1}$) is 75% of this. The RGR of both plants was almost one hundred per cent of that reported by Hodgson (1967) for *Vicia faba*. Whilst there are undoubtedly appreciable differences between the various species of legumes as well as strong environmental influences upon the RGR, another reason for the slightly diminished RGRs observed is undoubtedly due to the plants being inadequately spaced apart. This was necessitated by the limitations of space available within the glasshouse as well as by the size of boxes providing the shade. For this reason, all discussion in this communication will be confined to the early phases of growth, before mutual shading of the leaves became a serious problem.

As both nitrate-supplied and nodulated plants grown under forty-one percent sunlight showed no signs of nutrient deficiency, were pathogen, and insect free, and (where appropriate) were adequately nodulated, it was concluded that some exogenous factor(s) limited plant growth.

Growth under different light regimes

Both plants responded dramatically to light (Table 1). The response was linear up until 25% daylight when it began to level off, reaching a maximum

Table 1. Effect of light intensity on dry weight accumulation in *G. max* and *V. unguiculata* plants.

Light intensity (% daylight)	Dry weight (g ± S.E.)			
	G. max[1]		*V. unguiculata*[2]	
	NO_3^-	Nod	NO_3^-	Nod
6	0.17 ± 0.02	0.22 ± 0.01	0.10 ± 0.02	0.15 ± 0.02
16	0.39 ± 0.04	0.55 ± 0.09	0.20 ± 0.04	0.30 ± 0.07
25	0.67 ± 0.05	0.67 ± 0.10	0.40 ± 0.07	0.52 ± 0.16
41	0.80 ± 0.01	0.77 ± 0.01	0.68 ± 0.04	0.75 ± 0.01

[1] Dry weight measured 26 d after inoculationwith 31/b 143.
[2] Dry weight measured 28 d after inoculation with UMKL 76.

somewhere just above 40% daylight. This is in complete agreement with gas-exchange studies (W.J. Broughton & F. Lenz, unpublished). Both plants in the nodulated condition grew better in limiting light intensities than those supplied with nitrate. These differences, however, vanished with increasing sunlight. Light had a qualitatively similar effect on nodulation (Table 2).

Table 2. Effect of light intensity on nodulation of *G. max* and *V. unguiculata* plants.

Light intensity (% daylight)	Number of nodules (± S.E.)			
	G. max[1]		*V. unguiculata*[2]	
	uninoculated	inoculated	uninoculated	inoculated
6	0	0.3 ± 0.7	0	0
16	0	6.3 ± 3.6	0	8.0 ± 0.6
25	0	9.3 ± 6.6	0	20.7 ± 2.0
41	0	22.4 ± 12.5	0	37.0 ± 7.3

[1] Nodulation assessed 31 d after inoculation with 31/b 143.
[2] Nodulation assessed 28 d after inoculation with UMKL 76.

Nodules were not observed in *V. unguiculata* plants grown under six per cent daylight, while some *G. max* plants possessed an occasional nodule. As plants raised under the six per cent of sunlight regime increased in dry weight (albeit slowly *vide infra*), it seems that photosynthate for nodule development is denied to the roots under conditions of severe carbohydrate deprivation. We may conclude therefore, that light is the factor limiting plant growth under the conditions of the experiment.

Growth analyses

A fundamental condition for integrating the equation which allows calculation of the net assimilation rate (NAR) is that the relationship between the mass and area of the assimilating apparatus is known (Šesták *et al.*, 1971). This is plotted for both plants grown under different light intensities in Fig. 3 and 4. Not only is the relationship between leaf area and total plant dry weight linear, but the slope of the relationship in plants supplied with nitrate is identical to that for nodulated plants. This is *prima facie* evidence that nitrate or dinitrogen uptake affects plant growth in similar ways. Or, in other words the energy requirements of dinitrogen- or nitrate- uptake are not significantly different in these plants within the limits of experimental error (8.8% in *G. max* and 9.5% in *V. unguiculata*).

On this basis, the NAR was calculated using the linear equation of Šesták *et al.* (1971) (number 10.12). These data are shown in Table 3 (for *G. max*) while the data for *V. unguiculata* are shown in Fig. 5 and Table 4. An ontogenic drift in both RGR and NAR occurred with development such that both parameters increased with increasing plant age during the duration of the experiment. Similarly, the light compensation point drifted downwards from about fifteen per cent of daylight twelve to fourteen days after inoculation to about five per

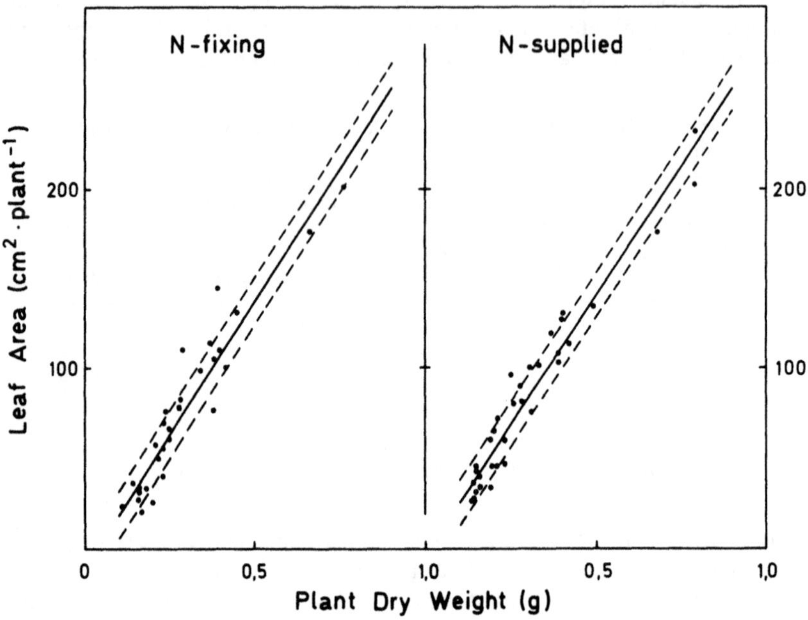

Fig. 3. Relationship between leaf area and total plant dry weight of nodulated or ammonium nitrate supplied *G. max* plants. The data are for all the plants from a particular experiment. Regression equations are: Nodulated – leaf area (cm^2) = 296.49 × dry weight (g) – 11.06 (r = 0.96, df = 34); nitrate supplied – leaf area (cm^2) = 287.63 × dry weight (g) – 1.99 (r = 0.97, df = 34). Therefore the leaf area ratio (LAR) of nodulated plants = 296.49 ± 13.03, and for nitrate supplied plants = 287.63 ± 11.79. Or, in other words, the differences between the treatments are not significant in terms of LAR.

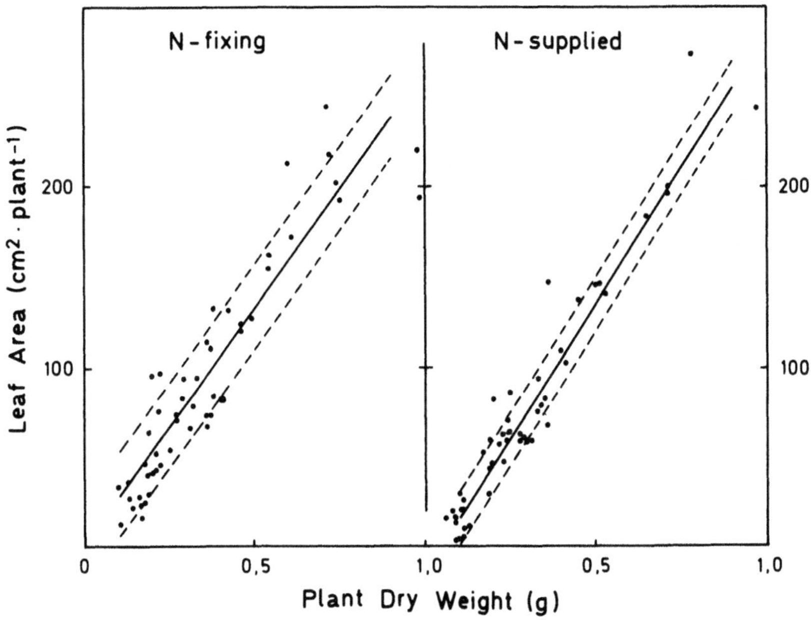

Fig. 4. Relationship between leaf area and total plant dry weight of nodulated or potassium nitrate supplied *V. unguiculata* plants. The data are for all the plants from a particular experiment. Regression equations are: nodulated – leaf area (cm^2) = 264.91 × dry weight (g) + 1.27 (r = 0.93, df = 46); nitrate supplied – leaf area (cm^2) = 299.2 × dry weight (g) – 13.91 (r = 0.97, df = 46). Therefore, the leaf area ratio (LAR) of nodulated plants = 264.91 ± 23.71 and for nitrate supplied plants = 299.22 ± 14.28. Or, in other words, the differences between the treatments are not significant in terms of LAR.

Table 3. Effect of light intensity and time on the relative growth rates (RGR) and net assimilation rates (NAR) of nitrate-supplied or nitrogen-fixing *G. max* plants.

Time period (d)	Light intensity (% daylight)	Treatment[1]	RGR ($mg \cdot g^{-1} \cdot d^{-1}$)	NAR ($mg \cdot dm^{-2} \cdot d^{-1}$)
12		Nod	19 ± 3	7.1 ± 1.3
to	25			
19		NO_3^-	24 ± 11	8.6 ± 2.4
12		Nod	53 ± 14	17.1 ± 2.5
to	41			
19		NO_3^-	66 ± 4	20.8 ± 1.6
19		Nod	95 ± 17	37.4 ± 7.9
to	41			
26		NO_3^-	90 ± 10	33.5 ± 4.3

[1] NO_3^- – plants watered daily with 2.5 mH NH_4NO_3.
Nod – plants inoculated with 31/b 143.

Table 4. Partitioning of the relative growth rate (RGR) during development of nitrate-supplied or nitrogen-fixing *V. unguiculata* plants. The data are for plants raised under 41% daylight.

Time period (d)	Treatment[1]	RGR ($mg \cdot g^{-1} \cdot d^{-1}$) (± S.E.)	Percent RGR in		
			Leaves	Stems	Roots
14	NO_3^-	23 ± 4	0	20	80
to					
21	Nod	30 ± 5	28	69	3
21	NO_3^-	64 ± 12	65	35	−3
to					
28	Nod	55 ± 6	67	33	−5

[1] NO_3^- – plants watered daily with 10 mM KNO_3.
Nod – plants inoculated with UMKL 76.

cent of daylight twenty-eight days later. This lower value is in agreement with that obtained by gas-exchange methods (W.J. Broughton & F. Lenz, unpublished). Thus it appears as if a gradual adaptation to low light intensities occurs, particularly in nodulated *V. unguiculata* plants. The net effect of this response to light is to raise both the RGR and NAR curves with respect to the light compensation point as time progresses, without changing the relative response to the form of applied nitrogen

Perhaps the most important observation of all however, is that there was no significant difference between the RGR and NAR of plants watered with nitrate or fixing nitrogen. This was true for both plants at all stages of growth up to one month after inoculation. It means of course that the response to light of these legumes assimilating nitrogen in the two different ways is identical.

An interesting sidelight to these observations concerns partitioning of the RGR in different parts of the plant. During the early establishment of an

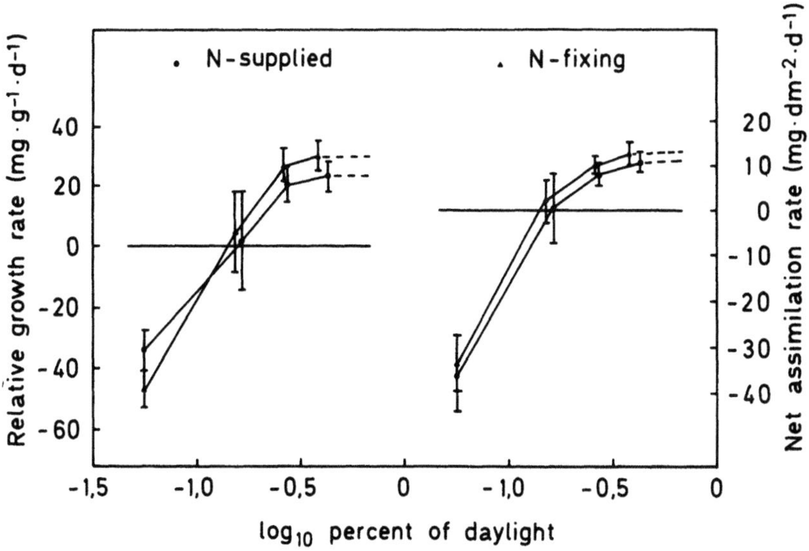

Fig. 5. Effect of light intensity on the relative growth rates (RGR) and net assimilation rates (NAR) of nitrate-supplied and nodulated *V. unguiculata* plants. The data are for the period 14 to 21 days after imposition of treatment and are shown with standard errors. N-supplied – plants watered daily with 10 mM KNO_3; N-fixing – plants inoculated with UMKL 76.

effective symbiosis, the roots of nodulated plants grow very slowly (Table 4). Much of the dry matter input during this period goes to the stems, with a smaller fraction going to the leaves. In nitrate-treated plants, the trend is reversed. Leaves do not grow. Four times more dry matter is accumulated in the roots than in the stems. At the next harvest, these differences between treatments had disappeared. Dry matter accumulation was confined to the shoot, with two-thirds of the gains occurring in the leaves and the rest in the stems. This equality between treatments persisted in the following harvests, although the relative distribution of the RGR between the various organs constantly changed.

DISCUSSION

Perhaps the greatest limitation in most experiments designed to compare the relative efficiencies of dinitrogen- and nitrate-uptake by legumes, is that one can never be sure that optimum conditions for both these processes were established. Certainly this criticism can be levelled at most of the work cited in the introduction. Great care was taken in the experiments described here to ensure equality between treatments. Four different nitrogen sources were evaluated in concentrations ranging from 2.5 mM to 160 mM, and the best treatments chosen for these experiments. Even then the method of application (100 ml daily) may have led to sub-optimal dry weight gains.

Similarly, much preliminary investigation went into establishing the best conditions for efficiently nodulating the host plants. *Rh. japonicum* strain 31/b 143 was selected from fifty other isolates capable of nodulating *G. max* in the Leonard jar trials of Faizah *et al.* (1979). A comparable investigation of inoculants for *V. unguiculata* was undertaken by Ikram & Broughton (1979) leading to the selection of UMKL 76. It could be argued however that these were not the best rhizobia available, that the inoculation delays (12 days for *G. max* and 7 days for *V. unguiculata*) were unsatisfactory (delayed inoculation stimulates nodulation – Nutman, 1949; Dart & Pate, 1959), or that the method of applying the inoculum was inferior.

It seemed to me that only two ways of overcoming such deficiencies in these experiments exist. One is to use non-nodulating isolines as employed by Weber (1966 a & b and Olsen *et al.*, 1975) but this approach carries the attendant danger of other (unseen) modifications to the plant. The other is to vary the energy input (light), and measure the response of nitrate-treated or nodulated-legumes. A comparable effect of light on both plants would be a direct indication of similar energy costs for the two processes. That this is indeed so in young *G. max* and *V. unguiculata* plants is the most important finding of this communication. It means either that there is no difference in the energy requirements for nitrate- or dinitrogen-assimilation, or that in one treatment the plants were inherently more photosynthetically efficient. Without detailed gas-exchange measurements there is no way to distinguish between these two possibilities.

Analysis of the plants' response to light was restricted to the first six weeks of growth for two reasons. First, it may take up to two weeks after inoculation to establish an effective symbiosis (see Broughton *et al.*, 1978) and there is obviously no point in analyzing the plants before a treatment effect was manifest. Second, inadequate spacing meant that mutual shading of one plant by another occurred as they grew larger, and this would have compounded these effects. Thus, by limiting the period of observation to the month following development of the first nodules, these disadvantages were eliminated. All inoculated plants grown under the higher light intensities possessed nodules by the first harvest, and there was no decline of the RGR during the period reported.

Fortuitously, observations on the partitioning of the RGR during development may have provided an insight into the nitrogen economy of these legumes. Some legumes, like *Lupinus albus* reduce most of the nitrate absorbed in the root system, while others like *T. subterraneum* reduce the majority of the absorbed nitrate in the leaves (Pate, 1973). As the theoretical free-energy relationships for both dinitrogen- and nitrate-reduction require relatively small (and almost equal) energy inputs as shown in equations one and two (from Delwiche, 1977), the site of reduction of these compounds would affect the organs' RGR.

Dinitrogen fixation

$$1/2\ N_2 + 3/2\ H_2O + H^+ \rightleftharpoons NH_4^+ + 3/4\ O_2 \qquad (1)$$
$$\Delta G'_{298}\ (\text{pH } 7) = +\ 75.73$$

Nitrate reduction

$$NO_3^- + 2\ H^+ + H_2O \rightleftharpoons NH_4^+ + 2\ O_2$$
$$\Delta G'_{298}\ (\text{pH } 7) = +\ 83.25 \qquad (2)$$

These data suggest therefore that most of the nitrate is reduced in the leaves of these legumes, while dinitrogen-reduction occurs in the roots.

Furthermore, the fact that peak nitrate reductase (EC 1.6.6.1.) activity does not coincide with that of nitrogenase (EC 1.7.99.2) activity in *G. max* (Harper & Hageman, 1972; Thibodeau & Jaworski, 1975) or *C. pubescens* (Broughton *et al.* 1978), that nitrate reductase activity displays substantial diurnal rhythms (*e.g.*, Nicholas, Harper, & Hageman, 1976), while the rhythm in nitrogenase activity is not so pronounced (Pate, 1976; Williams & Broughton, 1979), means that it is not possible to compare the efficiencies of dinitrogen- or nitrate-uptake by measuring carbon dioxide effluxes in the dark as Silsbury (1977) reported.

REFERENCES

Blackman, G.E. & G.L. Wilson. Physiological and ecological studies in the analysis of the plant environment. VII. An analysis of differential effects of light intensity on the net assimilation rate, leaf area ratio, and relative growth rate of different species. *Ann. Bot.* 15: *373–408* (1951).

Bouma, D. Effect of nitrogen nutrition on leaf expansion and photosynthesis of *Trifolium subterraneum* L. 2. Comparison between nodulated plants and plants supplied with combined nitrogen. *Ann. Bot.* 34: *1143–1153* (1970).

Broughton, W.J. & M.J. Dilworth. Control of leghaemoglobin synthesis in snake beans. *Biochem. J.* 125: *1075–1080* (1971).

Broughton, W.J., C.H. Hoh, C.A. Behm, & H.F. Tung. Development of the nitrogen fixing apparatus in the legumes, *Centrosema pubescens* Benth., and *Vigna unguiculata* L. Walp. *Planta* (Berl.) 139: *183–192* (1978).

Broughton W.J. & C.K. John. Rhizobia in tropical legumes. III. Experimentation and supply in Malaysia–1927–1976, in: Broughton, W.J., C.K. John, J.C. Rajarao & B. Lim (eds.), Soil microbiology and plant nutrition. University of Malaya Press, Kuala Lumpur (in press)

Dart, P.J. Root nodule symbiosis and tropical grain legume production. Proc. Intl. Inst. Tropical Agric. Grain Legume Improvement Workshop, Ibadan, Nigeria, pp. 185–197 (1973).

Dart, P.J. & J.S. Pate. Nodulation studies in legumes. III. The effects of delaying inoculation on the seedling symbiosis of barrel medic, *Medicago tribuloides* Desv. *Aust. J. Biol. Sci.* 12: *427–444* (1959).

Delwiche, C.C. Energy relations in the global nitrogen cycle. *AMBIO* 6: *106–111* (1977).

Faizah, A.W., W.J. Broughton & C.K. John. Rhizobia in tropical legumes. X. Growth in coir-dust soil compost. *Soil Biol. Biochem* (in press).

Faizah, A.W., S. Padmanabhan, W.J. Broughton & C.K. John. Rhizobia in tropical legumes. V. Problems involved in selecting inoculants for soyabeans, in: Broughton, W.J., C.K. John, J.C. Rajarao & B. Lim (eds.), Soil microbiology and plant nutrition, University of Malaya Press (in press).

Gibson, A.H. The carbohydrate requirements for symbiotoc nitrogen fixation: a 'wholeplant' growth analysis approach. *Aust. J. Biol. Sci.* 19: *499–515* (1966).

Gibson, A.H. The control of dinitrogen assimilation by nodulated legumes. pp. 13–22, in: Bieleski, R.L., A.R. Ferguson & M.M. Cresswell (eds.), Mechanisms of regulation of plant growth, Royal Society of New Zealand, Wellington (1974).

Gibson, A.H. The energy requirements of symbiotic nitrogen fixation, in: Broughton, W.J., C.K. John, J.C. Rajarao & B. Lim (eds.), Soil microbiology and plant nutrition, University of Malaya Press, Kuala Lumpur (in press).

Ham, G.E., R.J. Lawn & W.A. Brun. Influence of inoculation, nitrogen fertilisers and photosynthetic source-sink manipulations on field-grown soybeans, pp. 239–253, in: P.S. Nutman (ed.), Symbiotic nitrogen fixation in plants. Cambridge Univ. Press, Cambridge (1976).

Hanway, J.J. & C.R. Weber. Accumulation of N, P, and K by soyabean (*Glycine max* L. Merr.) plants. *Agron. J.* 63: *406–408* (1971).

Hardy, R.W.F., R.C. Burns & R.D. Holsten. Applications of the acetylene-ethylene assay for measurement of nitrogen fixation. *Soil Biol. Biochem.* 5: *47–81* (1973).

Harper, J.E. & R.H. Hageman. Canopy and seasonal profiles of nitrate reductase in soybeans. *Plant Physiol.* (Lancaster) 49: *146–154* (1972).

Havelka, U.D. & R.W.F. Hardy. Legume N_2 fixation a problem in carbon nutrition. pp. 456–475, in: Newton, W.E. & Nyman, C.J. (eds.), Proc. 1st Intl. Symp. Nitrogen Fixation, Washington State University Press, Pullman (1976).

Hodgson, G.L. Physiological and ecological studies in the analysis of the plant environment. XIII. A comparison of the effects of seasonal variations in light energy and temperature on the growth of *Helianthus annus* and *Vicia faba* in the vegetative phase. *Ann. Bot.* 31: *291–308* (1967).

Ikram, A. & W.J. Broughton. Rhizobia in tropical legumes. XVIII. Glasshouse and field observations with a variety of plants. *Plant & Soil* (in preparation).

Kunelius, H.T. & K.W. Clark. Influence of root temperature on the early growth and symbiotic nitrogen fixation of nodulated *Lotus corniculatus* plants. *Can. J. Plant Sci.* 59: *569–575* (1970).

Minchin, F.R. & J.S. Pate. The carbon balance of a legume and the functional economy of its root nodules. *J. exptl. Bot.* 24: *259–71* (1973).

Nicholas, J.C., J.E. Harper & R.H. Hageman. Nitrate reductase activity in soybeans (*Glycine max* L. Merr.). I. Effects of light and temperature. *Plant Physiol* (Lancaster) 58: *731–735* (1976).

Nutman, P.S. Physiological studies on nodule formation. II. Influence of delayed inoculation on the rate of nodulation in red clover. *Ann. Bot.* 13: *261–283* (1949).

Olsen, F.J., G. Hamilton & D.M. Elkins. Effect of nitrogen on nodulation and yield of soybean. *Exptl. Agric.* 11: *289–294* (1975).

Pate, J.S. Uptake, assimilation and transport of nitrogen compounds by plants. *Soil Biol. Biochem.* 5: *109–119* (1973).

Pate, J.S. Transport in symbiotic systems fixing nitrogen. pp. 278–303, in: Lüttge, U., & M.G. Pitman (eds.). Encyclopaedia of plant physiology. Transport in plants. vol. 2, part B. Tissues and organs. Springer Verlag, Berlin (1976).
Rachie, K.O. & L.M. Roberts. Grain legumes of the lowland tropics. *Adv. Agron.* 26: *1–132* (1974).
Šesták, Z., J. Čatský & P.G. Jarvis. Plant photosynthetic production. A manual of methods. pp. 343–391. Dr. W. Junk, The Hague (1971).
Silsbury, J.H. 1977. Energy requirement for symbiotic nitrogen fixation. *Nature* 267: *149–150*.
Summerfield, R.J., P.J. Dart, P.A. Huxley, A.R.J. Eaglesham, F.R. Minchin & J.M. Day. Nitrogen nutrition of cowpea (*Vigna unguiculata*). I. Effects of applied nitrogen and symbiotic nitrogen fixation on growth and seed yield. *Exptl. Agric.* 13: *129–142* (1977).
Thibodeau, P.S. & E.G. Jaworski. Patterns of nitrogen utilization in the soybean. *Planta* (Berl.) 127: *133–147* (1975).
Weber, C.R. Nodulating and non-nodulating soybean isolines. I. Agronomic and chemical attributes. *Agron. J.* 58: *43–46* (1969a).
Weber, C.R. Nodulating and non-nodulating soybean isolines. II. Response to applied nitrogen and modified soil conditions. *Agron. J.* 58: *46–49* (1966b).
Williams, W.M., & W.J. Broughton. Measurement of acetylene reduction in some tropical legumes. *Planta* (in preparation).

Author's address:
Department of Genetics and Cellular Biology
University of Malaya
Kuala Lumpur
Malaysia.

CONTROL OF NITRATE UPTAKE BY PHOTOSYNTHESIS IN *LEMNA PAUCICOSTATA* 6746

H. LÖPPERT & W. KRONBERGER*
(Vienna, Austria)

ABSTRACT

From the intracellular nitrate level and the negative cell potential of *Lemna paucicostata* 6746 it appears that nitrate is taken up against the gradient of its electrochemical potential. Energy for this active transport process is supplied by the H^+ gradient across the plasmalemma: discharging of this gradient with a permeant acid completely inhibited nitrate uptake.

The rate of nitrate influx, which was measured with nitrate-starved plants, exceeded the net ion movement considerably. This is explained by the finding that passive efflux of nitrate occurs with a rate, comparable to influx rate. While net nitrate uptake was dependent upon illumination, influx rate was the same in light and dark. DCMU inhibited only net uptake, whereas influx was not affected. It is therefore concluded that energy supply is sufficient even in the dark to maintain a high influx rate and that light-stimulation of net nitrate uptake does not occur for energetic reasons. It is assumed that net nitrate uptake is under kinetic control: efflux increases with increasing intracellular nitrate concentration and diminishes net uptake. If intracellular nitrate is not consumed by metabolism, efflux rate will increase to influx rate until in the stationary state net uptake is zero. Therefore the rate of removal of nitrate by photosystem-dependent assimilation controls net nitrate uptake.

INTRODUCTION

The alkalinization of the nutrient medium by *Lemna* in the case, nitrate is the sole nitrogen source, has been attributed to a $NO^-_3 - OH^-$ antiport (or a $NO^-_3 - H^+$ symport) Löppert *et al.*, 1977). The 1:1 stoichiometry of net nitrate uptake and alkalinization during steady-state conditions is in accordance with the requirements for disposal of OH^-, produced during nitrate assimilation to maintain constant intracellular pH. Furthermore, a $NO^-_3 - OH^-$ antiport would be favourable for energetic reasons. The gradient of the electrochemical potential of OH^- (or H^+) could then serve as driving force for nitrate uptake (Raven & Smith, 1976). The H^+ gradient could be maintained by active H^+ extrusion, a mechanism which has been assumed to be responsible for the highly negative electrical potential of plant cells (Raven & Smith, 1976: Raven & Smith,

* The authors have made equal contributions to the paper. Thanks are due to Prof. R. Kandeler for his encouraging interest and helpful discussions, and to Dr. R. Heldwein for critical comments during the preparation of the manuscript. Financial support from the 'Fonds zur Förderung der wissenschaftlichen Forschung' of the Republic of Austria is gratefully acknowledged.

1977; Novacky *et al.*, 1978; Löppert, 1978). In addition, the formation of the H^+ gradient could be achieved by metabolic OH^- production within the cell, as it occurs during nitrate assimilation (Raven & Smith, 1976). As both processes are dependent at least partly on functioning of the photosystems, the latter might participate in a control over nitrate uptake.

Besides this 'energetic control', photosystem-dependent assimilation of nitrate could exert a 'kinetic control' upon nitrate uptake. If it is assumed that nitrate can leak through the cell membranes, influx is diminished by this passive efflux. As the latter is a function of intracellular nitrate concentration, net nitrate uptake (= influx minus efflux) is dependent on the rate of intracellular nitrate consumption. Whether energetic control or kinetic control is responsible for regulation of nitrate uptake into *Lemna* will be discussed in this paper.

MATERIAL AND METHODS

Lemna paucicostata 6746 was cultured under 'long-day' conditions (16 hr light phase, irradiance: 20 $w.m^{-2}$) as described (Löppert *et al.*, 1977). The chief constituents of the culture medium (Pirson & Seidel, 1950) were: 4 mM KNO_3, 1.5 mM KH_2PO_4, 5.5 mM $CaCl_2$, 1.2 mM $MgSO_4$. For the nitrate uptake experiments, the concentration of KNO_3 was reduced to 1 mM.

The rate of nitrate uptake into the plants was calculated from the rate of the concentration change in the external solution. Samples of 1 ml were withdrawn and nitrate activity was measured with a nitrate-specific electrode (Orion Research Inc., Cambridge, MA.). The pH of the medium was set to 5.0 and was checked in intervals of 20 min.; if necessary, readjustment was achieved by addition of HCl.

For determination of the nitrate content of the plants, samples of about 200 mg fresh weight were withdrawn from the medium, rinsed with about 50 ml 1 mM $CaCl_2$ and were immediately frozen. Extraction was achieved by homogenization in 2 ml of a solution, containing: 16.7 g $Al_2(SO_4)_3.18\,H_2O$, 1.2 g H_3BO_3, 4.7 g Ag_2SO_4, 2.4 g NH_2SO_3H per liter. With this procedure, interferences from chloride and organic anions in the consecutive electrode measurement are eliminated (Orion Research 1977).

Membrane potential measurements were carried out according to the conventional electrophysiological methods. Experimental details have been described elsewhere (Löppert *et al.*, 1978; Löppert, 1978).

RESULTS

The intracellular nitrate content decreased by about an order of magnitude within 24 hr after removal of exogenous nitrate when the plants were illuminated during this time (Table 1). Nitrate-starved plants, obtained in this way did not show a significantly changed cell potential (Table 1). The potential values were obtained from measurements in the middle of the light phase; the dark values are steady-state values which established within 30 min. after transition from light to dark.

Plants harvested from complete culture medium ('high nitrate plants') and nitrate-starved plants ('low nitrate plants') showed a quite different behaviour

Table 1. Nitrate content and electrical cell potential of *Lemna*.

	intracellular nitrate (mmoles.g_{FW}^{-1})	vacuolar potential (mV)	
High nitrate plants	9.15 ∓ 1.3	in light	-208 ∓ 11
		in darkness	-196 ∓ 8
Low nitrate plants	0.95 ∓ 0.2	in light	-206 ∓ 9
		in darkness	-201 ∓ 12

in respect to nitrate uptake under various conditions. In the light, the rate of uptake by high nitrate plants was considerably smaller than that by low nitrate plants (Fig. 1). With the addition of Na isobutyrate it was intended to change the intracellular pH (for an explanation, see 'Discussion'). Isobutyrate not only inhibited nitrate uptake but even induced nitrate leakage from high nitrate plants. The uncoupler CCCP exerted a very similar effect (Fig. 1).

When high nitrate plants were transferred from light to dark, nitrate uptake was abolished (Fig. 2). In contrast, nitrate uptake into low nitrate plants did not respond significantly to the light-dark transition. Presence of 1 μM DCMU in the light could substitute for dark conditions (Fig. 2).

The values plotted in Fig. 1 and Fig. 2 represent nitrate uptake rates during the first 2 hr after the beginning of the experiments. Each value is the mean of at least three corresponding results, obtained with plants from different harvests.

DISCUSSION

The investigation of the uptake of an ion which is metabolized within the cell is complicated by the fact that metabolism generates a sink which could then be responsible for the movement of this ion. Nitrate uptake into *Lemna* however cannot be regarded as a simple source-sink movement as it occurs against the gradient of its electrochemical potential. This can be estimated from the data given in Table 1. At an external nitrate concentration of 1mM, $\triangle \mu$ is about 6 kcal.mole^{-1} for high nitrate plants and 4.6 kcal.mole^{-1} for low nitrate plants (for details of calculation, see 'Appendix').

From the observation of a NO^-_3 $-OH^-$ antiport (Löppert *et al.*, 1977), it could be argued that nitrate uptake is a secondary active process, the H^+ gradient across the plasmalemma serving as driving force. Nitrate uptake should then respond in a sensitive way to a manipulation of the H^+ gradient. This could be achieved with permeant acids; thereby blocking of energy transduction by uncouplers or inhibitors of phosphorylation is avoided. Weak acids are known to penetrate biological membranes by diffusion of the undissociated form (Collander, 1959). Upon reequilibration with their ionic form, protons are released at the plasmic side of the plasmalemma. In this way the H^+ gradient should be drastically diminished.The inhibition of nitrate uptake by Na isobutyrate (Fig. 1) is in accordance with this view and can be taken as a direct proof that the H^+ gradient is the primary energy source for nitrate uptake. The finding that isobutyrate induces a nitrate release from high nitrate plants will be mentioned below. A similar action might be expected for the uncoupler of

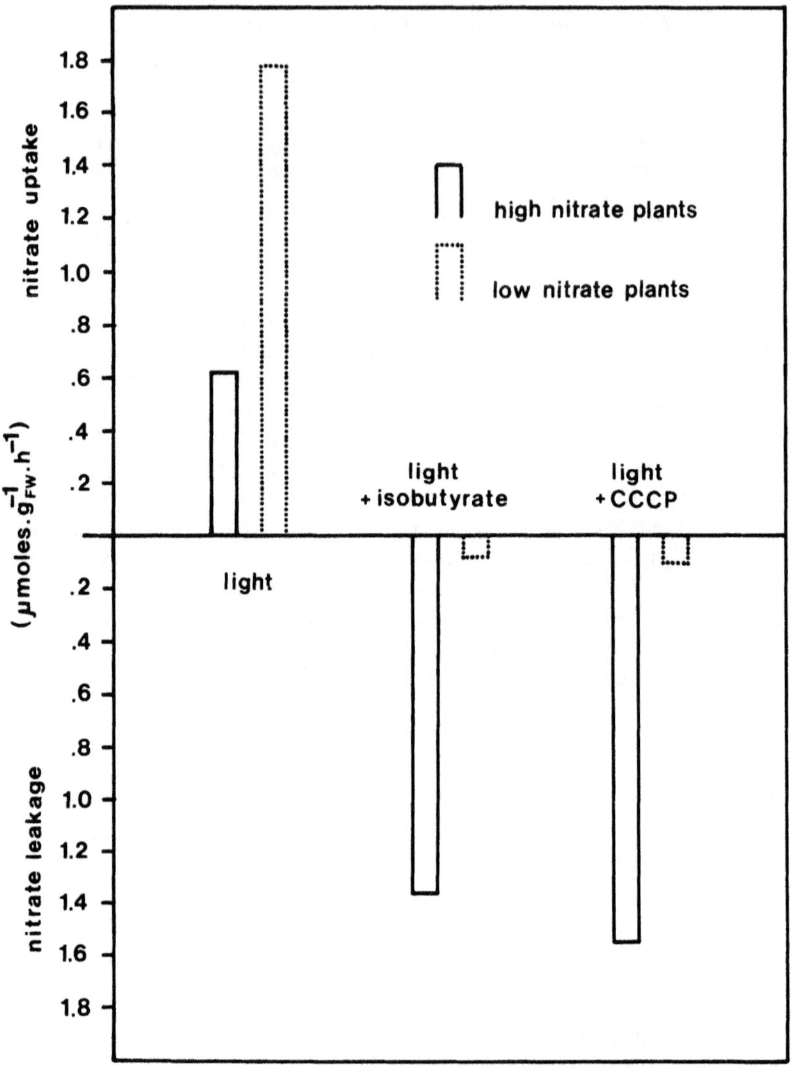

Fig. 1. The effect of culture condition, Na isobutyrate and CCCP on nitrate uptake. High nitrate plants are those, harvested from complete culture medium, low nitrate plants were obtained by starvation for 24 hr in the culture medium from which KNO_3 has been omitted. External NO^-_3 -concentration was 1mM at the beginning of the experiment; 5mM Na isobutyrate or 1 μM CCCP were present during the uptake period.

phosphorylation, CCCP. As it shuttles protons across biological membranes, it discharges the H^+ gradient and therefore abolishes nitrate influx (Fig. 1). However, in this case uncoupling of phosphorylation at energy-conserving membranes must be taken into consideration as well and therefore the inhibition of nitrate uptake could be partly due to a decreased energy supply for the maintainance of the H^+ gradient. In this context it is noteworthy that uncoupling of phosphorylation by isobutyrate is less probable under the present

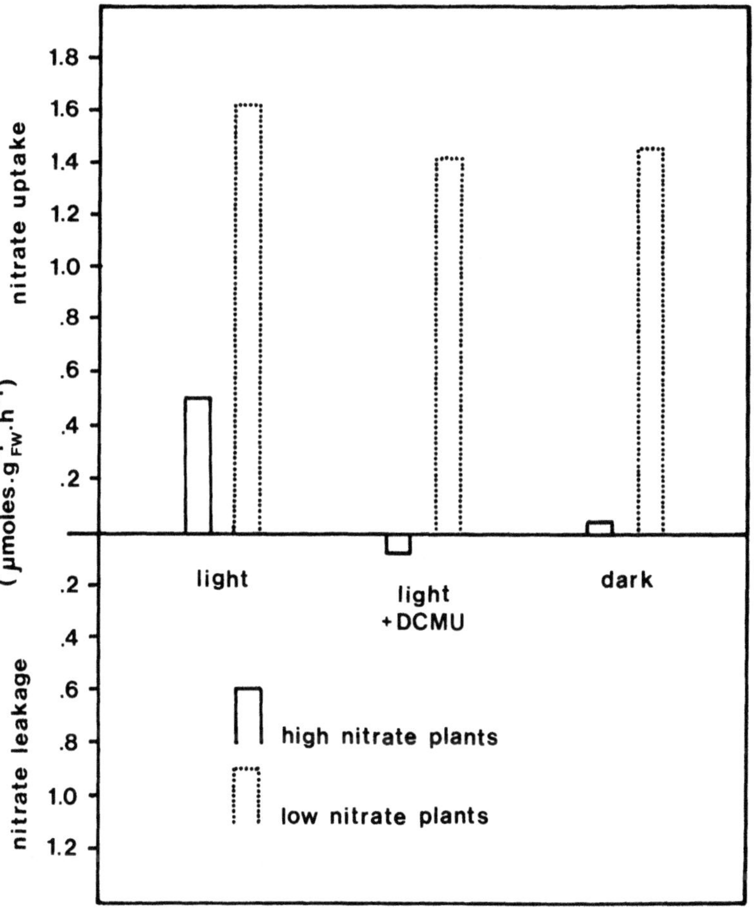

Fig. 2. The effect of illumination and DCMU on nitrate uptake into high nitrate and low nitrate plants. For explanation of 'high-nitrate' and 'low nitrate' plants, see Fig. 1. External NO^-_3 -concentration was 1mM at the beginning of the experiment, DCMU-concentration was 1 μM.

experimental conditions: oxygen uptake in the dark is not stimulated by this substance (Löppert, unpublished) in contrast to the well-known enhancement of respiration by CCCP.

If it is accepted that the H^+ gradient serves as driving force for nitrate uptake and that this gradient is discharged by the $NO^-_3 - OH^-$ antiport, reactions which lead to the reestablishment of this gradient are of primary importance for nitrate uptake. Two processes are to be envisaged: active extrusion of H^+ and intracellular OH^- production by metabolism. For the first one, support has been obtained from cell potential measurements (Löppert, 1978). Energy for this process is supplied by ATP as well as by a more direct – yet unknown – way from the photosystems. On the other hand, nitrate reduction itself has been regarded as predominantly responsible for intracellular alkalinization (Smith & Raven 1976). It was therefore argued that nitrate uptake should be at least partly dependent upon functioning of the photosystems. The considerable enhancement of nitrate uptake by light seemed to justify this assumption (Fig. 2). However, it could be shown that nitrate uptake is under kinetic, rather than under energetic control.

Net uptake of nitrate into *Lemna* is composed of influx and efflux. Under certain conditions it is possible to estimate the flux rates from concentration changes in the external solution. After inhibition of the influx with isobutyrate or CCCP, the rate of nitrate leakage can be taken as efflux rate. As the latter is near zero with low nitrate plants (Fig. 1), uptake of nitrate into these plants can be regarded approximately as influx. This ion movement is no longer dependent on functioning of the photosystems (Fig. 2). Therefore, even in the dark H^+ extrusion rate is sufficient for maintainance of the H^+ gradient and is not rate-limiting for nitrate influx. The light-dependence of net nitrate uptake is rather due to the kinetics of the 'pump' and 'leak' system. In the case, nitrate consumption by assimilation is stalled (dark or presence of DCMU in light), an increasing intracellular nitrate level produces an increasing influx until in the stationary state efflux equals influx, yielding a zero net uptake. This view is an contrast to the interpretation of the DCMU-effect on nitrate uptake into *Wolffia arrhiza* by Swader *et al.* (1975). They postulated that ATP, produced by cyclic photophosphorylation is less readily available for nitrate uptake.

The concept of a 'pump' and 'leak' system implicates the necessity to limit substrate accumulation within the cells of a tissue. In the case of nitrate uptake into *Lemna* it could be shown that the photosystem-dependent metabolism represents an additional flux (J_3 in Fig. 3) which provides for the removal of intracellular nitrate and thus finally controls the net movement of this ion.

ABBREVIATIONS

CCCP = Carbonylcyanide m-chlorophenyl hydrazone
DCMU = 3-(3,4-dichlorophenyl)-1, 1-dimethyl urea
FW = Fresh weight.

APPENDIX

The difference of the electrochemical potential is given by the equation

$$\Delta\mu = \mu^i - \mu^o = z.F\,(E_M - E_N)$$

where E_M is the cell potential and E_N the Nernst potential for nitrate. From the nitrate content of the plants, given in Table 1, and the assumption that the fresh weight is representative for the total cell volume, the Nernst potential for nitrate amounts to about 59 mV for high nitrate plants and zero for low nitrate plants at an external nitrate concentration of 1 mM. Inserting these values together with a mean cell potential of -200 mV into the above equation, $\Delta\mu$ becomes 6kcal.mole^{-1} for high nitrate plants and 4.6 kcal.mole^{-1} for low nitrate plants.

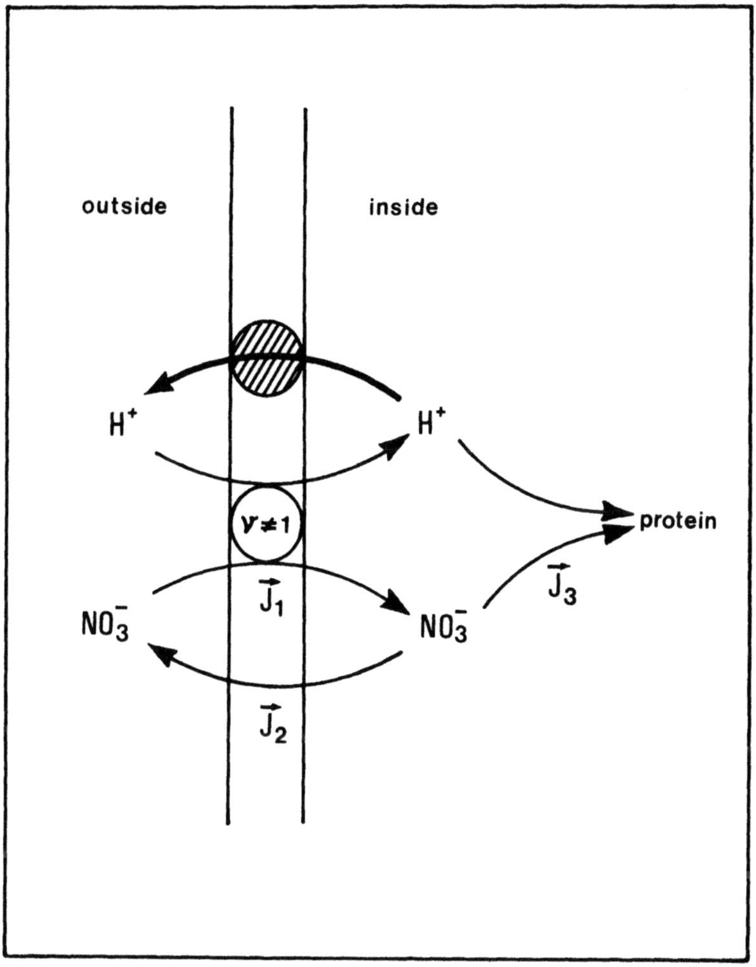

Fig. 3. A model for nitrate uptake and assimilation in *Lemna*. Nitrate influx is coupled to a H^+ influx with a stoichiometry deviating from 1:1. This has been concluded from potential measurements and is not discussed in this paper. A passive efflux (J_2) and the reaction flux of assimilation (J_3) are responsible for the limitation of intracellular nitrate concentration.

REFERENCES

Collander, R. Cell membranes: Their resistance to penetration and their capacity for transport. pp. 3–93, in: Steward, F.C. (ed.), Plant Physiology, A treatise. Academic Press, London, New York (1959).

Löppert, H., W. Kronberger & R. Kandeler. Correlation between nitrate uptake and alkalinization by *Lemna paucicostata* 6746. pp. 283–288, in: Thellier M., A. Monnier, M. Demarty & J. Dainty (eds.), Transmembrane ionic exchanges in plants. Editions du C.N.R.S. et de l'Université de Rouen, Paris, Rouen (1977).

Löppert, H., W. Kronberger & R. Kandeler. Phytochrome-mediated changes in the membrane potential of subepidermal cells of *Lemna paucicostata* 6746. *Planta* (Berl.), 138: *133–136* (1978).

Löppert, H. Evidence for electrogenic proton extrusion by subepidermal cells of *Lemna paucicostata* 6746. *Planta* (Berl.) 144: *311–315* (1979).
Novacky, A., C.I. Ullrich-Eberius & U. Lüttge. Membrane potential changes during transport of hexoses in *Lemna gibba* G1. *Planta* (Berl.), 138: *236–270* (1978).
Orion Research. Nitrate in plant tissue. Analytical methods guide May 1977: 30 (1977).
Pirson, A. & F. Seidel. Zell- und stoffwechselphysiologische Untersuchungen an der Wurzel von *Lemna minor* L. unter besonderer Berücksichtigung von Kalium und Kalziummangel *Planta* (Berl.) 38: *431–473* (1950).
Raven, J.A. & F.A. Smith. Cytoplasmic pH regulation and electrogenic H^+ extrusion. *Curr. Adv. Plant Sci.* 24: *649–660* (1976).
Raven, J.A. & F.A. Smith. Characteristics, functions and regulation of active proton extrusion. pp. 25–40, in: Marrè, E. and O. Ciferri (eds.), Regulation of cell membrane activities in plants. Elsevier/North-Holland Biomedical Press, Amsterdam (1977).
Smith, F.A. & J.A. Raven. H^+ transport and regulation of cell pH. pp. 317–346, in: Lüttge, U. and M.G. Pitman (eds.), Transport in plants II, part A. Springer, Berlin Heidelberg, New York (1976).
Swader, J.A., C.R. Stocking & Chin Ho Lin. Light-stimulated absorption of nitrate by *Wolffia arrhiza*. *Physiol. Plant*. 34: *335–341* (1975).

Authors' address:
Botanical Institute
Agricultural University
Gregor Mendel Str. 33
A–1180 Vienna
Austria.

INTEGRATION OF PHOTOSYNTHETIC CARBON METABOLISM AND NITROGEN METABOLISM ON A DAILY BASIS

BARRIE T. STEER
(Griffith, N.S.W., Australia)

ABSTRACT

In *Capsicum annuum cv.* California Wonder, within the photoperiod leaf nitrate reductase activity, levels of free amino acids, and the rate of labelling of amino acids from $^{14}CO_2$ have three maxima. Their timing is controlled by a circadian oscillator and by the light-on signal. Pyruvate kinase is activated when nitrate reductase activity is high.

Nitrate uptake by the whole plant and leaf polyribosome content have 24-hour patterns different to nitrate reductase activity. They exhibit an harmonic pattern with maxima in the middle of the photoperiod and minima in the middle of the dark period. Both of these patterns may be controlled by adenosine phosphates. For the first group, changes in the ATP/ADP ratio and the energy change during the photoperiod, and for the second, by changes in the total adenosine phosphate levels during the 24-hour cycle.

INTRODUCTION

This review is concerned with photosynthesis at the metabolite level and with processes controlling photosynthetic carbon metabolism. Much of the subject matter I will discuss will be somewhat removed from photosynthetic carbon assimilation *per se*, because we are seeking controls for functions which themselves influence carbon metabolism. Our published information on the integration of carbon and nitrogen metabolism in *Capsicum annuum* will be reviewed and some new information added, and new interpretations made where necessary. In addition, information from species other than *Capsicum annuum* will be presented for comparison.

The approach used in this work depends upon pattern comparison of within-day changes in a number of leaf and whole plant functions. This approach yields two sets of information. One, a description of the behaviour of plants on a daily basis, overall a poorly researched area, and, two, the possibility of identifying control mechanisms and the integration of metabolic sequences without using perturbative techniques.

MATERIALS AND METHODS

Seeds of *Capsicum annuum* L. *cv*., California Wonder were a gift of Arthur Yates and Co., Pty., Ltd., Sydney. Plants were grown as described in previous publications (Steer, 1973; 1974a).

Measurement of the adenosine phosphate content of leaves

Plants were grown in a constant environment cabinet with a 12-hour photoperiod and at 25° C. Samples comprising leaves number 7, 8 and 9 from three plants were harvested at intervals. Laminae were cut from the midvein and placed directly on to dry ice (−78° C) and stored at that temperature until extraction. The deep-frozen sample was placed in a mortar containing 7 ml 0.9N perchloric acid at −20° C. Sample weight was obtained in the frozen state and was near to 6 g fresh weight. Sample and acid were ground to a frozen slurry and then slowly allowed to thaw to 4° C. After centrifugation, the pellet was re-extracted with 0.2N perchloric acid, re-centrifuged and the supernatants bulked.

The acidic extract was carefully neutralized with KOH to pH 6.4 and then allowed to stand for one hour at 4° C. After centrifugation an aliquot of the supernatant was passed through a copper-Chelex 100 (200–400 mesh) column prepared according to Goldstein (1967). This step removed substances that interfered with subsequent adenosine phosphate estimations (Krauss & Reinbothe, 1977). Tests showed that all adenosine phosphates could be recovered from the column in the void volume.

Spectrophotometric estimations of ATP, ADP, AMP and glucose-6-phosphate were made in the column eluate using standard enzymic methods (Bergmeyer, 1965). Tests with standards showed that close to 95% recovery was obtained using this procedure.

RESULTS AND DISCUSSION

The daily pattern in distribution of assimilated carbon

The variation within the photoperiod of the distribution of ^{14}C assimilated from short exposure of leaves on intact plants to $^{14}CO_2$ is shown in Fig. 1 (from Steer, 1973). The main difference within the photoperiod is in the variable flow of carbon into the amino acid pools. The supply of reduced nitrogen necessary for amino acid synthesis in the leaves, as assessed by nitrate reductase activity (NRA), was shown to have the same pattern as the ^{14}C labelling of amino acids (Steer, 1973; 1974b). Three peaks of enzyme activity are evident in the photoperiod (Steer, 1974b; 1976) and these are shown in a stylized form in Fig. 2. The relationship between NRA and ^{14}C in amino acids during the major nitrate reductase (NR) peak in the middle of the photoperiod is shown in Fig. 3. There is good agreement between *in vitro* enzyme activity and ^{14}C content of amino acids. Thus in these leaves there is a close relationship between the supply of reduced nitrogen and the flow of recently fixed carbon in photosynthesis.

What control mechanisms are involved in this relationship? What controls this daily pattern of NRA? These two questions are now examined.

Carbon-nitrogen controls

Measurements of some compounds intermediate to 3-phospho glyceric acid and pyruvic acid on the one hand and 3PGA and fructose-6-phosphate on the other

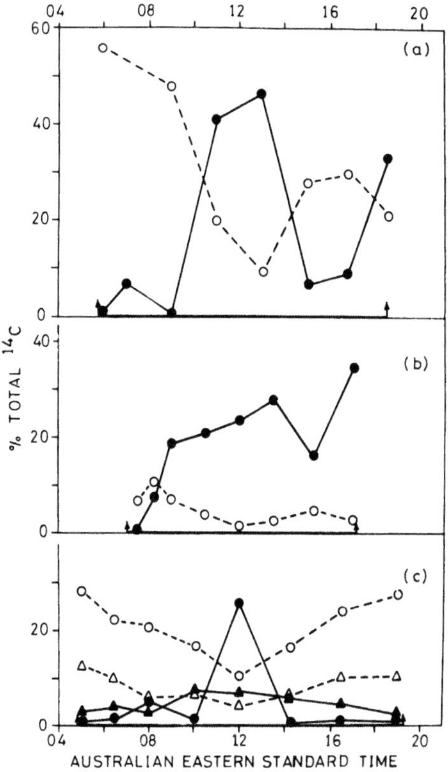

Fig. 1 Diurnal variations in the ^{14}C content of metabolites after five minutes exposure to $^{14}CO_2$ at different times in the photoperiod. Values are percentage of total ^{14}C recovered. a) *Capsicum annuum cv.* California Wonder. b) *Cucumis melo cv.* Hales Best. c) *Citrus sinensis cv.* Washington Navel. ● total acids; ○ sucrose; ▲ fructose; △ malate. Daylength marked by arrows (from Steer, 1973; reproduced with permission from *Plant Physiology)*.

(Steer, 1974a) indicated that at times in the photoperiod when amino acid synthesis was appreciable, pyruvate kinase was activated. Thus when nitrate reductase activity was high, pyruvate kinase activity was high. Pyruvic kinase from seeds is activated by K^+ (Miller & Evans,1957; Tomlinson & Turner, 1973) and to a lesser extent by NH_4^+. An effect of NH_4^+ on photosynthetic carbon metabolism in Chlorella (Kanazawa, Kirk & Bassham, 1970) and in Alfalfa leaf discs (Platt, Plaut & Bassham, 1977) has been interpreted as an activation of pyruvic kinase by NH_4^+.

There is some uncertainty about the pathways of carbon flow to amino acids. For example, alanine and glutamate are synthesised in *Vicia faba* from different carbon precursor pools (Kent, 1972, 1977), via 3PGA and serine respectively. Are the measured changes in phosphoenolpyruvate and pyruvate concentrations (Steer, 1974a) from at least two pools of each with two separate pyruvic kinases? Is the activation of pyruvic kinases the control link common to both alanine and glutamate synthesis pathways, the levels of which both increase in parallel when total amino acid synthesis is occurring?

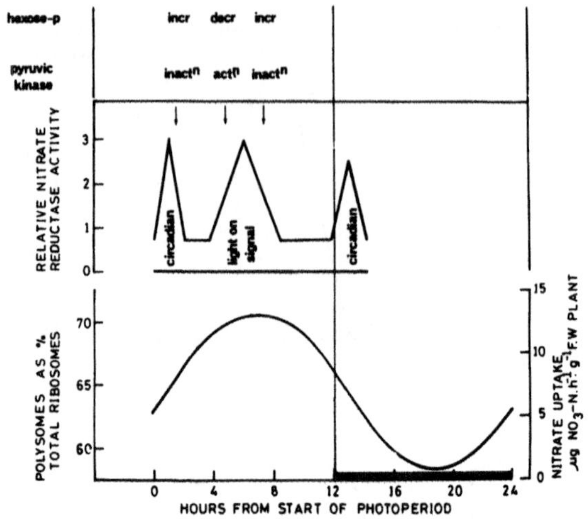

Fig. 2. Stylized diagram of the patterns of leaf nitrate reductase activity, leaf polyribosome content and whole plant nitrate uptake rates, with times of pyruvate kinase activation and inactivation, and changes in fructose biphosphate and fructose 6-phosphate (hexose-p) pool sizes. (after Steer, 1974a, b; Pearson & Steer, 1977; Steer & Blackwood, 1978).

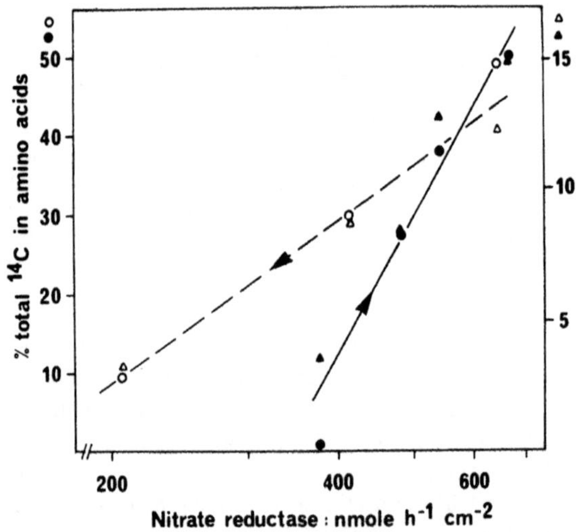

Fig. 3. Correlation between *in vitro* nitrate reductase activity and percentage total recovered ^{14}C in amino acids from two experiments. Fitted lines are for the increase and decrease of enzyme activity during the mid-photoperiod peak.

In the crossover analyses (Steer, 1974a) of intermediates that pointed to pyruvate kinase as a controlling step in the synthesis of amino acids, there was an indication that aldolase was at a control step also (Steer, 1974a). This was most clearly indicated during the decrease in the mid-photoperiod peak of NRA.

Leaf aldolase participates in a number of metabolic sequences: the reductive pentose phosphate cycle in the chloroplasts, the cytoplasmic glycolytic sequence and sucrose synthesis from the triose phosphates. Based on the properties of cytoplasmic and chloroplastic aldolases (Anderson, 1972; Anderson & Pacold, 1972), it has been suggested that there may be little effective compartmentation of these different functions and little control by effector molecules. Implicit in this is either that aldolase does not operate as a controlling enzyme in these pathways or there is a complex system of substrate control via the modulation of triose phosphate isomerase and the levels of triose phosphates (Anderson, 1971).

Citrate is a competitive inhibitor of pea-leaf aldolase (Anderson & Pacold, 1972) and the citrate content of *Capsicum annuum* leaves is significantly less later in the photoperiod than soon after light-on. However, aldolase in the 20 to 40% ammonium sulphate fraction from *Capsicum* leaves did not show a response to citrate (up to 2.3 mM). Thus there is no direct evidence that citrate levels control aldolase activity in *Capsicum* leaves.

While aldolase is indicated as a controlling enzyme only late in the photoperiod, what is consistent throughout the photoperiod is that at times when NRA is increasing, the hexose phosphate content decreases and when NRA is falling, hexose phosphate content increases. This increase when NR is inactivated is consistent with the demonstration in *Chlorella* (Chaparro *et al.*, 1976) that glucose-6-phosphate (G6P) increases when NR is inactivated by amines but measurements of (G6P) levels in *Capsicum* during the 24-hour cycle does not bear this out. Levels of G6P, like total adenosine phosphate (see Fig. 4), are highest near mid-photoperiod and fall during the later part of the photoperiod. The level remains low at night showing no increase when NRA is low. At the moment the change in hexose phosphate levels cannot be explained.

Control of leaf nitrate reductase activity

It was shown (Steer, 1976) that the timing of the two small peaks of NRA was controlled by a circadian oscillator. The major, mid-photoperiod, peak was not controlled in the same way but some evidence was presented (Steer, 1974b) that the timing of this peak depended upon a response to the light-on signal.

Other aspects of nitrogen metabolism have been investigated in order to seek for mechanisms controlling NRA in the leaves. Where diurnal variations in leaf amino acid content have been found in other species: in tomato (Noguchi & Tamaki, 1972) and in pea (Bauer *et al.*, 1977) and mechanisms have been suggested, the rate of arrival of nitrate in the leaves is usually seen as controlling leaf NRA and hence the supply of reduced nitrogen for amino acid synthesis. For many species this is likely. For example, maize leaf NRA responds rapidly to the nitrate flux in xylem sap (Shaner & Boyer, 1976). *Capsicum* is different in that nitrate uptake by whole plants has been shown not to have a direct control of NRA (Pearson & Steer, 1977). Other controls of leaf NRA impose patterns on amino acid synthesis that are not found in species such as maize.

The nitrate uptake rate in *Capsicum annuum* has a pattern that is described by a Fourier harmonic function (Pearson & Steer, 1977) with a maximum in the middle of the photoperiod and a minimum in the middle of the dark period (Fig. 2). The same pattern has been found for another solanaceous species, *Solanum*

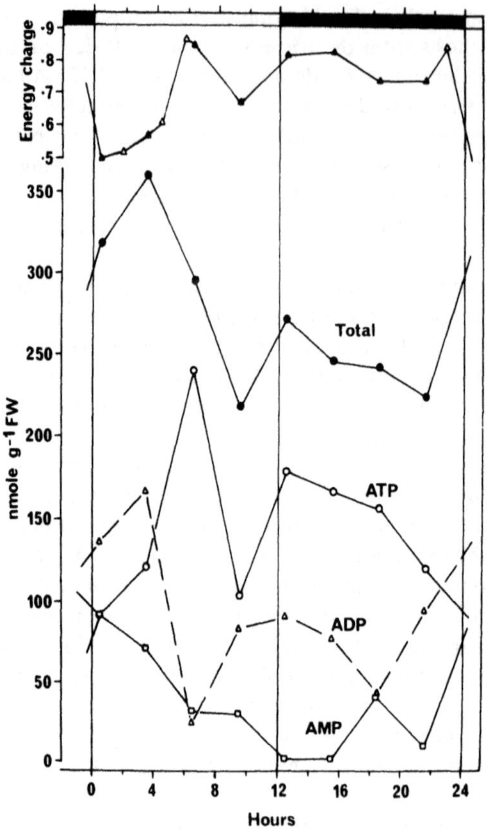

Fig. 4. ATP (○), ADP (△), AMP (□), total adenosine phosphate (●) contents and energy charge values (△) from leaves of *Capsicum annuum* throughout a 24-hour cycle. Energy charge values are from two experiments (▲△) but the other plots are from one experiment only. The second experiment gave lower absolute adenosine phosphate levels but showed the same trends as illustrated. The dark period is shown by the black bar.

melongena cv. Supreme, but not for *Zea mays cv.* Stylepak nor *Cucumis melo cv.* Hales Best (see also Fig. 1b) both of which exhibit a steady rate of nitrate uptake for four hours before and during the photoperiod.

The polyribosome content of *Capsicum* leaves shows a pattern in the 24-hour cycle that is also described by a Fourier harmonic function (Steer & Blackwood, 1978) (Fig. 2). In many cases, the different polyribosome content of organs gives a relative measure of their capacity for protein synthesis. If this is true for *Capsicum* leaves, then protein synthesis exhibits a simple pattern throughout the photoperiod with less synthesis at the beginning and the end than in the middle of the photoperiod. There is no evidence for a complex pattern in protein synthesis similar to that for leaf NRA and amino acid pool sizes.

Nitrate uptake and polyribosome content of leaves are at either end of the nitrogen synthetic chain, and the similar patterns suggest a common controller. The energy status of the plant may play that role. The adenosine phosphates in leaves have been measured during a 24-hour cycle. The results (Fig. 4) suggest that the total nucleotide content may be implicated. Separately AMP, ADP, ATP and energy charge (Atkinson, 1968) do not have patterns at all like a harmonic. Separate data for chloroplasts and cytoplasm are not available for *Capsicum* but information from other species suggests that chloroplastic and cytoplasmic pools of adenosine phosphates are in equilibrium (Keys & Whittingham, 1969; Heber & Santarius, 1970; Sellami, 1976) and photophosphorylation in the light and oxidative phosphorylation in the dark can keep the overall energy charge of leaf cells fairly high throughout the 24-hour cycle (Sellami, 1976). Published information on the change in energy charge after changes from light to dark or dark to light has been obtained within ten minutes of the change in conditions (Keys & Whittingham, 1969; Heber & Santarius, 1970; Sellami, 1976), but it seems likely that the rapid changes in energy charge at light-on or light-off are followed by a slow re-establishment to the pre-change value by about one hour (Sellami, 1976). Thus the fairly constant energy charge values in *Capsicum* leaves during the late photoperiod and the dark period are not at variance with the published information.

While the pattern of energy charge is not like that of nitrate uptake nor the polyribosome content, it does have similarities to the pattern of NRA. Within 30 minutes of the beginning of the photoperiod the energy charge has fallen to 0.50 (Fig. 4), largely as a result of an increase in AMP content that occurs rapidly at light-on. The gradual depletion of AMP throughout the photoperiod (Fig. 4) and an increase in ATP and decrease in ADP (Fig. 4) results in an increase in the energy charge (Fig. 4) and of the ATP/ADP ratio (Fig. 5) of the leaves by the sixth hour. It is at this time that the major, mid-photoperiod, peak of NRA occurs (Steer, 1974b). Thus the occurrence of this peak of NRA may be controlled by the changes in the levels of adenosine phosphates during the first

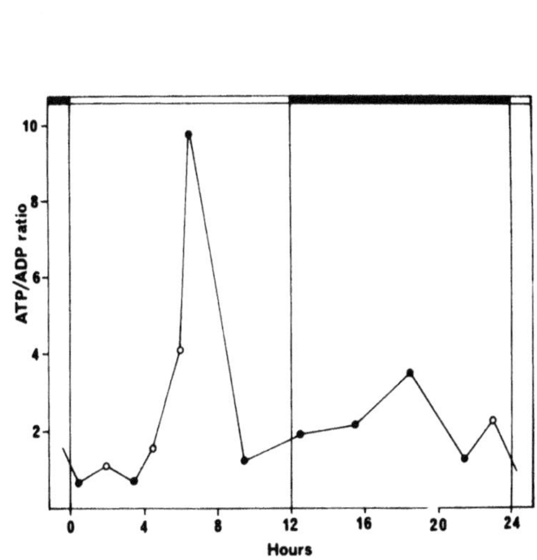

Fig. 5. ATP/ADP ratios from leaves of *Capsicum annuum* during the 24-hour cycle. From two experiments (●○).

half of the photoperiod and is consistent with the known inhibition of NRA by ADP (Nelson & Ilan, 1969; Eaglesham & Hewitt, 1971).

However, the small peak of NRA early in the photoperiod occurs at a time when the energy charge is low (contrast the similar peak at the end of the photoperiod at a time of high energy charge) and, presumably, must be induced by systems other than energy charge and be isolated in another cellular compartment.

The ATP/ADP ratio of *Capsicum* leaves is fairly constant throughout the 24-hour cycle except during the mid-photoperiod (Fig. 5). This supports the interpretation that the low energy charge early in the photoperiod is due to the rise in AMP content. This dominant effect of AMP content on energy charge in *Capsicum* is similar to that in the band mutant of *Neurospora crassa* (Delmer & Brody, 1975) where a circadian rhythm in AMP content imposes a circadian rhythm in energy charge. It is unlikely that the AMP content of *Capsicum* leaves exhibits a circadian rhythm but rather has a direct link to the light-on signal in a manner similar to the mid-photoperiod peak of NRA.

An anomaly is apparent when the supply of carbon for amino acid synthesis during the same period is considered. Both 3-phosphoglycerate kinase and pyruvate kinase activity will be favoured by a low ATP/ADP ratio, opposite in response to NRA. Duggleby & Dennis (1973) have shown that cotton seed pyruvic kinase is controlled by the levels of adenosine phosphates such that changing the energy charge from 0.5 to 0.85 invokes a 60 to 70% decrease in the maximum velocity of the enzyme. If *Capsicum* leaf enzyme responds in the same way then there is a discrepancy between conclusions to be drawn from the energy charge data of Fig. 4 and that drawn from crossover analysis of phosphoenolpyruvate and pyruvate levels in the leaves (Steer, 1974a). However, the information on control of leaf compared to seed pyruvic kinases is scarce and as Tomlinson & Turner (1973) have pointed out, enzymes from different sources do show different responses to effector compounds so that leaf pyruvate kinase may be less sensitive to changes in energy charge than is the enzyme from seeds.

Circadian rhythms

Almost all of this work has been based on the 24-hour cycle with normal light and dark periods. Without observations made in continuous light regimes, we are not able to say whether the patterns in nitrate uptake, leaf polyribosome contents and total adenosine phosphates in *Capsicum annuum* possess the characteristics of circadian rhythms, except for the NRA peaks early and late in the photoperiod (Steer, 1976) (Fig. 2).

In *Acetabularia mediterranea* and *Phaseolus multiflorus* there is evidence for circadian rhythms in RNA synthesis (Vanden Driessche & Bonotto, 1969; Rückebeil, 1961) with maxima in the light period and minima in the late dark period. In *Acetabularia* (Vanden Driessche, 1970) there is evidence for a rhythm in chloroplastic ATP contents although it is out of phase with RNA synthesis. Chloroplastic ATP is highest during the first three hours of the photoperiod and falls to a minimum at the sixth hour. The energy charge in

Chenopodium rubrum seedlings exhibits a circadian rhythm in continuous darkness (Wagner *et al*., 1974).

Observations on net photosynthesis and transpiration rates in peanut (Pallas *et al*., 1974) have shown that both peak near the middle of the photoperiod and have minima 12 hours later. In continuous light the free-running period is about 26 hours. Stomatal aperture seems to be the major controlling factor and rhythms in stomatal functions have been demonstrated before (*e.g.*, Stälfelt, 1965; Martin & Meidner, 1972).

Pallas (1973) measured net photosynthetic rate and transpiration in a 14-hour photoperiod for four species including *Capsicum annuum cv*. California Wonder. A 30% difference in transpiration and photosynthetic rate was evident within the photoperiod. Steer (1976) measured apparent transpiration as leaf/air temperature differences and found some evidence for an entrained pattern of transpiration rate when plants were kept in continuous light. We considered (Pearson & Steer, 1977) that this pattern in transpiration was not sufficiently like the harmonic in nitrate uptake to suggest a common control, basing our argument on the occurrence of a rapidly-damped cycle in transpiration in the first two hours of the photoperiod (Barrs & Klepper, 1968), that does not appear in the nitrate uptake pattern, and on the fairly steady temperature difference during the dark period. However, in the dark this can tend towards a curvilinear pattern. Ignoring the cycling during the first two hours of the photoperiod Fourier harmonic functions fitted to the relative transpiration rates accounts for 79.6 and 91.3% (from Figs. 5a and 5c of Steer, 1976), of the time-of-day variance. Thus transpiration rates are similar to nitrate uptake and on this criterion we may add transpiration to the list of functions exhibiting harmonic variations during the 24-hour cycle in *Capsicum annuum*.

While nitrate uptake rates do not explain leaf NRA and may be separated temporally in manipulative experiments (Pearson & Steer, 1977), the possibility remains that nitrate uptake and transpiration may have a common, parallel, control, or that in a control sequence, an entrained transpiration pattern may control the rate of nitrate uptake.

CONCLUSIONS

Fig. 6 summarizes the results obtained from *Capsicum annuum* and groups them into the two daily patterns. In a recent review, Wagner (1977) suggests that the energetic integration of the whole plant takes place through the symplast. This concept is useful for *Capsicum annuum* where the uptake of nitrate by the roots has the same pattern as the polyribosome content of the leaves. The total adenosine phosphate content of the leaves has approximately the same daily patterns as these other functions and it seems possible that the rates of nitrate uptake and the rates of protein synthesis are limited by the pool sizes of the total adenosine phosphates.

The reconciliation of the daily pattern of nitrate reduction in the leaves with the pattern of nitrate uptake and protein synthesis can also be made at the level of the adenosine phosphates expressed as energy charge or ATP/ADP ratios.

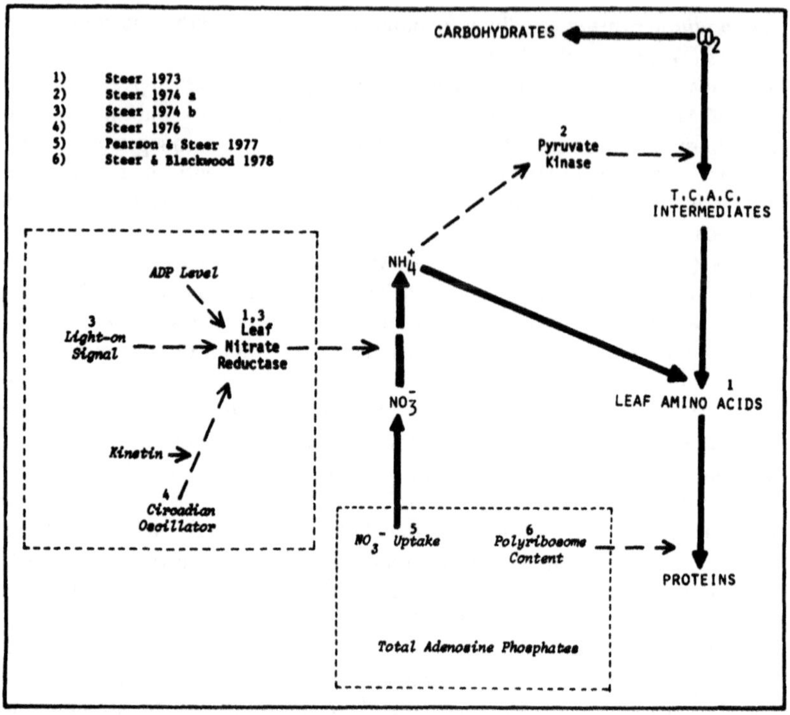

Fig. 6. Scheme of suggested control sequences. Processes that exhibit the two daily patterns are grouped together in the dotted boxes. Numbers refer to the published articles listed in the upper left-hand corner.

Further investigations are needed on the rapid increase in AMP content immediately after light-on, and on the large increase in ATP/ADP ratio for three hours in the middle of the photoperiod. The presence of isozymes of adenylate kinase (Criss *et al.*, 1970) with different equilibrium constants (Bomsel & Pradet, 1968) in plant cells would provide one mechanism controlling changes in energy charge and ATP/ADP ratios within a photoperiod.

Although the behaviour of the adenosine phosphates within the 24-hour cycle seem to control the flow of carbon to amino acids and proteins, and the integration with nitrogen metabolism, light does not seem to influence the processes simply through photophosphorylation. For example, the rapid change in AMP content early in the photoperiod and the changes in total adenosine phosphates suggest additional mechanisms linked to light signals that remain to be investigated.

REFERENCES

Anderson, L.E. Chloroplast and cytoplasmic enzymes. II. Pea leaf triose phosphate isomerases. *Biochim. Biophys. Acta* 235: *237–244* (1971).

Anderson, L.E. Similarities between chloroplast and cytoplasmic fructose-1,6-diphosphate

aldolases. *Proc. II. Intern. Congress Photosyn. Res.* 3: *1855–1860* (1972).
Anderson, L.E. & I. Pacold. Chloroplast and cytoplasmic enzymes. IV. Pea leaf fructose-1,6-diphosphate aldolases. *Plant Physiol.* 49: *393–397* (1972).
Atkinson, D.E. The energy charge of the adenylate pool as a regulatory parameter. Interaction with feedback modifiers. *Biochemistry* 7: *4030–4034* (1968).
Barrs, H.D. & B. Klepper. Cyclic variations in plant properties under constant environmental conditions. *Physiol. Plant.* 21: *711–730* (1968).
Bauer, A., A.A. Urquhart & K.W. Joy. Amino acid metabolism of pea leaves. Diurnal changes and amino acid synthesis from 15N-nitrate. *Plant Physiol.* 59: *915–919* (1977).
Bergmeyer, H.U. Methods of Enzymatic Analysis. Academic Press, New York (1965).
Bomsel, J.-L. & A. Pradet. Study of adenosine 5'-mono-, di-, and triphosphates in plant tissues, IV. Regulation of the level of nucleotides, *in vivo*, by adenylate kinase: theoretical and experimental study. *Biochim. Biophys. Acta* 162: *230–242* (1968).
Chaparro, A., J.M. Maldonado, J. Diez, A.M. Relimpio & M. Losada. Nitrate reductase inactivation and reducing power and energy charge in *Chlorella* cells. *Plant Sci. Lett.* 6: *335–342* (1976).
Criss, W.E. G. Litwack, H.P. Morris & S. Weinhouse. Adenosine Triphosphate: Adenosine monophosphate phosphotransferase isozymes in rat liver and hepatomas. *Cancer Res.* 30: *370–375* (1970).
Delmer, D.P. & S. Brody. Circadian rhythms in *Neurospora crassa*: Oscillation in the level of an adenine nucleotide. *J. Bacteriol.* 121: *548–553* (1975).
Duggleby, R.G. & D.T. Dennis. Pyruvate kinase, a possible regulatory enzyme in higher plants. *Plant Physiol.* 52: *312–317* (1973).
Eaglesham, A.R.J. & E.J. Hewitt. Kinetics and inhibition by adenosine phosphates and nitrite of nitrate reductase from *Spinacia oleracea* L. *Biochem. J.* 122: 18P–19P (1971).
Goldstein, G. Ligand-exchange chromatography of nucleotides nucleosides, and nucleic acid bases. *Anal. Biochem.* 20: *477–483* (1967).
Heber, U. & K.A. Santarius. Direct and indirect transfer of ATP and ADP across the chloroplast envelope. *Z. Naturforsch* 25b: *718–728* (1970).
Kanazawa, T., M.R. Kirk & J.A. Bassham. Regulatory effects of ammonia on carbon metabolism in photosynthesizing *Chlorella pyrenoidosa. Biochim. Biophys. Acta.* 205: *401–408* (1970).
Kent, S.S. Photosynthesis in the higher plant *Vicia faba*. II. The non-Calvin cycle origin of acetate and its metabolic relationship to the photosynthetic origin of formate. *J. Biol. Chem.* 247: *7293–7302* (1972).
Kent, S.S. On the metabolic relationship between the Calvin cycle and the tricarboxylic acid cycle. IV. A plant survey for anomalous acetyl coenzyme A. *Plant Physiol.* 60: *274–276* (1977).
Keys, A.J. & C.P. Whittingham. Nucleotide metabolism in chloroplast and non-chloroplast components of tobacco leaves. *Prog. Photosyn. Res.* 1: *352–358* (1969).
Krauss, G. & H. Reinbothe. Purification and fractionation of free nucleotides from *Euglena gracilis* Z by a combined procedure of ligand-exchange and anion-exchange gel chromatography. *Anal. Biochem.* 78: *1–8* (1977).
Martin, E.S. & H. Meidner. The phase-response of the dark stomatal rhythm in *Tradescantia virginiana* to light and dark treatments. *New Phytol.* 71: *1045–1054* (1972).
Miller, G. & H.J. Evans. The influence of salts on pyruvate kinase from tissues of higher plants. *Plant Physiol.* 32: *346–354* (1957).
Nelson, N & I. Ilan. Inhibition of nitrate reductase from tomato leaves by adenosine-5'-diphosphate. *Plant Cell Physiol.* 10: *143–148* (1969).
Noguchi, M. & E. Tamaki. Studies on nitrogen metabolism in tobacco plants. A. Part II. Diurnal variation in the amino acid composition of tobacco leaves. *Arch. Biochem. Biophys.* 98: *197–205* (1962).
Pallas, J.E. jr. Diurnal changes in transpiration and daily photosynthetic rate of several crop plants. *Crop Sci.* 13: *82–84* (1973).
Pallas, J.E. jr., Y.B. Samish & C.M. Willmer. Endogenous rhythmic activity of

photosynthesis, transpiration, dark respiration, and carbon dioxide compensation point of peanut leaves. *Plant Physiol.* 53: *907–911* (1974).

Pearson, C.J. & B.T. Steer. Daily changes in nitrate uptake and metabolism in *Capsicum annuum. Planta* (Berl.), 137: *107–112* (1977).

Platt, S.G., Z. Plaut & J.A. Bassham. Ammonia regulation of carbon metabolism in photosynthesizing leaf discs. *Plant Physiol.* 60: *739–742* (1977).

Rückebeil, A. Untersuchungen zur tagesperiodizität der 32P-Einlagerung in die Verbindungen der Nukleinsäurefraktionen. *Z. Botan.* 49: *1–22* (1961).

Sellami, A. Evolution des adenosine phosphates et de la charge energetique dans les compartiments chloroplastique et non-chloroplastique des feuilles de blé. *Biochim. Biophys. Acta.* 423: *524–539* (1976).

Shaner, D.L. & J.S. Boyer. Nitrate reductase activity in maize (*Zea mays* L.) leaves. I. Regulation by nitrate flux. *Plant Physiol.* 58: *499–504* (1976).

Stalfelt, M.G. The relation between the endogenous and induced elements of the stomatal movements. *Physiol. Plant.* 18: *177–184* (1965).

Steer, B.T. Diurnal variations in photosynthetic products and nitrogen metabolism in expanding leaves. *Plant Physiol.* 51: *744–748* (1973).

Steer, B.T. Control of diurnal variations in photosynthetic products. I. Carbon metabolism. *Plant Physiol.* 54: *758–761* (1974a).

Steer, B.T. Control of diurnal variations in photosynthetic products. II. Nitrate reductase activity. *Plant Physiol.* 54: *762–765* (1974b).

Steer, B.T. Rhythmic nitrate reductase activity in leaves of *Capsicum annuum* L. and the influence of kinetin. *Plant Physiol.* 57: *928–932* (1976).

Steer, B.T. & G.C. Blackwood. Within-day changes in the polyribosome content and in synthesis of proteins in leaves of *Capsicum annuum* L. *Plant Physiol.* 62: *907–911* (1978).

Tomlinson, J.D. & J.F. Turner. Pyruvate kinase of higher plants. *Biochim. Biophys. Acta* 329: *128–139* (1973).

Vanden Driessche, T. & S. Bonotto. The circadian rhythm in RNA synthesis in *Acetabularia mediterranea. Biochim. Biophys. Acta.* 179: *58–66* (1969).

Vanden Driessche, T. Circadian variation in ATP content in the chloroplasts of *Acetabularia mediterranea. Biochim. Biophys. Acta* 205: *526–528* (1970).

Wagner, E., L. Stroebele & S. Frosch. Endogenous rhythmicity and energy transduction. V. Rhymicity in adenine nucleotides and energy charge in seedlings of *Chenopodium rubrum. J. Interdiscipl. Cycle Res.* 5: *77–88* (1974).

Wagner, E. Molecular basis of physiological rhythms. *Symp. Soc. Expt. Biol.* 31: *33–72* (1977).

Author's address:
C.S.I.R.O.
Division of Irrigation Research
Griffith, N.S.W. 2680
Australia.

CARBON ASSIMILATION IN RELATION TO TRANSPIRATION AND FLUXES OF AMMONIA*

G.D. FARQUHAR
(Canberra City, A.C.T., Australia)

ABSTRACT

The sensitivity of rate of transpiration to change of stomatal conductance is the definition of one environmental gain. The sensitivity of rate of assimilation of CO_2 to change of conductance is another. The ratio of the two sensitivities is called the gain ratio. The minimum amount of water transpired in a day, given a certain amount of carbon fixed, occurs when stomatal conductance varies temporally and spatially in such a way that the gain ratio is maintained constant (Cowan & Farquhar, 1977).

Opening of stomata in response to illumination and closure in high vapour pressure deficits is consistent with keeping the gain ratio constant. Spatial uniformity of gain ratio demands that stomatal conductance in leaves of differing ages (*i.e.*, differing positions) on a plant (at a particular calendar date) must be in harmony with the microenvironments and assimilatory metabolisms of those leaves.

The rate of assimilation of CO_2 is strongly influenced by the nitrogen nutrition of the plant because the levels of ribulose biphosphate carboxylase are so sensitive to nitrogen content. There are also many reports of stomatal conductance reflecting the nitrogen status of plants. Stomatal conductance and the rate of transpiration and CO_2 assimilation increase with nitrogen content in such a way that the gain ratio itself changes little. A similar correlation may occur as each declines during senescence.

Nitrogen assimilation and photorespiration involve fluxes through ammonia which are large and comparable. However the exchange of ammonia between leaf and atmosphere is normally small.

INTRODUCTION

In this paper I intend to link water loss with carbon gain by plants, and examine possible effects of nitrogen status on this linkage. I then wish to discuss recent observations relating to ammonia and carbon fluxes in higher plants.

The loss of water and assimilation of CO_2 in higher plants is mainly *via* the stomata. The resultant link between the rates of transpiration, E, and assimilation, A, of a leaf is given by

* I wish to thank S.C. Wong for the data used in Fig. 2. His results encouraged me to compute the effects of carboxylase levels on optimal stomatal behaviour. I thank S. von Caemmerer for assistance with those computations. I also acknowledge Drs. V.P. Aneja and J.S. Paul for providing recent data relating to ammonia in plants.

$$E = \frac{g(e_i - e_a)}{P}$$

$$A = \frac{0.6\, g(c_a - c_i)}{P} \qquad (1)$$

Where E and A are molar flux densities, P is total air pressure, e_i is the vapour pressure in the intercellular spaces in the leaf and e_a is the ambient vapour pressure, c_a and c_i are the partial pressures of CO_2 in the ambient air and in the intracellular spaces, respectively. The variable in common is the leaf conductance to water vapour, g, whose magnitude is dominated by stomatal properties. The factor 0.6 arises because CO_2 diffuses more slowly in air than does water vapour. Strictly the equations should take into account the presence of cuticular transpiration from the leaf and a boundary layer around it. These complications have been adequately dealt with elsewhere (Cowan & Farquhar, 1977).

We pose the following problem: how to minimize the water lost during a day for a given amount of carbon gained, *i.e.*,

$$\text{minimize } \int^T E\, dt$$
$$\text{subject to } \int^T A\, dt = \text{constant, } A_T \text{, say.}$$

The rates of transpiration, E, and assimilation, A, are complex functions of the environmental variables irradiance, temperature, windspeed and the physiological variable leaf conductance. That is to say, the saturation vapour pressure at leaf temperature, e_i, and the partial pressure of CO_2 in the intercellular spaces, c_i, depend on the environmental variables and on leaf conductance, g. Using the calculus of variations it may be shown (Cowan & Farquhar, 1977) that the solution to the above problem is for leaf conductance to vary in such a manner that the ratio of $\partial E/\partial g$ to $\partial A/\partial g$ is kept constant, at λ, say. $\partial E/\partial g$ has previously been termed an environmental gain (Farquhar & Cowan, 1974) and $(\partial E/\partial g)/(\partial A/\partial g)$ which may be written more simply as $\partial E/\partial A$, may be called the gain ratio. Instead of asking how stomata should behave in order to minimize total transpiration for a given amount of dry matter accumulated, we can ask how to maximize the carbon gained for a fixed amount of water lost, *i.e.*,

$$\text{maximize } \int^T A\, dt$$
$$\text{subject to } \int^T E\, dt = \text{constant, } E_T \text{, say.}$$

The solution is again that leaf conductance should vary in such a manner that the gain ratio is constant. Cowan & Farquhar (1977) showed that certain aspects of stomatal behaviour such as closure in darkness and decreased conductance in dry air are consistent with keeping the gain ratio constant.

In the preceding section temporal variation in environment and stomatal response were considered. Analogous considerations lead to the requirement that the gain ratio be uniform in space and this demands that stomatal conductance in leaves of differing ages (*i.e.*, differing positions) on a plant (at a particular calendar date) must be in harmony with the microenvironments and assimilatory metabolisms of those leaves. To the extent that these leaves represent different stages of 'development' our theory of optimal behaviour is already relevant to this symposium.

Yoshida & Coronel (1976) have shown that stomatal conductance and photosynthetic rate are correlated in rice, regardless of the age of the leaf, when various levels of nitrogen nutrition are given to the plants. Is the observation of a correlation between conductance and assimilation rate over a range of levels of carboxylase activity consistent with the optimization theory I have briefly described? I used a model of the CO_2 assimilatory metabolism to test the effects of differing levels of carboxylase activity on optimal stomatal behaviour. The model is similar to that of Hall & Björkman (1975) as modified by Cowan & Farquhar (1977) except that the enzyme velocity depends on the total concentration of free RuP_2 and RuP_2 bound in intermediate enzyme complexes (Farquhar, 1978) which in turn depends *inter alia* on irradiance. The results are shown in Fig. 1. The thin lines represent changes of assimilation rate, A, with conductance, g, at various maximum velocities of the carboxylase, Vc_{max}. The *loci* of constant gain ratio are shown as thick lines, which are approximately linear and pass through the origin. Optimal conductance is correlated with assimilation rate as carboxylase levels vary. In Canberra, S.C. Wong (unpublished data) has recently varied the levels of nitrogen nutrition applied to *Zea mays* and monitored the resulting changes in chlorophyll content and phosphoenolpyruvate carboxylase and RuP_2 carboxylase levels. He measured leaf conductance, g, and assimilation rate, A, in a large number of these plants under standard conditions. The relationship between g and A is indicated by the small dots in Fig. 2. He made more detailed measurements on four of these plants whose conductances and assimilations are indicated by the larger dots. From measurements of the response of assimilation rate to changes of intercellular partial pressure of CO_2 he was able to calculate how assimilation rate would vary with conductance in these plants – the lines in Fig. 2. The similarity between Figs. 1 and 2 is obvious. The curves relating A to g at particular enzyme levels may be expected to differ between figures because those in Fig. 1 are based on a model of C_3 photosynthesis whereas *Zea mays* is a C_4 plant. The linearity of the relationship between A and g, particularly in the data of Wong, is such that intercellular partial pressure of CO_2, c_i, is maintained nearly constant (see Eq.(1)). Of interest here is the implication that stomata tend to keep the gain ratio constant, despite large changes in the levels of carboxylating enzymes. The correlation between stomatal conductance and assimilation rate that occurs with varying enzyme levels is consistent with our hypothesis of optimal stomatal behaviour.

It is of considerable interest that Davis *et al.* (1977) observed a continual decrease with age in the leaf conductance of fully expanded bean leaves (*Phaseolus vulgaris*), approaching zero as the leaf approached abscission. The rate of change of conductance with time varied among the leaves within a plant, but the differences were zero when the time scale was expressed as the fractional lifespan of each leaf. The effect of leaf age was independent of plant age: there was no influence of the stage of plant development on the relation between stomatal conductance and leaf age. Davis *et al.* (1977) suggested that changes in enzyme capacity with age may also be more consistent when plotted versus percentage time from maturity to death. Čatský *et al.* (1976) demonstrated that

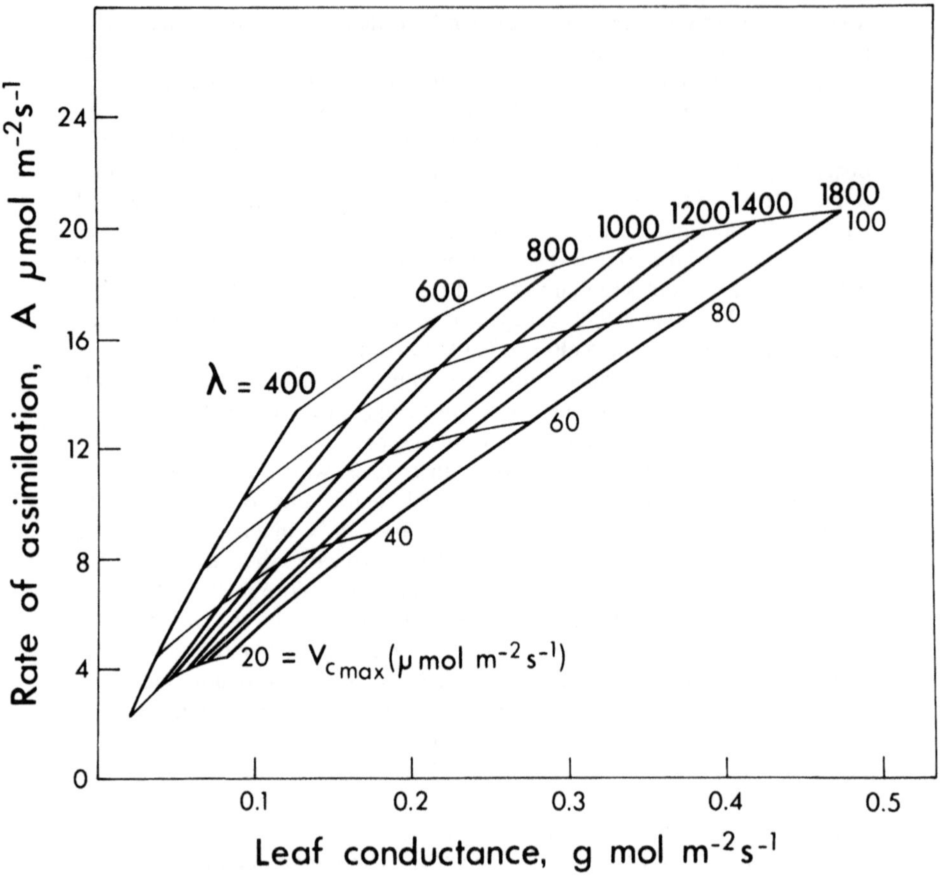

Fig. 1. Rate of assimilation, A, *vs.* leaf conductance, g. The thin lines show how A responds to g at various levels of maximum carboxylase level, Vc_{max}. The thick lines are the *loci* of constant gain ratio as Vc_{max} varies, for a number of values, λ, of that constant.

the slope of the relationship between assimilation rate and intercellular CO_2 concentration declined as leaves of *P. vulgaris* senesced. Several authors (*e.g.*, Sinclair *et al.*, 1977) have devized equations to predict this slope, the so-called mesophyll conductance, which indicate a dependence on the activity of RuP_2 carboxylase/oxygenase. The experiments of Medina (1970) Björkman (1968) and Neales *et al.* (1971) are consistent with this.

It appears then that stomatal conductance declines as proteolysis of RuP_2 carboxylase/oxygenase occurs during leaf senescence. Such behaviour by stomata is consistent with the results presented in Figs. 1 and 2. Although these results are encouraging, it would be surprising if the gain ratio remained constant through the life history of the plant – the marginal costs and benefits of water and carbon will not be fixed. Cowan & Farquhar (1977) suggested that stomata should operate to keep $\partial E/\partial A$ constant, at λ, over periods of the order of a day but that λ itself might decline in conditions of reduced soil water supply

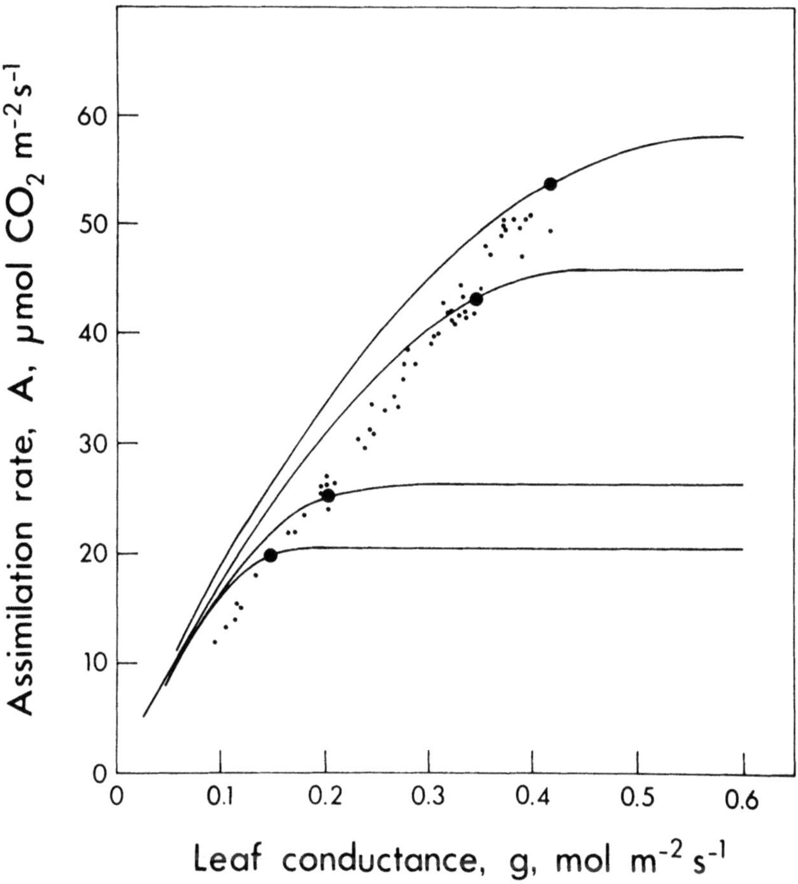

Fig. 2. Rate of assimilation, A, *vs.* leaf conductance, g, in *Zea mays*. Unpublished data of S.C. Wong. The data represent values of A and g at 2 mE m^{-2} s^{-1} irradiance, 320 μbar ambient partial pressure of CO_2 and 27° C. Each dot represents a single plant and the larger dots represent plants on which measurements were made of A vs. c_i, the intercellular partial pressure of CO_2. From these later measurements the responses of assimilation rate to conductance were calculated and are also illustrated here.

and decline somewhat with plant age. The reduction of stomatal opening with rapid stress is well known, and leads to a reduced gain ratio. Only detailed experimentation will establish whether the decline of stomatal conductance with leaf age is more than is necessary to keep the gain ratio constant and whether such a simple hypothesis could be useful over different stages of plant development.

I shall now turn to another aspect of nitrogen metabolism in plants – the fluxes of ammonia within the plant and between it and the atmosphere.

The nitrogen and carbon contents of plant dry matter are typically 2 and 40% by weight, respectively. When the rate of assimilation of CO_2 is 25 μmol m^{-2} s^{-1} the rate of assimilation of nitrogen should be of the order of 1 μmol m^{-2} s^{-1}, when both are expressed on leaf area bases. Much of the flux of nitrogen will be as ammonia/ammonium, either directly or after reduction of nitrate. The ammonia is then fixed into glutamine *via* glutamine synthetase. Paul *et al.* (1978) have found that an increased ammonia concentration causes a rapid rise in glutamine which draws carbon off the tricarboxylic-acid (TCA) cycle as a-ketoglutarate. Reactions which feed into the TCA cycle are stimulated at the expense of sucrose synthesis. The results of Paul *et al.* establish the capability of ammonia to act as a powerful regulatory agent in the carbon metabolism of higher plants.

Recent studies in Canberra on the photorespiratory carbon oxidative (PCO) cycle have revealed a massive release and subsequent refixation of ammonia (Woo *et al.*, 1977) which I estimate as representing a flux through ammonium of 2 μmol m^{-2} s^{-1}, on a leaf area basis. The ammonia is refixed into glutamine, again *via* glutamine synthetase. Thus it appears that there are two large fluxes of ammonia *via* the same enzyme, although not necessarily at identical sites, since the enzyme occurs in the root (Miflin & Lea 1977). Fair *et al.* (1972) suggested that nitrogen metabolism was linked to the photorespiratory activity of the plant and observed substantial transient increases of the CO_2 compensation point of barley leaves, lasting for some days, when plants were transferred from a nitrate nitrogen medium to an ammonium nitrogen medium. They also found similar results in C_4 plants (Grossman & Cresswell, 1973), and most interestingly, that the resulting compensation point responded linearly to oxygen in a similar manner to that usually found in C_3 plants. The foregoing indicate some important relationships between large fluxes of CO_2 and ammonia within the plant. To what extent are these 'visible from the outside'? A number of experiments have revealed that leaves assimilate gaseous ammonia when the ambient partial pressures are high (30 −20000 nbar) (Porter *et al*, 1972; Hutchinson *et al.*, 1972; Aneja, 1977). This may be of considerable importance in control of pollution by ammonia. An equation may be written for the molar flux density, J, of ammonia into the leaf through stomata

$$J = \frac{1.03\ g\ (n_a - n_i)}{P}$$

where n_a and n_i are the partial pressures of ammonia in the ambient air and in the intercellular spaces, respectively. Hutchinson *et al.* (1972) and Aneja (1977) found that the rates of assimilation were proportional to the stomatal conductance and ambient partial pressure, implying that n_i was negligible. However, normal ambient levels of ammonia in unpolluted areas are small, typically 1 − 8 nbar (Junge, 1956). Farquhar *et al.* (1978) have recently demonstrated that non-senescent *Zea mays* plants do not take up ammonia from the trace amounts found normally. This implies that there was either no metabolism of ammonia, which is unlikely, or that reactions involving ammonia/ammonium in the leaf resulted in finite partial pressures in the intercellular spaces, similar to those found in the atmosphere. Proteolysis occurs in senescing leaves and this may account for the observation (Farquhar *et al.*, 1978) that senescing leaves of *Zea mays* release ammonia to the atmosphere at a rate of approximately 0.6 nmol m^{-2} s^{-1}. Recent advances in the techniques for rapidly measuring small amounts of ammonia (Aneja, 1977) may allow us to test whether C_3 plants release ammonia by diffusion into the intercellular spaces from the organelles involved in photorespiration. In the scheme of Woo *et al.* (1977) the CO_2 released in photorespiration is the result of the mitochondrial decarboxylation of glycine and is accompanied by release of ammonia. It may also be possible to determine whether the post-illumination burst of CO_2 (Tregunna *et al.*, 1964) is accompanied by one of ammonia.

REFERENCES

Aneja, V.P. Dynamic studies of ammonia uptake by selected plant species under flow reactor conditions. North Carolina State University, Ph.D Thesis (1977).

Björkman, O. Carboxydismutase activity in shade-adapted and sun-adapted species of higher plants. *Physiol. Plant.* 21: *1–10* (1968).

Čatský, J., I. Tichá & J. Solárová. Ontogenetic changes in the internal limitations to bean-leaf photosynthesis. 1. Carbon dioxide exchange and conductances for carbon dioxide transfer. *Photosynthetica* 10(4): *394–402* (1976).

Cowan, I.R. & G.D. Farquhar. Stomatal function in relation to leaf metabolism and environment. *Symp. Soc. Exp. Biol.* 31: *471–505* (1977).

Davis and McCree (*Crop Sci.* 18: 280–282 (1978)) have recently demonstrated in leaves of *Phaseolus vulgaris* that stomatal conductance and assimilation rate declined in parallel with age, resulting in a constant intercellular CO_2 concentration.

Davis, S.D., C.H.M. van Bavel & K.J. McCree. Effect of leaf aging upon stomatal resistance in bean plants. *Crop Sci.* 17: *640–645* (1977).

Fair, P., J. Tew & C.F. Cresswell. The effect of age and leaf position on carbon dioxide compensation point Γ and potential photosynthetic capacity, photorespiration and nitrate assimilation in *Hordeum vulgare* L. *J.S. Afr. Bot.* 38: *81–95* (1972).

Farquhar, G.D. Models describing the kinetics of ribulose biphosphate carboxylase-oxygenase. *Arch. Biochem. Biophys.* (in press).

Farquhar, G.D. & I.R. Cowan. 1974. Oscillations in stomatal conductance. The influence of environmental gain. *Plant Physiol.* 54: *769–772* (1978).

Farquhar, G.D., R. Wetselaar & P. Firth. Ammonia volatisation from senescing leaves of maize. *Science* (in press).

Grossman, D. & C.F. Cresswell. Influence of nitrogen supply in nutrient media on carbon dioxide compensation point (Γ) of C-4 photosynthetic plants. *S. Afr. J. Sci.* 69: *244–246* (1973).

Hall, A.E. & O. Björkman. Model of leaf photosynthesis and respiration. pp. 55–72, in: Gates, D.M. & R.B. Schmerl (eds.), Perspectives of Biophysical Ecology, Ecological Studies 12. Springer-Verlag, New-York (1975).

Hutchinson, G.L., R.J. Millington & D.B. Peters. Atmospheric ammonia: absorption by plant leaves. *Science* 175: *771–772* (1972).

Junge, C.E. Recent investigations in air chemistry. *Tellus* 8: *27–139* (1956).

Medina, E. Relationships between nitrogen level, photosynthetic capacity, and carboxy-dismutase activity in *Atriplex patula* leaves. *Carnegie Inst. Wash. Yearbook* 69: *655–662* (1970).

Miflin, B.J., & P.J. Lea. The pathway of nitrogen assimilation in plants. *Phytochemistry* 15: *873–885* (1976).

Neales, T.F., K.J. Treharne & P.F. Wareing. A relationship between net photosynthesis, diffusive resistance and carboxylating enzyme activity in bean leaves. pp. 89–96, in: Hatch, M.D., C.B. Osmond & R.O. Slatyer (eds.), Photosynthesis and Photorespiration. Wiley Interscience, New York (1971).

Paul, J.S., K.L. Cornwell & J.A. Bassham. Effects of ammonia on carbon metabolism in photosynthesizing isolated mesophyll cells from *Papaver somniferum*. *Planta* (Berl.), 142: *49–54* (1978).

Porter, L.K., F.G. Viets, jr. & G.L. Hutchinson. Air containing nitrogen -15 ammonia: foliar absorption by corn seedlings. *Science* 175: *759–761* (1972).

Sinclair, T.R., J. Goudriaan & C.T. de Wit. Mesophyll resistance and CO_2 compensation concentration in leaf photosynthesis models. *Photosynthetica* 11(1): *56–65* (1977).
Tregunna, E.B., G. Krotkov & C.D. Nelson. Further evidence on the effects of light on respiration during photosynthesis. *Can. J. Bot.* 42: *989–997* (1964).
Woo, K.C., J.A. Berry, C.B. Osmond & G.H. Lorimer. Photorespiration and the metabolism of ammonia. Abstract 4th Inter. Congress on Photosynthesis, Reading, U.K. (1977).
Yoshida, S. & V. Coronel. Nitrogen nutrition, leaf resistance, and leaf photosynthetic rate of the rice plant. *Soil Sci. Plant Nutr.* 22(2): *207–211* (1976).

Author's address:
Department of Environmental Biology
Research School of Biological Sciences
The Australian National University
P.O. Box 475
Canberra City, A.C.T. 2601
Australia.

SESSION VI

PHOTOSYNTHESIS AND PESTICIDES PHOTOSYNTHESIS AND PATHOGEN INFECTION

INFLUENCE OF PESTICIDES ON PHOTOSYNTHESIS OF CROP PLANTS*

DAVID C. FERREE
(Wooster, Ohio, U.S.A.)

ABSTRACT

Certain older inorganic pesticide materials such as bordeaux, lime sulfur and oil have been found to reduce leaf photosynthesis. Tests of 33 newer organic pesticides revealed that 62% had no influence on photosynthesis nor on dark respiration. Pesticides that decrease photosynthesis generally have their greatest impact one to three days following application with a gradual recovery thereafter. However, with some materials the influence persisted for 3 weeks or more. Method and number of applications as well as combination with non-pesticides occasionally influenced photosynthesis. It is suggested that new pesticides be screened for their impact on photosynthesis before they are developed commercially.

Since the biochemical processes of the cellular and subcellular levels of the target organism and the host are often quite similar, the possibility of direct effects of protective chemicals on the host plant are great. These effects may be subtle in nature and go unobserved unless an effort is made to study their effect on the vital photosynthetic and respiratory mechanisms of the plant. If the impact of each cultural practice on productive efficiency of the crop plant is known, a combination of practices can be made to achieve optimum yield and economic input.

The various types of pesticide chemicals can influence the growth and development of crop plants in many ways. A 1968 compilation by the National Academy of Sciences (Subcommittee on chemicals affecting fruit and vegetable physiology (1968), enumerates the effects of protective chemicals on seed germination, vegetative storage organs, maturation, chemical composition, appearance and other aspects of growth and fruiting of many different agricultural crops. Casida & Lykken (1969) reviewed metabolism of pesticide chemicals in higher plants. The effect of herbicides on crop and target species has been reviewed (Ashton & Crafts, 1973; Good, 1962). The present paper is an attempt to review the effect of acaricides, fungicides, and insecticides on photosynthesis of crop plants. The interaction of pesticide and nonpesticide groups and influence of method of application are also treated.

* Approved for publication as Journal Article No. 114–178 of the Ohio Agricultural Research and Development Center, Wooster, Ohio 44691.

Inorganic compounds

Many of the early pesticide chemicals reduced growth and yield of the crop plants they were used to protect. Several investigators (Agnew & Childer, 1939; Berry, 1939, 1940; Brody & Childers, 1939; Clore, 1936; Heinicke, 1939; Murphy, 1940; Pickett & Birkeland, 1941; Sirois *et al.*, 1964) have reported a marked reduction in photosynthesis of apple leaves due to lime sulfur and other forms of sulfur. Hoffman (1935) indicated that lime-sulfur injury to apple leaves was largely due to a decrease in photosynthesis rather than the slight increase in respiration. Pickett & Birkeland (1941) found the combination of lime sulfur and lead arsenate altered the internal structures, particularly the palisade cells of apple leaves which could partially explain their influence on photosynthesis.

The form of sulfur and weather conditions surrounding application influence their effect on photosynthesis. Agnew & Childers (1939) found sulfur formulations, that remained in suspension, had less effect on photosynthesis of apple leaves than sprays with sulfur in solution. Lime sulfur caused greater reduction in photosynthesis than colloidal sulfur or sulfur dusts (Agnew & Childers, 1939; Hyre, 1939; Christopher *et al.*, 1944). The reduction in photosynthesis because of sulfur compounds was greater at higher temperatures, low humidity and high light intensity (Berry, 1939; Heinicke, 1939; Hyre, 1939). The formation of H_2S from lime sulfur and colloidal sulfur has been demonstrated (Abbot, 1945; Berry, 1940; Turrell, 1950; Turrell & Chervenak, 1949. Berry (1940) postulated that the conversion of H_2S is rapid and maximum effect will occur soon after application. Turrell (1950) and Turrell & Chervenak (1949) suggested that the heat liberated during oxidation may play an important role in causing injury along with a lowered pH due to formation of sulfuric acid in lemon leaves. The resulting pH after sulfur application was often lower than the isoelectric points of the proteins involved.

Copper based sprays, especially Bordeaux mixture, reduced photosynthetic activity and transpiration of treated apple and cherry trees (Clore, 1936; Menzel, 1935; Payeur, 1949; Murphy, 1939). Southwick & Childers (1940 and 1941) concluded that the influence of Bordeaux mixture was primarily physiological rather than mechanical shading as postulated by earlier workers. Transpiration of tomato leaves was increased by Bordeaux mixture (Childers, 1936), but photosynthesis or transpiration of pecan foliage was unaffected (Loustalot, 1944). Hyre (1939) compared the effects of several early pesticides on apple foliage and found Bordeaux caused a slight reduction in photosynthesis, wettable sulfurs were intermediate and lime sulfur caused the greatest reduction.

Organic pesticides

Spray oils have been shown to decrease photosynthesis in apples (Schroeder, 1935; Ayers & Barden, 1975; Ferree & Hall, 1975), pears (Sugiyama, 1941), bananas (Reidhart, 1961), citrus (Wedding *et al.*, 1952), McMillan & Reidhart, 1961), parsnip and mustard (Helson & Minshall, 1962), Heliconia (Reidhart, 1970), and tomatoes (Gudin *et al.*, 1976). Respiration was increased by spray oils in citrus (McMillan & Reidhart, 1964) and beans and barley (Green, 1936), but it was decreased in apple (Oberle *et al.*, 1944). Yield reductions as the result of oil sprays have been reported with apple (Spotts *et al.*, 1975) and citrus (Dean *et al.*, 1978).

The chemical and physical properties of spray oil and the amount applied influence the effect on photosynthesis. Schroeder (1935) found higher viscosity oils caused a greater decrease in photosynthesis, but reported little difference in effect between oils of 80 and 110 viscosity. The adverse effects of spray oil are also positively correlated with unsulfonatable residue numbers (Green & Johnson, 1931; Tutin, 1932) iodine values (Tutin, 1932) and degree of refinement (Green, 1936). Baker (1970) found toxicity increased with increasing amounts of low boiling compounds, unsaturated compounds, aromatics and acids, the more toxic the oil.

Riehl and Wedding (1959) reported that the difference in rate of recovery of photosynthesis associated with the difference in paraffinicity was greater for oils of comparable viscosity than for those of comparable molecular weight. McMillan and Reidhart (1964) tested pure hydrocarbons and found those having a distillation range of 419 to 487° F exhibited a faster rate of penetration and caused a greater decrease in photosynthesis than those with a higher distillation range. Ferree *et al.* (1976) found no affect on photosynthesis when recommended rates (.32 ml/liter) of corn, cotton seed and soybean oil were sprayed on apple leaves. Several investigators (Schroeder, 1935; Ayers & Barden, 1975; Ferree & Hall, 1975; McMillan & Reidhart, 1964) have shown that the higher the application rate the greater the magnitude and duration of photosynthesis inhibition.

MacMillan & Reidhart (1964) applied various oils to the dorsal or ventral surface of citrus leaves and found penetration only from the lower surface. Since stomata occur only on the dorsal surface, they postulated that oil penetrated through the stomata and not the cuticle. Reidhart (1970) in working with Heliconia which has stomata distributed on both surfaces, found that application of an oil emulsion to only the lower surface resulted in slightly increased CO_2 uptake through the upper surface. The results of these studies corroborate the hypothesis that the inhibition of photosynthesis by spray oils results from mechanical interference of gaseous exchange (Helson & Minshall, 1962; Reidhart, 1961; Wedding *et al.*, 1952). Baker (1970) also supports this hypothesis but indicates that other possible explanations would be the disruption of chloroplast membranes and inhibition caused by accumulation of end-products.

Following the introduction of DDT, Moon & Harley (1946) measured fruit growth on mature apple trees and found no difference between DDT sprayed trees and the control. They concluded that leaf efficiency was not impaired. Pickett (1948) found that DDT caused an increase in respiration of peach leaves, particularly of the younger leaves with basal leaves showing little effect. Harris (1952) found DDT reduced photosynthesis of cucumbers by 50% but found it had little effect on beans. Pickett *et al.* (1951) could not determine any effect on photosynthetic activity in peach leaves that had been sprayed 3 times with DDT and chlordane. Lawler & Rogers (1967) found DDT caused a 75% reduction in the Hill reaction of chloroplasts collected from selected cultivars of barley 4 days after the plants were sprayed. Wurster (1968) reported that DDT reduced photosynthesis in 4 species of fresh water algae.

Other organic insecticides have been evaluated for their impact on photosynthesis as they became available. Pickett *et al.* (1951) found the palisade tissue of chlordane sprayed apple leaves was thicker and had a higher photosynthetic rate than leaves sprayed with parathion. Johnson *et al.* (1950) found Parathion caused greater palisade development in sour cherry leaves than benzene hexachloride or arsenate of lead. Cole & Plapp (1974) found that parathion and piperonyl butoxide (an insecticide synergist) caused a decrease in photosynthesis, particularly when multiple doses were applied to *Chlorella pyrenoidosa*. In this study, DDT was a weak photosynthesis inhibitor and the least inhibition was caused by dieldrin and carbaryl. Heinicke & Foott (1966) found diazinon resulted in the greatest reduction in photosynthesis of apple leaves followed by ethion and guthion. Maximum retardation of photosynthesis occurred immediately after treatment and decreased with time, but was still significant after 14 days.

Westwood *et al.* (1960) found that Aramite reduced photosynthesis in apple leaves, but dicofol, tetraditon and parathion did not. Sulphenone and Aramite resulted in decreased dry weight production and some visual injury while tedion produced a reduction in dry weight, although the trees appeared healthy. These workers also reported that growth rate of Bartlett pear fruit was reduced by 6 sprays of toxaphene or dieldrin but not by 3 sprays. Blackleaf 40 or malathion did not affect fruit growth rate. In beans parathion and disulfoton depressed photosynthesis and transpiration for 15 days after application (Bogdanow, 1962). Parathion increased photosynthesis and respiration in rice, wheat, and barley but did not affect the yield of grain (Ito & Yamamoto, 1959).

Leigh (1963) found that a soil application of phorate delated emergence of cotton and reduced the number of plants per acre and increased plant height and leaf size. Snyder & Sheridan (1974) reported that the insecticide Zectran decreased both respiration and photosynthesis in 2 species of algae while 2 others were relatively unaffected. In long term growth experiments, all 4 algae species exhibited a toxicity threshold between 1 and 10 ppm Zectran. Ayers & Barden (1975) tested 24 commonly used insecticides and acaricides and found 8 of these significantly affected photosynthesis of apple leaves with 5 causing a decrease and 3 an increase. Oil and emulsifiable concentrate formulation generally reduced photosynthesis. Suzuki and Uchiyama (1977) found that parathion blocks the electron transfer from photosystem II to photosystem I and consequently inhibits the Hill activities of common chloroplasts. They suggested that the inhibitory effect of other organophosphorous insecticides was due to the same mechanism. The inhibitory effect of the insecticides tested was weak compared with that of DCMU (3–3, 4-dichlorophenyl) – 1, 1-dimethyluren or Atrazine. The mechanism of action of the inhibitory effect on photosynthesis of fungicides and other forms of insecticides needs further study.

Organic fungicides

Organic fungicides can also influence photosynthesis of plants, but available evidence would indicate a less profound influence than insecticides. Captan had little effect on respiration of pea seedlings, but the utilization of radio-labeled sugars and organic acids was drastically altered (Dugger *et al*., 1957). Bosian *et al*. (1960) found that captan reduced the rate of photosynthesis in grapes while Dithane increased it. Engelhard (1969) tested the influence of fungicides on flower quality of chrysanthemums and found Dithane M-45, Dithane Z-78, orthocide caused no damage. Daconil caused no injury to foliage, but slightly injured blooms and botran was injurious to both.

Ross *et al*. (1969) and Ross & Longley (1963) found no significant difference in yield, trunk growth, amount of bloom, fruit size or color in mature apple trees whether sprayed with captan, dodine, or dichlone. However, Winter (1962) reported reduced yields of apples when sprayed with dichlone, phenylmercury acetate and sulfur compared to trees sprayed with captan. Donoho (1964) showed that spraying trees with Phygon XL plus DDT or Guthion plus captan reduced fruit sets and yields, but not to the extent or as consistently as Phix plus DDT. Although the crop was reduced by these treatments, return bloom or size was not affected. Burth (1973) reported that ziram, captan and thiuram caused enlargement of apple leaf surface as well as increase fresh weight and dry matter content of leaves. Thiuram increased chlorophyll content of leaves while the following materials caused a decrease in chlorophyll: wettable sulphur, chinomethionate, dinocap, dichlofuanide and mancozeb. Sirois *et al*. (1964) reported that Niacide-M and glyodin reduced the rate of CO_2 assimilation in apple trees. The initial suppression in assimilation rate was followed by rates higher than the original rate with glyodin causing the greater increase. Wentzler & White (1949) found 8 sprays of Crag 341 or Fermate applied to apple trees increased the chlorophyll content of the leaves. Pieniazek & Christopher (1944) found Fermate reduced photosynthesis of apple leaves slightly at low temperatures and the reduction increased at higher temperatures. The reduction did not approach that experienced with lime sulfur.

Shishiyama *et al*. (1965) reported that the antibiotic cycloheximide inhibited the synthesis of both chlorophyll-a and DNA but the content of RNA was found to be increased more than two-fold in treated onion leaves. Dyrene was favorable to the synthesis of chlorophylls in newly developed leaves. Zineb did not appear to produce physiological changes in onion leaves. Absorption curves of chlorophylls extracted from the leaves treated with fungicides were identical with that from non-treated leaves.

Many of the newer fungicides are systemic in nature and their influence on the physiological processes of plants has been of concern to many investigators. Kristeva & Kristev (1971) found benomyl ((methyl-l-butyl carbamoyl)-2-benzimidazol carbamate) that was sprayed on apple leaves decreased apparent photosynthesis and caused a 50% inhibition of the Hill reaction in the chloroplasts. Benomyl did not influence the rate of respiration nor change the activity of isocitric dehydrogenase. However, with potted trees, Ayers & Barden (1975) and Ferree & Hall (1975) found no influence of benomyl on net photosynthesis of attached apple leaves.

Solel *et al*. (1973), using dosages of benomyl and carbendazin (methyl benzimidase-2-yl carbamate) 5- to 10-fold higher than those used commercially, found the materials were translocated basipetally *via* the phloem in apple and cucumber. However, under field conditions most movement was likely by apoplastic migration carried by water and moving toward transpiring surfaces. Skene (1972) found benomyl to possess cytokinin-like activity in soybean callus and the radish cotyledon assay but the response was weak compared to kinetin. Thomas (1947) tested seven systematic benzimidazole fungicides and found four showed cytokinin activity in a modified *Amaranthus* betacyanin bioassay. Schreiber & Hock (1975) found the fungicides benomyl and thiabendazole incorporated into soil, greatly depressed the growth of seedlings of American elm, marigold, buckthorn, sycamore and to a lesser degree silver maple. Pepper, tomato, and bean were less sensitive than other species grown in benomyl-treated soil, while corn and pea growth were not affected. The growth of turfgrass was stimulated by benomyl.

Prusky, *et al.* (1973) worked with a number of other experimental systemic fungicides on wheat. They found that Hoe-6052 (2-methyl-5, 6-dihydro-4-pyran-3-carbonoxide-anilide) and EL-273 (a-(2,4-d, chlorophenyl)-a-phenyl-5-pyrimidinemethanol) affected the respiration of wheat while BAS 3050 (0-methylbenzoic acid anilide) did not. Hoe-6052 also decreased photosynthesis while EL-273 and BAS 3050 did not. Hoe-6052 reduced the yield of a rust resistant cultivar of wheat and did not increase the yield of a rust susceptible cultivar even though it controlled the rust. Carlson (1970) reported that the systemic fungicide Vitavax (2,3-dihydro-5-carboxanilido-6-methyl-1, 4, oxathin) caused an immediate inhibition of photosynthesis and respiration of barley leaves. Treated leaves returned to apparently normal photosynthesis and respiration within 5 days; however, their dry weight to fresh weight ratio remained lower than non-treated leaves even 20 days after treatment. The chlorophyll content of treated leaves was slightly higher than that of non-treated during the first 7 days, and markedly higher 20 days after treatment.

Suzuki (1966) found the systemic bactericide, dihydrostreptomycin, caused no significant differences in sugar, starch, or chlorophyll content or photosynthesis or respiration between sprayed and control rice leaves. Suzuki and Sumina (1966) found that 2 or 3 sprays of dihydrostreptomycin did cause a reduction in the percentage of grains of standard weight, but not in the number of panicles per plant nor number of spikelets per panicle in rice. The authors suggested that the bactericide may have an inhibitory effect on translocation of sugar from flag leaf to grain and starch accumulation in the grain. Ferree & Hall (1978) found the systemic insecticide, aldicarb, caused a decrease in photosynthesis of apple leaves when application exceeded 5 g per 2.9 liter/pot. Rates above 14 g/pot reduced transpiration and induced leaf chlorosis and defoliation.

Interaction of pesticides

The various protective chemicals not only can influence the physiology of the plant directly, but one group of chemicals can influence the behavior of another. Chang *et al.* (1971) found that in half of 72 combinations between 8 insecticides and 9 herbicides that insecticides inhibited the metabolism of herbicides in isolated leaf tissues. In another study they (Chang *et al.*, 1971) found the reverse was also true with herbicides linuorn and propanil inhibiting the metabolism of insecticides dyfonate and malathion in bean tissues. Tonks (1969) found that when insecticides were combined with growth retardants and applied to chrysanthemums the effectiveness of the retardant was reduced or nullified. Ferree & Hall (1978) applied the plant regulators daminozide and ethephon to apple trees that also received 5 applications of plictran or oxamyl. The plant regulators alone had no influence on photosynthesis or dark respiration, and they did not alter the decrease caused by the acaricide and insecticide. Ferree *et al.* (1976) tested several commonly used spray adjuvants for their affect on photosynthesis of apple leaves and found Triton CS-7 increased, and X100 decreased net photosynthesis. The following adjuvants had no influence on photosynthesis: X77, NuFilm, 17, Nufilm P, Regulaid, Triton B-1956.

Methods of pesticide application

The method of application can effect the influence of a pesticide on the photosynthetic process. Several investigations (Ayers & Barden, 1975; Ferree & Hall, 1975) have found that quickly dipping leaves in an agitated solution containing spray oil caused a greater reduction in photosynthesis than a spray application. Ayers & Barden (1975) compared a spray application of oil with a deep dip (vertical) and shallow dip (horizontal) and found the shallow dip caused a greater reduction in photosynthesis than the deep dip at 1, 6 and 11 days after treatment. They suggested the response may have been due to a greater deposit of oil in the non-agitated shallow dip. The severity of the photosynthetic inhibition reported by Prusky *et al.* (1973) may partially have been due to application of the fungicides by dipping. Ferree & Hall (1975) made spray applications of oil to apple trees with a laboratory aspirator, air blast sprayer and low volume sprayer and found no difference in photosynthesis with a single application of benomyl and oil. The same materials did cause a reduction in net photosynthesis of fully expanded apple leaves when 3 weekly applications were made and 5 weekly applications decreased photosynthesis of both expanded and newly expanding leaves with the latter showing the greatest reduction.

Spotts *et al.* (1975) reported that multiple application of benomyl and oil throughout the season caused a yield reduction of apple. Sharma *et al.* (1977) found multiple applications of dicofol caused a greater reduction in photosynthesis of apple leaves than a single spray with a maximum decline occurring after the second spray. Multiple sprays of dodine had no influence on photosynthesis. Ayers & Barden (1975) found a single application of plictran increased photosynthesis in apple. However, Ferree & Hall (1978) found 5 applications decreased net photosynthesis in 2 apple cultivars. One to 3 applications of diazinon, methomyl and oxamyl had no affect on photosynthesis and transpiration but 5 aplications reduced photosynthesis, but had no effect on transpiration. Application by aerosol, spray, soil surface incorporation and fumigation using commercially available pesticide materials adaptable to each had no influence on photosynthesis.

The evident physiological effects (chlorosis, phytoxicity, *etc.*) of new pesiticide chemicals on economic crops are usually identified during routine advanced screening and product development. However, the subtle and interacting effects on photosynthesis, transpiration and respiration and ultimately yield may go unobserved. These effects may be manifested under a specific environment or may be expressed only after repeated use. Enough evidence exists for proven reductions in photosynthesis and ultimately yield that a systematic screening of new pesticides for photosynthetic impact is strongly encouraged. When the effects of each cultural practice on the physiology of the crop are known, the blend of practices that result in the greatest productive efficiency can be made. If more than one pesticide will control a given pest, the pesticide of choice should have the least impact on photosynthesis and yield.

REFERENCES

Abbott, C. The toxic gases of lime-sulfur. *J. Econ. Ent.* 38: *618–620* (1945).

Agnew, E.L. & N.F. Childers. The effect of two mild sulfur sprays on the photosynthetic activity of apple leaves. *Proc. Amer. Soc. Hort. Sci.* 37: *379–383* (1939).

Ashton, F.M. & A.S. Crafts. Mode of action of herbicides. John Wiley & Sons, N.Y. pp. *504* (1973).

Ayers, J.C. & J.A. Barden. Net photosynthesis and dark respiration of apple leaves as affected by pesticides. *J. Amer. Soc. Hort. Sci.* 100(1): *24–28* (1975).

Baker, J.M. The effects of oils on plants. *Environ. Pollut.* 1: *27–44* (1970).

Berry, W.E. Spray injury studies. Progress report I. Some observations on the probable cause of lime-sulfur injury. Ann. Rep. Long Ashton Res. Sta. pp. *124–144* (1939).

Berry, W.E. Spray injury studies. Progress report II. The effects of time and temperature on the production of hydrogen sulfide during atmospheric decomposition of lime sulfur. Ann. Rep. Long Ashton Hort. Res. Sta. pp. *52–56* (1940).

Bogdanow, V. Changes in the transpiration and photosynthesis of bean plants under the influence of the insecticides Ekaten and parathion. *Izv. Nauch, Inst. Zart. Rast.* 3: *175–183* (1962).

Bosian, G.M., M. Paetzholdt, & A. Ensgrather. The effect of plant protection chemicals on the photosynthesis of the vine. Proc. *4th Int. Congr. Crop. Prot.* (Hamburg) 2: *1515–1522* (1960).

Brody, H.W. & N.F. Childers. The effect of dilute liquid lime-sulfur sprays on the photosynthesis of apple leaves. *Proc. Amer. Soc. Hort. Sci.* 36: *205–209* (1939).

Burth, U. The effect of organic fungicides in the vegetative performance of apple trees. *Archiv. fuer Gartenbau* 21: *567–585* (1973).

Carlson, L.W. Effects of Vitavax on chlorophyll content, photosynthesis and respiration of barley leaves. *Can. J. Plant Sci.* 50: *627–630* (1970).

Casida, J.E. & L. Lykken. Metabolism of organic pesticide chemicals in higher plants. *Ann. Rev. of Plant Physiol.* 20: *607–636* (1969).

Chang, F.Y., L.W. Smith & G.R. Stephenson. Insecticide inhibition of herbicide metabolism in leaf tissues. *J. Agr. Food Chem.* 19: *1183–1186* (1971).

Chang, F.Y., G.R. Stephenson, & L.W. Smith. Influence of herbicides on insecticide metabolism in leaf tissue. *J. Agr. Food Chem.* 19: *1187–1190* (1971).

Childers, N.F. Some effects of sprays on growth and transpiration of tomatoes. *Proc. Amer. Soc. Hort. Sci.* 33: *532–535* (1936).

Christopher, E.P., S.A. Pienizek & C. Jennings. Lime as a 'Safner' in lime sulphur and lead arsenate sprays. *Proc. Amer. Soc. Hort. Sci.* 44: *101–104* (1944).

Cole, D.R. & F.W. Plapp, Jr. Inhibition of growth and photosynthesis in *Chlorella pyrenoidosa* by a polychlorinated biphenyl and several insecticides. *Environ. Ent.* 3: *217–220* (1974).

Clore, W.J. The effect of Bordeaux, copper and calcium sprays upon carbon dioxide intake of Delicious apple leaves. *Proc. Amer. Soc. Hort. Sci.* 33: *177–179* (1936).

Dean, H.A., H. Tannahill & J.R. Bush. Effects of postbloom and summer oils on yield, fruit size, and growth of four varieties of oranges, 1970–74. *J. Econ. Entomol.* 71: *211–216* (1978).

Dugger, W.M., Jr., T.E. Humphreys, & B. Calhoun. Influence of n-(trichloromethylthio)-4-cyclohexane-1,2 dicarboximide (Captan) on the metabolism of pea and corn seedlings. *Plant Physiol.* 32(suppl): *vii* (1957).
Donoho, C.W., Jr. Influence of pesticide chemicals on fruit set, return bloom, yield and fruit size of the apple. *Proc. Amer. Soc. Hort. Sci.* 85: *53–59* (1964).
Englehard, A.W. Effects of fungicides and insecticides on flower quality of commercial chrysanthemums in Florida. *Proc. Fla. State. Hort. Sco.* 82: *381–385* (1969).
Ferree, D.C. & F.R. Hall. Influence of benomyl and oil on photosynthesis of apple leaves. *Hort. Science* 10: *128–129* (1975).
Ferree, D.C., F.R. Hall, & R.A. Spotts. Influence of spray adjuvants and multiple applications of benomyl and oil on photosynthesis of apple leaves. *Hort. Science* 11: *391–392* (1976).
Ferree, D.C. & F.R. Hall. Effects of growth regulators and multiple application of pesticides on net photosynthesis and transpiration of greenhouse apple trees. *J. Amer. Soc. Hort. Sci.* 103: *61–64* (1978).
Good, N.E. Inhibitors of photosynthesis as herbicides. *World Rev. Pest. Con.* 1: *19–28* (1962).
Green, J. & A.H. Johnson. Effect of petroleum oil on the respiration of bean leaves. *Plant Physiol.* 6: *149–159* (1931).
Green, J.R. Effects of petroleum oils on the respiration of bean plants, apple twigs and leaves and barley seedlings. *Plant Physiol.* 11: *101–113* (1936).
Gudin, C., W.J. Syratt & L. Boize. The mechanisms of photosynthetic inhibition and the development of scorch in tomato plants treated with spray oils. *Ann. Appl. Biol.* 84: *213–219* (1976).
Harris, C.S. Effects of certain insecticides and related chemicals on photosynthesis in cucumbers and beans. *Proc. Amer. Soc. Hort. Sci.* 60: *335–340* (1952).
Heinicke, A.J. The influence of sulfur dust on the rate of photosynthesis of an entire apple tree. *Proc. Amer. Soc. Hort. Sci.* 36: *202–204* (1939).
Heinicke, D.R. & J.W. Foott. The effect of several phosphate insecticides on photosynthesis of Red Delicious apple leaves. *Can. J. Plant. Sci.* 46: *589–591* (1966).
Helson, V.A. & W.H. Minshall. Effects of petroleum oils on the carbon dioxide uptake in the apparent photosynthesis of parsnip and mustard. *Can. J. Bot.* 40: *887–896* (1962).
Hoffman, M.B. The effect of lime-sulphur spray on the respiration rate of apple leaves. *Proc. Amer. Soc. Hort. Sci.* 33: *173–176* (1935).
Hyre, R.A. The effect of sulfur fungicides on the photosynthesis and respiration of apple leaves. Cornell Exp. sta. Memo 222. 40 pp. (1939).
Ito, S. & F. Yamamoto, Effects of EPN emulsion and dusts on the physiological behavior of crops. *Tokyo Nagyo Daigaka Nogaku Shuho* 5: *73–89* (1959).
Johnson, E.M., A.L. Kenworthy & A.E. Mitchell. Influence of spray materials on the structure of sour cherry leaves (*Prunus cerasus* L, var. Montmorency). *Proc. Amer. Soc. Hort. Sci.* 55: *195–198* (1950).
Kristeva, M. & K. Kristev. Respiratory and photosynthetic rates in apple leaves treated with the systemic fungicide benomyl. *Act Phytopath. Acad. Scien. Hungaricae* 6: *365–369* (1971).
Lawler, P.D. & L.J. Rogers. DDT effects on photosynthesis in certain varieties of barley. *Nature* 215: *1515* (1967).
Leigh, T.F. The influence of two systemic organophosphates on growth, fruiting and yield of cotton in California. *J. Econ. Entomol.* 56: *517–522* (1963).
Loustalot, A.J. Apparent photosynthesis and transpiration of pecan leaves treated with Bordeaux mixture and lead arsenate. *J. Agr. Res.* 68: *11–19* (1944).
McMillan, R.T. & J.M. Riedhart. The influence of hydrocarbon on photosynthesis of citrus leaves. *Proc. Fla. State Hort. Soc.* 77: *15-21* (1964).
Menzel, K.C. Investigation on the harmful effects of copper sprays. *Angew Bot.* 17: *225–263* (1935).
Moon, H.H. & C.P. Harley. Effect of DDT spray on apple leaf efficiency. *Proc. Amer. Soc. Hort. Sci.* 47: *11–14* (1946).
Murphy, L.M. The effect of certain fungicides on the photosynthetic activity of sour cherry leaves. *Proc. Amer. Soc. Hort. Sci.* 37: *375–378* (1939).

Oberle, G.O., G.W. Pearce, P.J. Chapman, & A.W. Avens. Some physiological responses of deciduous fruit trees to petroleum oil sprays. *Proc. Amer. Soc. Hort. Sci.* 45: *119–130* (1944).

Payeur, J.B. Effects of fungicides and insecticides on apple foliage. *Rev. ol' Oka* 12: *180–189* (1948).

Pickett, B.S. Respiration of peach leaves as influenced by some spray materials. *Proc. Amer. Soc. Hort. Sci.* 51: *196–198* (1948).

Pickett, W.F. & C.J. Birkeland. Common spray materials alter the internal structure of apple leaves. *Proc. Amer. Soc. Hort. Sci.* 38: *158–162* (1941).

Pickett, W.F., A.S. Fish, Jr. & K.S. Shan. The influence of certain organic spray materials on the photosynthetic activity of peach and apple foliage. *Proc. Amer. Soc. Hort. Sci.* 57: *111–114* (1951).

Pieniazek, S.A. & E.P. Christopher. The effect of some new spray materials on the rate of apparent photosynthesis of apple leaves. *Proc. Amer. Soc. Hort. Sci.* 44: *105–106* (1944).

Prusky, D., A. Dinoor & Y. Eshel. Symptomless effects of experimental fungicides on wheat. *Phytopath.* 64: *812–813* (1973).

Reidhart, J.M. Influence of petroleum oil on photosynthesis of banana leaves. *Trop. Agr. Trin.* 38: *23–27* (1961).

Reidhart, J.M. Effects of an oil emulsion on absorption of carbon dioxide by Heliconia. *Proc. Fla. State Hort. Soc.* 83: *370–372* (1970).

Riehl, L.A. & R.T. Wedding. Effect of naphthenic and paraffinic petroleum composition at a comparable molecular weight or viscosity on photosynthesis of Eureka Aemon leaves. *J. Econ. Ent.* 52: *883–884* (1959).

Ross, R.G. & R.P. Longley. Effect of fungicides on McIntosh apple trees. *Can. J. Plant Sci.* 43: *497–502* (1963).

Ross, R.G., A.D. Crowe & R.P. Longley. Performance of McIntosh apple trees sprayed with captan, dodine, or pre-cover and cover sprays of dichlone. *Can. J. Plant Sci.* 49: *655–658* (1969).

Schreiber, L.R. & W.K. Hock. Effects of benomyl and thiabendazole on growth of several plants species. *J. Amer. Soc. Hort. Sci.* 100: *309–313* (1975).

Schroeder, R.A. The effect of some summer oil sprays upon the carbon dioxide absorption of apple leaves. *Proc. Amer. Soc. Hort. Sci.* 33: *170–172* (1935).

Sharma, D.P., D.C. Ferree, & F.O. Hartman. Multiple application of dicofol and dodine sprays on net photosynthesis of apple leaves. *Hort. Science* 12: *154–155* (1977).

Shishiyama, J., M. Fukutomi & S. Akai. Effect of some fungicides on the synthesis of chlorophylls, deoxyribonucleic acid and ribonucleic acid in onion leaves. *Phytopath.* 55: *844–847* (1965).

Sirois, D., M.T. Hilborn & G.R. Cooper. Influence of certain fungicides on apparent photosynthesis of an entire apple tree. *Me. Agr. Exp. Sta. Bull.* 629: *p. 18* (1964).

Skene, K.G.M. Cytokinin-like properties of the systemic fungicide benomyl. *J. Hort. Sci.* 47: *179–182* (1972).

Synder, C.E. & R.P. Sheridan. Toxicity of the pesticide zectran on photosynthesis, respiration, and growth in four algae. *J. Phycol.* 10: *137–139* (1974).

Solel, Z., J.M. Schooley & L.V. Edgington. Uptake and translocation of benomyl and carbendazim (Methyl benzimidazol-2-yl carbamate) in the symplast. *Pestic. Sci.* 4: *713–718* (1973).

Southwick, F.W. & N.F. Childers. The influence of Bordeaux mixture on the rate of photosynthesis and transpiration of apple leaves. *Proc. Amer. Soc. Hort. Sci.* 69: *41–47* (1940).

Southwick, F.W. & N.F. Childers. Influence of Bordeaux mixture and its component parts on transpiration and apparent photosynthesis of apple leaves. *Plant Physiol.* 16: *721–754* (1941).

Spotts, R.A., F.R. Hall & D.C. Ferree. Yield reduction in apples related to benomyl oil spray applications. *Plant Dis. Reptr.* 59: *541–543* (1975).

Subcommittee on Chemicals Affecting Fruit and Vegetable Physiology. Effect of pesticides on fruit and vegetable physiology. Principles of Plant and Animal Pest Control. Volume 6, pp. 90. Nat'l. Acad. of Sci. Pub. 1968, Wash. D.C.

Sugiyama, T. Effects of spraying on photosynthesis. *J. Hort. Sci. Jap.* 12: *24–33* (1941).
Suzuki, T. Effect of dihydrostreptomycin spray upon the photosynthesis, respiration and chlorophyll content in rice. *Ann. Phytopath. Soc. Jap.* 32: *60–61* (1966).
Suzuki, T. & M. Uchiyama. Inhibition of the photosynthetic system in spinach chloroplasts by the organophosphate insecticide parathion. *Ectoxicology and Environ. Safety* 1: *263–269* (1977).
Suzuki, T. & I. Sumina. Effect of dihydrostreptomycin spray upon the grain development and carbohydrate content of rice. *Ann. Phytopath. Soc. Jap.* 32: *52–57* (1966).
Thomas, T.H. Investigations into cytokinin-like properties of benzimidazol-derived fungicides. *Ann. Appl. Biol.* 76: *237–241* (1974).
Tonks, N.V. Stem lengths and bloom weights of greenhouse pot chrysanthemums treated with combinations of growth retardant and systemic insecticides. *Can. J. Plant. Sci.* 49: *219–222* (1969).
Turrell, F.M. & M. Chervenak. Metabolism of radioactive elemented sulfur applied to lemons as an insecticide. *Bot. Gaz.* 111: *109–122* (1949).
Turrel, F.M. A study of the physiological effects of elemented sulfur dust on citrus fruit. *Plant Physiol.* 25: *13–62* (1950).
Tutin, F. A note on the toxicity of mineral oil sprays to vegetation. *J. Pomol. Hort. Sci.* 10: *65–70* (1932).
Wedding, R.T., L.A. Riehl & W.A. Rhoads. Effect of petroleum oil spray on photosynthesis and respiration in citrus leaves. *Plant Physiol.* 27: *269–278* (1952).
Wentzler, J.E. & D.G. White. The effects of nitrogenous fungicides and insecticides on the chlorophyll content of apple leaves. *Proc. Amer. Soc. Hort. Sci.* 54: *81–85* (1949).
Wurster, C.F., Jr. DDT reduces photosynthesis by marine phytoplankton. *Science* 159: *1474–1475* (1968).
Westwood, M.N., L.P. Batjer & H.D. Billingsley. Effects of several organic spray materials on the fruit growth and foliage efficiency of apple and pear. *Proc. Amer. Soc. Hort. Sci.* 75: *59–67* (1960).
Winter, H.F. The comparative effects of various fungicide programs on fruit numbers and yields of apple trees. *Plant Dis. Rpt.* 46: *560–564* (1962).

Author's address:
Department of Horticulture
Ohio Agricultural Research and Development Center
Wooster, Ohio 44691
U.S.A.

CO_2 EXCHANGES IN PLANTS INFECTED BY OBLIGATELY BIOTROPHIC PATHOGENS

P.G. AYRES
(Lancaster, U.K.)

ABSTRACT

An examination is made of the effects of obligately biotrophic fungi, which only survive on photosynthetically active hosts, on CO_2 exchanges and photosynthate production in plants. Laboratory investigations are reviewed and special attention is paid to the author's research into powdery mildew diseases.

Gross photosynthesis first decreases 2 to 4 days after infection and falls more than 50% in the 7 to 10 days the pathogen requires to complete its life cycle. Often, net photosynthesis is initially stimulated transiently as photorespiration commences to decline before gross photosynthesis. Stimulation of dark respiration is invariably noted in later stages of the infection cycle. Compensatory stimulation of photosynthesis may occur, when chlorophyll is retained in 'green islands' at infection sites, or, more importantly, in uninfected regions of infected leaves or plants, but little is known of its extent.

Neither increased stomatal resistance to CO_2 diffusion, nor loss of pigments, appears to be the initial cause of reduced photosynthesis. The causal event is probably loss of chloroplast polyribosomes. No detailed study has been made of the carbon pathway in infected plants; there are indications that fixation, and subsequent metabolism, may occur through routes less used in comparable healthy plants.

INTRODUCTION

Obligately biotrophic fungi cause many of the world's most common and economically damaging plant diseases, and have a uniquely important physiological relationship with their host where photosynthesis is concerned. Defined in the past as 'fungi which cannot be grown axenically', today, because of the success of modern culturing techniques, they can only be described as 'fungi which grow and reproduce in nature in association with a living, photosynthesizing plant' (modified after Scott, 1972). The salient features of obligately biotrophic pathogens, which contrast them with necrotrophic pathogens, are that they are relatively host-specific, and they do not cause massive destruction of host tissues; thereby, they protect longevity and the photosynthetic capability of infected tissues. Also, they alter the host's physiology so that a continuous supply of solutes is translocated to the infected area (Lewis, 1973). Whereas the effect of a necrotroph on photosynthesis and growth is broadly equivalent to the effect resulting from an equal amount of defoliation by, for example, excision of photosynthetic tissues, the effect of a biotroph is more complex as will be discussed. Continued growth of the plant, and its capacity to defend itself actively, as well as the growth of the pathogen, depend upon production of photosynthate and, therefore, to understand the development of the host-parasite relationship this paper examines respiratory as well as synthetic processes that contribute to the net production of photosynthate.

Since obligate biotrophs are functionally defined, the limits of the group of diseases that they cause are not precise. However, most investigations of CO_2 exchanges in diseased plants have concerned diseases caused by filamentous fungi which are indisputably obligate biotrophs, downy mildews (Peronosporales: Phycomycetes), powdery mildews (Erysiphales: Ascomycetes) or, rusts and smuts (Uredinales: Basidiomycetes), so this review will concentrate on these groups, paying special attention to the reviewer's own work and interests in powdery mildew diseases. Each disease requires different environmental conditions (the significance of this in relation to photosynthesis is discussed under Methodology) and the course of infection is different, so generalizations must be made with caution.

METHODOLOGY

Techniques used for measuring CO_2 exchange rates in diseased plants are broadly similar to those used with healthy plants, and, as exemplified by work on cereal powdery mildews (*Erysiphe graminis*), have progressed from manometric measurements of gas exchange (Allen, 1942; Scott & Smillie, 1966), to measurement of $^{14}CO_2$ uptake over short periods (Edwards, 1970) using detached tissue pieces and, often, unphysiologically high concentrations of CO_2, to the use of infra red gas analyzers (IRGA) monitoring CO_2 fluxes in attached leaves (Aust, Domes & Kranz, 1977; Woolacott & Ayres, unpublished) (Table 1). Experiments have been conducted in the laboratory because pathologists, in addition to facing the normal problems caused by experimentation in the field, must control conditions so that uniformly high levels of infection are achieved. Often, such conditions do not allow maximal development of the photosynthetic capacity of the plant. Commonly, experiments have been conducted on young plants, grown in small pots, and maintained under artificial light of low intensity – conditions which lead to the production of shade leaves, with a high relative water content. Leaves become heavily infected and their physiology is more severely disturbed than that of older leaves grown under higher light intensities and with some water stress. Such considerations are important if quantitative rather than qualitative nature of disease-induced changes is considered.

Measurements of gas exchange on intact leaves have increasingly been linked to studies of the activities of isolated chloroplasts and their associated enzymes. Here there are special problems because disease has effects which may alter the isolability of chloroplast systems, *e.g.*, on chloroplast structure, or on phenol metabolism, which are not always appreciated.

Table 1. Major investigations into the effect of *Erysiphe graminis* on photosynthesis in barley and wheat.

Investigation	Plant Age	Plant Tissue	Light intensity, lux Growth	Light intensity, lux Measurement	Method of Measurement	Maximal rate of photosynthesis in healthy tissue at 22°C
Allen, 1942	10 to 12d	Wheat. Segments of 1st leaf	1,500	5,000–6,000	5% CO_2 in manometer	10.2 mg $CO_2/dm^2/hr$ (recalculated)
Scott & Smillie, 1966	5 to 7d	Barley. Segments of 1st leaf	20,000	Not stated	1% CO_2 in manometer	0.2 ml O_2 per g f. wt.
Paulech & H.-Horvatovicova 1970	10 to 14d	Barley. Attached	10,000 (Estimated)	11,000 (Estimated)	Gravimetric	11.2 mg $CO_2/dm^2/hr$
Edwards 1970	10d	Barley. Discs from 1st leaf	12,000	35,000	$^{14}CO_2$ feeding (0.04%)	16.2 mg $CO_2/dm^2/hr$*
	,,	,,	,,	,,	$^{14}CO_2$ feeding (1.0%)	114.0 mg $CO_2/dm^2/hr$*
Aust, Domes & Kranz, 1977	7d	Barley. Detached 1st leaf	38,000	42,000	I.R.G.A.	50.0 mg $CO_2/dm^2/hr$
Woolacott & Ayres, unpublished	28d	Barley. Attached 3rd & 4th leaves	10,000	40,000	I.R.G.A.	25.6 mg $CO_2/dm^2/hr$

* Respiratory loss not accounted for.

RESULTS AND DISCUSSION

CO_2 fluxes

All experimental evidence leads to the conclusion that in susceptible hosts infection causes a localized inhibition of net photosynthesis and stimulation of dark respiration (dark respiration proceeds at approximately the same rate in light as in darkness and CO_2 is derived mainly from glycolytic, Kreb's cycle, and pentose phosphate pathway activity) (Fig. 1, A to F). What has concerned most workers is the mechanism of these changes (to be considered later) and the transient stimulation of net photosynthesis that occurs in many diseases soon after infection.

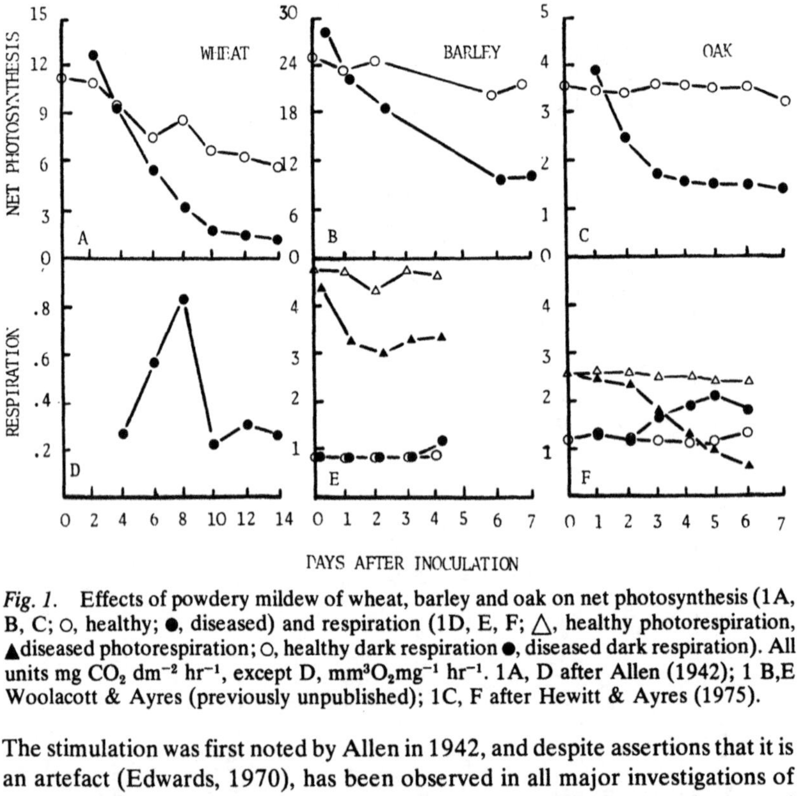

Fig. 1. Effects of powdery mildew of wheat, barley and oak on net photosynthesis (1A, B, C; ○, healthy; ●, diseased) and respiration (1D, E, F; △, healthy photorespiration, ▲diseased photorespiration; ○, healthy dark respiration ●, diseased dark respiration). All units mg CO_2 dm^{-2} hr^{-1}, except D, $mm^3O_2mg^{-1}$ hr^{-1}. 1A, D after Allen (1942); 1 B,E Woolacott & Ayres (previously unpublished); 1C, F after Hewitt & Ayres (1975).

The stimulation was first noted by Allen in 1942, and despite assertions that it is an artefact (Edwards, 1970), has been observed in all major investigations of photosynthesis in mildewed cereals, despite the different methods and conditions employed (Table 1) – Paulech & Haspelova-Horvatovicova's gravimetric analysis of photosynthesis could not be expected to show such a short-lasting effect. Stimulations were in the order of 10% and occurred between approximately the 8th and 48th hr after inoculation. Similar stimulations have been found in oak mildew, caused by *Microsphaera alphitoides* (Hewitt & Ayres, 1975), and yellow rust of wheat, caused by *Puccinia striiformis* (Doodson, Manners & Myers, 1965), but were not noted in powdery mildew of pea (*Erysiphe pisi*) (Ayres, 1976), or in white blister (downy mildew) disease of radish, caused by *Albugo condida* (Black, Gordon & Williams, 1968).

Until recently photorespiration (CO_2 only released in the light) has been ignored, or has not been measured, in studies of diseased plants but now we at Lancaster have evidence that the major cause of the transient stimulation in net photosynthesis is a temporally associated decrease in photorespiration (Fig. 1, E and F). It must be admitted the magnitude of the former is greater than that of the latter, and we are left with a small, yet unexplained, increase in gross photosynthesis resulting from infection. Edwards (1970) found a transient stimulation of net photosynthesis only occurred in an atmosphere of 1% CO_2; he concluded that mildew inhibits photorespiration since similar effects were produced by HPMS (α-hydroxy-2-pyrimidinemethane sulphonic acid) a known inhibitor of photorespiration. The reason neither HPMS, nor mildew, stimulated photosynthesis in an atomosphere of 0.04% CO_2 was possibly that uptake was measured less than 10 min. after labelled CO_2 was fed to the leaves, *i.e.*, before significant amounts of labelled carbon were released by photorespiration. In mildewed pea, photorespiration rather than photosynthesis was stimulated by infection (Ayres, 1976), possibly peroxisomes were affected before chloroplasts in this disease in which events following inoculation proceed more rapidly than in other powdery mildews.

So & Thrower (1976) measured uptake of $^{14}CO_2$ (concentration not given) by leaves of rusted *Vigna sequipedalis* over short periods to eliminate photorespiratory effects and found no transient stimulation. However, the inoculation procedure itself appears to have reduced photosynthesis, in which case, light infection caused a small stimulation of gross photosynthesis of about 2%, while heavy infection produced only a decrease. Decreased photorespiratory activity has been reported in wheat stem-rust (*Puccinia graminis tritici*) (Kiraly & Farkas, 1957) and flax rust (*Melampsora lini*) (Kakkar, 1966) diseases.

The increase in dark respiration found in all diseases, and the slow decline of photorespiration compared with gross photosynthesis in mildewed pea, mean an increasing proportion of a diminished amount of newly fixed photosynthate is immediately released through respiratory processes. The contribution that the fungus makes to measured respiratory increases is not certain, because host and parasite cannot be physically separated, but differences in their biomasses have caused most observers to conclude it is small. The mechanism whereby host respiratory activity increases is not agreed, but in a recent review Daly (1976) concluded that the most likely cause was increased activity of the hexose monophosphate (pentose phosphate) pathway which supplies carbon skeletons for synthesis of materials, *e.g.*, phenolics, involved in defence reactions of the plant. Daly suggested another of the more probable explanations is that in the later stages of infection there may occur some degree of uncoupling of electron transfer from oxidative phosphorylation. Both possibilities mean that photosynthate which would otherwise be available for growth of the plant is utilized in a non-productive way.

It must be noted that some pathogens can fix CO_2 directly. Zaki & Mirocha (1965) found bean leaves at the 'fleck' stage of rust disease, just prior to production of uredospores, fixed $^{14}CO_2$ in the dark at rates 3.5 to 4.5 faster than those of healthy leaves. Fixation probably occurred in fungal tissue because

Stallnecht & Mirocha (1971) later demonstrated that uredospores actively fixed $^{14}CO_2$ in the dark using malic enzyme. More recently, increased dark fixation in *Senecio squalidus* leaves infected by *Albugo tragopogonis* was demonstrated autoradiographically; increased fixation was located in host cells associated with young lesions (Thornton & Cooke, 1970). Cucumber leaves infected by *Pseudoperonospora cubensis* fixed 2.7 to 3.7 times more $^{14}CO_2$ in the dark than healthy leaves (Inaba & Kajiwara, 1974). The contribution of the fungus was unknown, but in this disease, where dark fixation was only about $\frac{1}{200}$th of light fixation, and others caused by obligate biotrophs, dark fixation is only a small item on the plant's carbon balance sheet.

As mentioned earlier, obligate biotrophs can alter translocation patterns in their host. Typically, export of photosynthate from the infected site, or leaf, is reduced, while import into it, from other healthy tissues, increases. Auxins and cytokinins also accumulate at the infection site, whether as a cause or a result of translocatory changes is not known. When infection densities are low, 'green islands' comprised of tissue with near normal levels of pigment and photosynthetic activity are left at infection sites as surrounding areas undergo senescence (Harding, Williams & McNabola, 1968). Green islands are important in consideration of questions of disease resistance and susceptibility for they are areas of retarded senescence, and, presumably, provide extra photosynthate for the parasite, but, because their total area is small, they probably have little effect on the carbon economy of the whole plant.

In healthy plants, photosynthetic efficiency is partly dependent on the translocation of photosynthate away from sites of synthesis (Liu, Wallace & Ozbun, 1973). Thus, in diseased plants, another consequence of increased export of photosynthate out of healthy tissues may be that their photosynthetic activity is thereby enhanced; this compensatory activity may partially offset the effects of disease. Last (1973) found, 7 to 10 days after heavy powdery mildew infection of barley leaves, photosynthesis was greater than was attributable to a simple loss of leaf area (Fig. 2). Livne (1964) found that heavy infections of

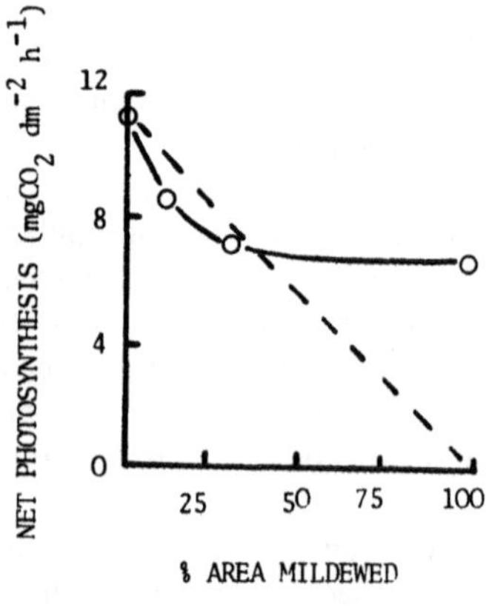

Fig. 2. Relationship between area of barley leaf mildewed and amount of photosynthesis (—— actual, – – – expected). After, Last (1973).

primary unifoliate bean leaves with rust (*Uromyces phaseoli*) stimulated rates of $^{14}CO_2$ fixation in uninfected trifoliate leaves on the same plant by more than 50%, while So & Thrower (1976) found that very light rust infections of second trifoliates of *V. sesquipedalis* slightly stimulated $^{14}CO_2$ fixation by uninfected third trifoliates. Heavy infection of the second trifoliate leaf reduced the rate of photosynthesis to 27.9% of the control 17 days after inoculation, and, remarkably, depressed rates in first and third trifoliates. In heavily infected plants, leaves 4 to 6 were slower to emerge than normal and, over a period of 20 days, the quantity of CO_2 fixed by an entire plant with one diseased leaf was only 72% of that fixed by a healthy plant. Infection of additional leaves would increasingly depress total photosynthesis by the plant.

Plant growth is progressively stunted by disease at a rate determined by the age-at-infection (young plants are most easily infected and suffer most stunting), severity of infection, and environmental effects on both host and pathogen. Thus, in mildewed barley, infection of young plants grown in wet soil inhibited leaf expansion (and ultimately yield) more severely than infection of older plants grown in drier soil (Table 2, Ayres, unpublished results). The degree to

Table 2. Leaf area of barley grown at three soil water levels and infected at three different stages with powdery mildew (Ayres, unpublished).*

	Age when sampled, days	Uninfected control	Non-desiccated leaf area, dm^2 $plant^{-1}$ Age at inoculation, days 64	45	26
Soil	26	2.74	2.75	2.81	3.19
	45	8.54	8.11	9.11	4.08
Wet	64	5.17	5.14	4.74	0.06
	83	1.48	0.71	0.03	0.00
	26	2.11	2.28	2.20	2.67
Moist	45	7.19	7.10	5.62	2.88
	64	3.70	3.52	2.58	0.45
	83	0.62	0.47	0.17	0.03
	26	1.70	2.11	1.79	1.80
Dry	45	4.91	5.27	4.79	2.63
	64	2.57	2.92	1.80	0.60
	83	0.54	0.55	0.16	0.12

* Each value is the mean of four replicates.

which compensatory photosynthetic activity occurs in healthy tissues of infected plants under field conditions is not known, but, because of its relevance to agriculture, merits investigation.

The causes of reduced photosynthesis

Few investigations have considered whether photosynthesis in diseased leaves is limited physically, by an increase in stomatal resistance to CO_2 diffusion, to the extent that the rate of CO_2 movement into the leaf limits photosynthesis. We know nothing of stomatal behaviour in plants infected by downy mildews and only a little about rust infections. In bean rust (*Ŭ. phaseoli*) Sempio, Majernik & Raggi (1966) found infection increased stomatal opening in the early stages of infection and inhibited it in the later stages, but Duniway & Durbin (1971) reported rust reduced average stomatal apertures in the light from 6.0 μm to 1.5 μm between the 3rd and 4th day after infection. Possibly, such changes could contribute to a reduction in photosynthesis in the later stages of infection.

Martin, Stuckey & Ellingboe (1975) found powdery mildew of wheat seedlings reduced stomatal opening in the light within 6 hr of infection, and possibly within 3 hr. Majernik (1965) found mildew reduced transpiration rates in 1st leaves of barley seedlings one day after infection, but he did not measure stomatal opening. Neither work measured concomitant changes in photosynthesis. Our own experience of barley mildew, involving older leaves, suggests that reduced opening is not apparent until 3 days after inoculation, and then the change is small. Woolacott & Ayres (unpublished results) found that in mildewed barley the CO_2 compensation point (which measures photosynthetic efficiency independently of stomatal behaviour) began to rise (from 30.5 ± 2.1 p.p.m. to 35.8 ± 0.4 p.p.m. CO_2) at the same time after inoculation that reduced levels of gross photosynthesis were detected, *i.e.,* 54 hr after infection. In pea, stomatal opening only began to be inhibited 4 days after mildew infection (Fig. 3). In mildewed oak, where stomata occur only on the lower uninoculated

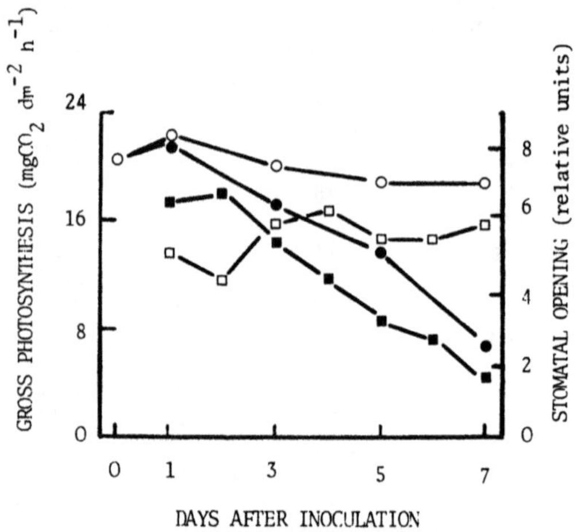

Fig. 3. Effects of powdery mildew of pea on gross photosynthesis (○ healthy, ● infected) and stomatal opening (□ healthy, ■ infected). Gross photosynthesis measured in an atmosphere of 2% O_2. After Ayres (1976).

epidermis, reduced stomatal opening was not found until six days after infection, and four days after the first reduction in photosynthesis (Hewitt & Ayres, 1975). Failure of stomata to open fully in mildewed or rusted leaves may result from some impairment of the activity of guard cell chloroplasts; thus, it may be an effect rather than a cause of failure of photosynthesis.

A popular subject for investigation has been the possibility that failure of photosynthesis derives from the loss of chlorophyll which sooner or later follows infection. However, the evidence is equally divided for and against the idea. Thus, Allen (1942) found photosynthesis per mole of chlorophyll in mildewed barley first increased, then later declined. This evidence for a decline was supported by Scott & Smillie (1963), although Pao & Godavari (1974) recalculated Scott & Smillie's results and concluded that photosynthesis per mg chlorophyll was 50% higher in mildewed leaves than in healthy leaves. Paulech & Haspelova-Horvatovicova (1970) found that total chlorophyll per mg dry matter declined sooner than photosynthesis (measured as an increase in dry matter per unit area). They also found the chlorophyll a/chlorophyll b ratio increased from about the sixth day after infection. Carotenoid pigments broke down at the same time.

In mildewed oak, photosynthesis declined long before total chlorophyll levels were reduced by infection. Again it was noted that infection reduced the ratio of chlorophyll a/chlorophyll b, from 3.30 in healthy leaves to 1.90 in infected leaves, six days after infection (Hewitt, 1976); the significance of this change is not clear. So & Thrower (1976) found there was a significant correlation (0.1%) between the decline of photosynthesis and chlorophyll levels between 0 and 14 days after rust infection of *V. sesquipedalis*, but Doodson, Manners & Myers (1965) found that 4 to 20 days after infection of wheat by *P. striiformis* photosynthesis declined more rapidly than chlorophyll levels.

There are several reports that obligate biotrophs deleteriously affect aspects of the light reaction of photosynthesis. Montalbini & Buchanan (1974) isolated chloroplasts from leaves of *Vicia faba* infected by rust (*Uromyces fabae*) and found a small decrease, 10 to 30%, in noncyclic but not in cyclic, photophosphorylation; on the basis of stoichiometry (electrons transferred to ATP formed), they suggested this was due to an inhibition of NADP reduction rather than uncoupling. Similar changes were found in chloroplasts isolated from powdery mildew-infected (*E. polygoni*) sugar beet leaves (Magyarosy, Schurmann & Buchanan, 1976). The changes noted do not seem to be the major cause of alterations in photosynthetic activity for, although they were first noted 2 days after infection of bean, they did not progress as rates of photosynthesis declined; in beet, the chloroplasts were examined 30 days after infection and no study was made of the course of changes with time.

The earliest effect of infection on chloroplasts known, and, therefore, probably the most significant, is that 1 day after mildew infection of barley, *i.e.*, 2 days before chlorophyll levels decline, chloroplast polysomes break down in susceptible, but not in resistant, leaves, and there is a decrease in the chloroplastic/cytoplasmic RNA ratio (Dyer & Scott, 1972). The fungus is confined to the epidermis of the host, so presumably acts through a diffusable intermediary. Chloroplasts are the major site of protein synthesis in the leaf, thus, many of the changes described before could originate from this loss of chloroplast ribosomes so early after infection. In sections of mildewed oak leaves examined under the electron microscope several chloroplasts near infection sites showed markedly fewer ribosomes than their more distant counterparts only 1 day after infection; the same chloroplasts showed some disorganization of the lamella system and a degree of vacuolation (Hewitt, 1976). As infection progressed, damage spread through the leaf along a line perpendicular to the leaf surface and away from the nearest haustorium. This pattern of development lead Hewitt to suggest that the failure of photorespiration to remove oxygen, and of photosynthesis to generate normal amounts of NADPH for the reduction of chlorophyll – protecting carotenoids, might result in photo-oxidation of chlorophylls, so establishing a chain reaction in the leaf.

Scott & Smillie (1966) reported decreases in the activity of some enzymes involved in photosynthesis, and suggested that the failure of photosynthesis to reduce NADP may lead to increased activity of the NADP-requiring pentose phosphate pathway. As Daly (1976) points out, the very fact that often the same enzymes and co-factors are involved in both photosynthesis and respiration makes it difficult to assess their role in the photosynthetic decline. The possibility remains that infection alters metabolic pathways in a qualitative as well as a quantitative way. Waygood *et al.* (1974) found increased phosphoenolpyruvate (PEP) carboxylase activity per unit protein in leaves of wheat two weeks after infection by stem rust – when uredosori and green island were present. They suggested that in green islands the more efficient Hatch & Slack Carbon cycle takes over from the normal Benson & Calvin cycle, and that similar changes may occur in mildewed barley where (as mentioned previously) their recalculation of Scott & Smillie's data showed increased photosynthetic activity per mg chlorophyll after infection. In contrast, Magyarosy, *et al.* (1976) found that when mildewed sugar beet leaves were fed $^{14}CO_2$ for 5 min. they formed fewer organic acids that did healthy controls. The difference was attributable to decreased activity of malate dehydrogenase (–37%) and PEP carboxylase (–50%) in mildewed leaves. Infected tissues, in which non-cyclic photophosphorylation is inhibited, formed more alanine and glutamate plus aspartate, but less 3-phosphoglyceric acid and sucrose, than healthy tissues. In sporulating leaves of rusted bean (*U. phaseoli*) fed with $^{14}CO_2$, aspartic and glutamic acids and their amides had a higher activity and constituted a larger percentage of the total pool of free amino acids than in healthy leaves (Raggi, 1975). Diminution in the labeling of glycine and, more particularly, serine was ascribed to a decrease in photorespiratory rate. In flax infected by rust, *Melamspora lini* (Kakkar, 1966), and in stem rusted wheat (Kiraly & Farkas, 1957), glycolate oxidase activity is reduced by infection. Similarly, in powdery mildewed oak (Hewitt & Ayres, 1977) glycolate oxidase activity per unit total chlorophyll declined, possibly reaching a level that limited photorespiration. In oak it was also found that activity of NAD glyoxylate reductase increased two days after infection. Tolbert (pers. comm.) has suggested that the enzyme functions *in vivo* as a hydroxypyruvate reductase – its rise after infection signifies a swing from glycine/serine synthesis from P-glycolate to serine/glycine synthesis from P-glycerate.

REFERENCES

Allen, P.J. Changes in the metabolism of wheat leaves induced by infection with powdery mildew. *Am. J. Bot.* 29: *425–534* (1942).

Aust, H.J., W. Domes & J. Kranz. Influence of CO_2 uptake of barley leaves on incubation period of powdery mildew under different light intensities. *Phytopathology* 67: *1469–1472* (1977).

Ayres, P.G. Patterns of stomatal behaviour, transpiration and CO_2 exchange in pea following infection by powdery mildew (*Erysiphe pisi*). *J. exp. Bot.* 27: *354–363* (1976).
Black, L.L., D.T. Gordon & P.H. Williams. Carbon dioxide exchange by radish leaves infected with *Albugo candida* measured with an infra red CO_2 analyser. *Phytopathology* 58: *173–178* (1968).
Daly, J.M. The carbon balance of diseased plants: Changes in respiration, photosynthesis and translocation. pp. 450–479, in: Heitefuss, R. & P.H. Williams (eds.), Encyclopedia of Plant Physiology, Vol. 4. Springer-Verlag, Berlin (1976).
Doodson, J.K., J.G. Manners & A. Myers. Some effects of yellow rust (*P. striiformis*) on C^{14} assimilation and translocation in wheat. *J. exp. Bot.* 16: *304–317* (1965).
Duniway, J.M. & R.D. Durbin. Some effects of *Uromyces phaseoli* on the transpiration rate and stomatal response of bean leaves. *Phytopathology* 61: *114–119* (1971).
Dyer, T.A. & K.J. Scott. Decrease in chloroplast content of barley leaves infected with powdery mildew. *Nature* 236: *237–238* (1973).
Edwards, H.H. Biphasic inhibition of photosynthesis in powdery mildewed barley. *Plant Physiol.* 45: *594–597* (1970).
Harding, H., P.W. Williams & S.C. McNabola. Chlorophyll changes, photosynthesis and ultrastructure of chloroplasts in *A. candida* induced 'green islands' on detached *Brassica juncea* cotyledons. *Can. J. Bot.* 46: *1229–1234* (1968).
Hewitt, H.G. The effects of infection by *Microsphaera alphitoides* upon the physiology and growth of Quercus robur. Ph.D. thesis University of Lancaster (1976).
Hewitt, H.G. & P.G. Ayres. Changes in CO_2 and water vapour exchange rate in leaves of *Quercus robur* infected by *Microsphaera alphitoides* (powdery mildew). *Physiol. Plant Pathol.* 7: *127–137* (1975).
Hewitt, H.G. & P.G. Ayres. Changes in glycolate oxidase and glyoxylate reductase in *Quercus robur* infected by *Microsphaera alphitoides*. *Physiol. Plant.* 40: *31–34* (1977).
Inaba, T. & T. Kajiwara. Fixation of $^{14}CO_2$ in the dark by cucumber infected with the cucumber downy mildew fungus. *Ann. Phytopath. Soc. Japan.* 40: *295–298* (1974).
Kakkar, R.K. Decline in glycolic acid oxidase activity in flax leaves infected with *Melampsora lini* (Pers.). *Phytopath Z.* 55: *172–176* (1976).
Kiraly, Z. & G.L. Farkas. Decrease in glycolic acid oxidase activity in wheat leaves infected with *Puccinia graminis* var. *tritici*. *Phytopathology* 47: *277–278* (1957).
Last, F.T. Metabolism of barley leaves inoculated with *Erysiphe graminis* Mérat. *Ann. Bot. N.S.* 27: *685–690* (1963).
Lewis, D.H. Concepts in fungal nutrition and the origin of biotrophy. *Biol. Rev.* 48: *261–278* (1973).
Liu. P. D.J. Wallace & J.L Ozbun. Influence of translocation on photosynthetic efficiency of *Phaseolus vulgaris* L. *Plant Physiol.* 52: *412–415* (1973).
Magyarosy, A.C., P. Schürmann & B.B. Buchanan. Effect of powdery mildew infection on photosynthesis by leaves and chloroplasts of sugar beets. *Plant Physiol.* 57: *486–489* (1976).
Majernik, O. Water balance changes of barley infected by *Erysiphe graminis* D.C. f. sp. *hordei* Marchal. *Phytopath. Z.* 53: *145–153* (1965).
Martin, J.J., R.E. Stuckey, G.R. Safir & A.H. Ellingboe. Reduction of transpiration from wheat caused by germinating conidia of *Erysiphe graminis* f. sp. *tritici*. *Physiol. Plant Pathol.* 7: *71–77* (1975).
Montalbini, P. & B.B. Buchanan. Effect of a rust infection on a photophosphorylation by isolated chloroplasts. *Physiol. Plant Pathol.* 4: *191–196* (1974).
Livne, A. Photosynthesis in healthy and rust affected plants. *Plant Physiol.* 39: *614–621* (1964).
Paulech, C. & A. Haspelova-Horvatovicova. Photosynthesis, plant pigments and transpiration in healthy barley and in barley infected by powdery mildew. *Biologia* (Bratislava) 25: *477–487* (1970).
Raggi, V. Incorporation of $^{14}CO_2$ in amides and amino acids in rusted bean leaves. *Can. J. Bot.* 53: *1244–1251* (1975).
Scott, K.J. Obligate parasitism by phytopathogenic fungi. *Biol. Rev.* 47: *537–572* (1972).
Scott, K.J. & R.M. Smillie. Metabolic regulation in diseased leaves. I. The respiratory rise in barley leaves infected with powdery mildew. *Plant Physiol.* 41: *289–297* (1966).

Sempio, C., O. Majernik & V. Raggi. Water loss and stomatal behaviour of bean (*Phaseolus vulgaris* L.) infected by *Uromyces appendiculatus* (Pers.) Link. *Biologia* (Bratislava) 21: *99–103* (1966).
So, M.L. & L.B. Thrower. The host-parasite relationship between *Vigna sesquipedalis* and *Uromyces appendiculatus*. I. Development of parasitic colonies and the pattern of photosynthesis. *Phytopath. Z.* 85: *320–332* (1976).
Stallnecht, G.F. & C.J. Mirocha. Fixation and incorporation of CO_2 in ribonucleic acid by germinating uredospores of *Uromyces phaseoli*. *Phytopathology* 61: *400–405* (1971).
Thornton, J.D. & R.C. Cooke. Accumulation of dark-fixed carbon compounds in pustules of *Albugo tragopogonis*. *Trans. Br. Mycol. Soc.* 54: *483–485* (1970).
Waygood, E.R., L.Y. Pao & H.R. Godavari. Stimulation of phosphoenolpyruvate carboxylase activity in rust-infected wheat leaves. *Experientia* 15: *986–988* (1974).
Zaki, A.I. & C.J. Mirocha. Carbon dioxide fixation by rust-infected bean plants in the dark. *Phytopathology* 55: *1303–1308* (1965)

Author's address:
Department of Biological Sciences
University of Lancaster
Bailrigg, Lancaster LA1 4YQ
United Kingdom.

THE EFFECT OF FOLIAR PATHOGENS ON THE LEAF PHOTOSYNTHETIC CARBON DIOXIDE UPTAKE OF BARLEY

DAVID HABESHAW
(Edinburgh, U.K.)

ABSTRACT

The changes in photosynthetic carbon dioxide uptake in barley during the progress of infections with foliar fungal pathogens were followed in four disease and barley cultivar combinations. The combinations were chosen to produce compatible or susceptible reactions and necrotic or resistant reactions and the results showed that leaf symptom responses were not necessarily indicative of equivalent responses in the changes in net photosynthetic carbon dioxide uptake by the leaf during the course of the disease. The use of photosynthetic measurements to characterize host/pathogen interactions and the implications of the observed photosynthetic changes in relation to the observed levels of resistance of the barley cultivars used in the experiment were discussed.

INTRODUCTION

Of all the stresses to which crop plants are subjected the commonest must be the invasion or attempted invasion by a fungal pathogen. The response of the plant to this challenge is complex depending as it does on three groups of factors. These are the environmental factors of climate and nutrition, the genetic make-up of the plant and the genetic make-up of the pathogen. If the environmental effects are excluded there remain potentially as many different host/pathogen responses as there are combinations of the genetic variants of both organisms. This genetic variability is easily seen if the visual symptoms within a single disease are examined. Not only are the differences in the rate of progress of the infection very marked but the type and extent of the lesions on the leaf are very variable. While this is of interest to the pathologist discriminatina between races of fungal pathogens or the degree of resistance in crop cultivars, the degree to which these variable leaf symptoms represent real differences in the physiological reaction of the plant to the disease organism deserves consideration. Dr. Lennard of the Botany Department of the East of Scotland College of Agriculture has been investigating the foliar diseases of barley with particular reference to resistance to cereal mildew and *Rhynchosporium* leaf blotch. Symptoms may be graded into a 'compatible' or 'susceptible' response in which the fungus spreads without forming localized lesions and a 'necrotic or resistant' response in which the fungus forms localized lesions with restricted areas of necrotic leaf tissue. The availability of standardized material infected in a reproducible manner from Dr. Lennard's experiments made possible an investigation of the photosynthetic behaviour of barley leaves during infection, with the aim of determining the extent to which the visual symptoms of disease, used as the basis of classification in disease studies, represented the variety of physiological responses actually present.

This was considered important for the programme of research in the department whose overall aim was the identification and characterization of plant-/pathogen interactions which contribute to some measure of resistance in non-immune varieties. The wide range of visual symptoms presently observed may represent fundamentally different host/pathogen responses each requiring a separate consideration in assessing the possible outcome of infection in a crop. If, however, the range of visual symptoms masks a much smaller number of possible physiological responses then the problem of estimating the consequences of infection in individual plants and plant populations is simplified.

The work reported here represents a preliminary investigation into the possibility of using measurements of leaf photosynthesis during infection as a means of relating leaf symptoms to physiological aspects of the host/pathogen interaction.

MATERIALS AND METHODS

Experimental material

Two diseases were investigated, barley mildew and *Rhynchosporium* leaf blotch. The mildew organism, *Erysiphe graminis* DC ex Merat f. sp. *hordei* Marshal, was a mixed race inoculum producing a compatible/susceptible reaction on *Hordeum sativum cv.* Golden Promise and a moderately resistant/necrotic response on *cv.* Maris Mink. The leaf blotch organism, *Rhynchosporium secalis* (Oudem.) J.J. Davis UK race 1, produced a susceptible response in Maris Mink and a moderately resistant/necrotic response in Golden Promise. Both varieties exhibited necrosis following *R. secalis* infections.

The barley plants were grown in Leavington 1 potting compost in 15 cm pots in a cool glasshouse. Supplementary lighting was available for use in the winter months. Inoculation was carried out, when the tip of the 4th leaf was visible, by gently shaking infected plants over the plants to be infected in the case of mildew, and by spraying the plants to be infected with a spore suspension obtained from 14-day-old cultures on lima bean agar, in the case of leaf blotch. After inoculation all the plants were covered with a transparent plastic hood to prevent any further cross or reinfection and the plants were kept covered until used for experiments.

In all experiments only fully expanded 2nd leaves were used since it was found that net carbon dioxide uptake differed considerably between leaves of different ages and positions on the plant. Restricting the leaves used for an experiment to one age and position of leaf on the plant increased the reproducibility considerably. However, even within such a restricted group the net photosynthetic carbon dioxide uptake varied during the 12–14 days necessary to complete an experiment, so for every batch of infected leaves taken for analysis a control batch of leaves of the same age was taken. The results were than expressed in all cases as a percentage of the value in the control leaves.

Experimental details

Plants were grown three to four to a pot until the second leaf had fully emerged. All the pots in a single sowing batch were randomly divided into two treatment groups. One group was inoculated, the other was left unsprayed or sprayed with water depending on the disease being used. All the pots were covered and left in the glasshouse. At intervals over the next 12–14 days pots of plants were removed at random from the control and treatment groups. Three leaves from each pot were taken and their photosynthetic carbon dioxide uptake under standard conditions was estimated using a flexible transparent grid ruled in 0.5 cm^2 squares. Then independent counts were made on each leaf and their average taken. If sugar estimations were to be carried out then the leaf was cut into 5 mm long segments and stored, frozen at −40° C, until required.

Determination of net photosynthetic CO_2 uptake

Leaves were cut from the plant 1–2 cm below the ligule and the cut ends immediately placed in distilled water. From here they were transferred to a leaf chamber of the design shown in Fig. 1. The leaf being cut about 5 mm above the ligule just prior to insertion through the slit in the plastic water pouch. The net carbon dioxide uptake was determined under standard conditions of light and temperature using a M.S.A. Lira 200 infra-red gas analyzer in the circuit shown in Fig. 2. The air supply was taken from a compressed air cylinder previously calibrated for CO_2 content against a standard gas mixture. The air supply was dried before entering the gas analyzer sample and reference cells by being passed over silica gel and calcium chloride in drying tubes. The only unusual item in the layout was the spore trap upstream of the sample cell which was necessary to prevent fungal spores from passing into the instrument. Illumination was provided by a bank of four 85 W fluorescent tubes supplemented by a 750 W tungsten filament projector bulb. The light sources were movable and enabled the light at the leaf surface to be adjusted between 6,000 and 10,000 lx depending on their distance from the leaf chamber. Light levels were measured by a small silicon photocell mounted in the leaf chamber and calibrated against an EEL portable photometer before each experiment. The temperature was not controlled, but the temperature of the air passing across the leaf was recorded and reached a constant value of between 24° C and 27° C after the apparatus had been running for some time. Variations in temperature during a single experiment were less than 1°C after the apparatus had warmed up and so no corrections were made for temperature. Flow rates through the apparatus were adjusted to keep carbon dioxide depletion rate below 20 ppm in its passage through the leaf chamber. In general the depletion rate was kept as low as was consistent with obtaining a reliable measurement.

The procedure for making a measurement of net photosynthetic uptake was as follows. A leaf was placed in the leaf chamber which was then switched in circuit with the analyzer and the flow rate through the chamber adjusted to give a suitable recording of CO_2 depletion on the analyzer. The gas flow was then switched through an identical leaf chamber containing no leaf and the zero position of the recorder noted. Once this had achieved a steady value the flow

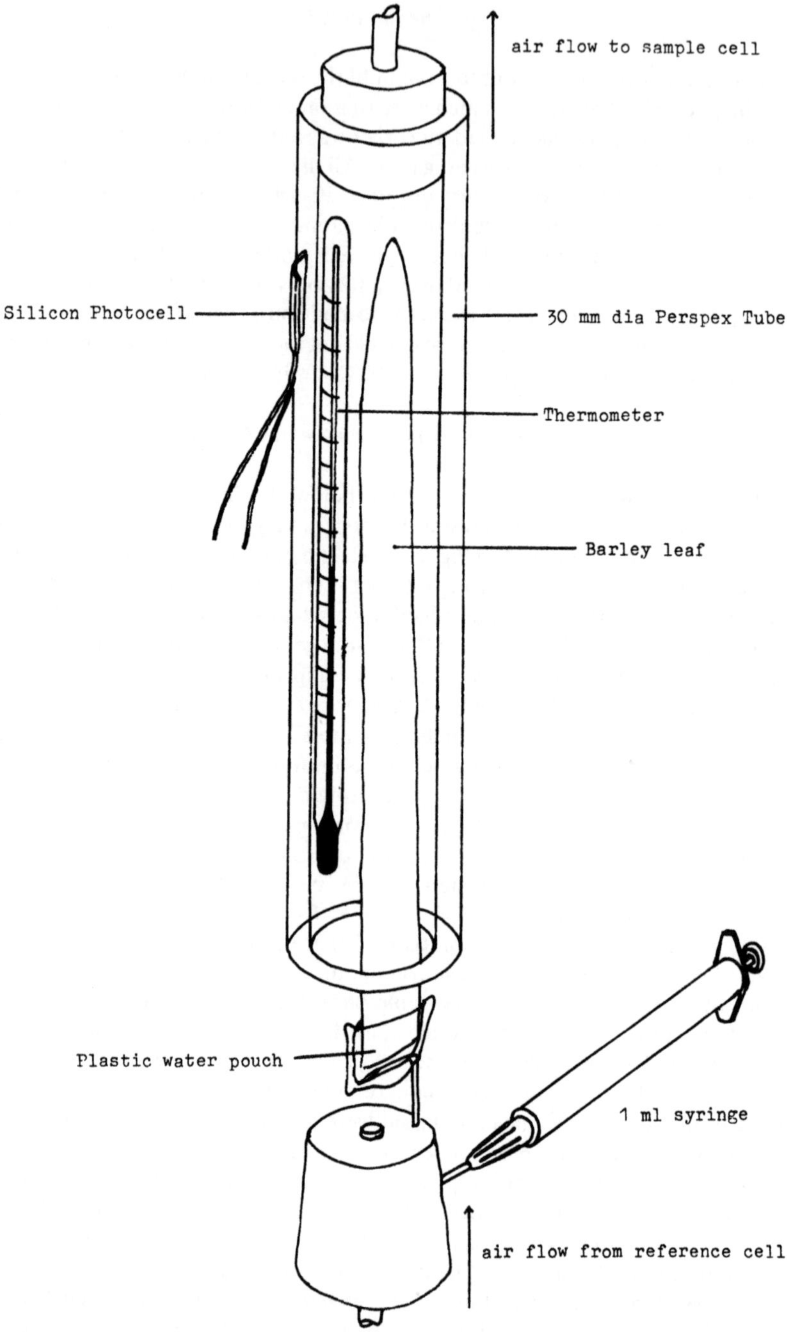

Fig. 1. Diagrammatic drawing of leaf chamber used in experiments with the MSA L.I.R.A. Model 200 gas analyzer.

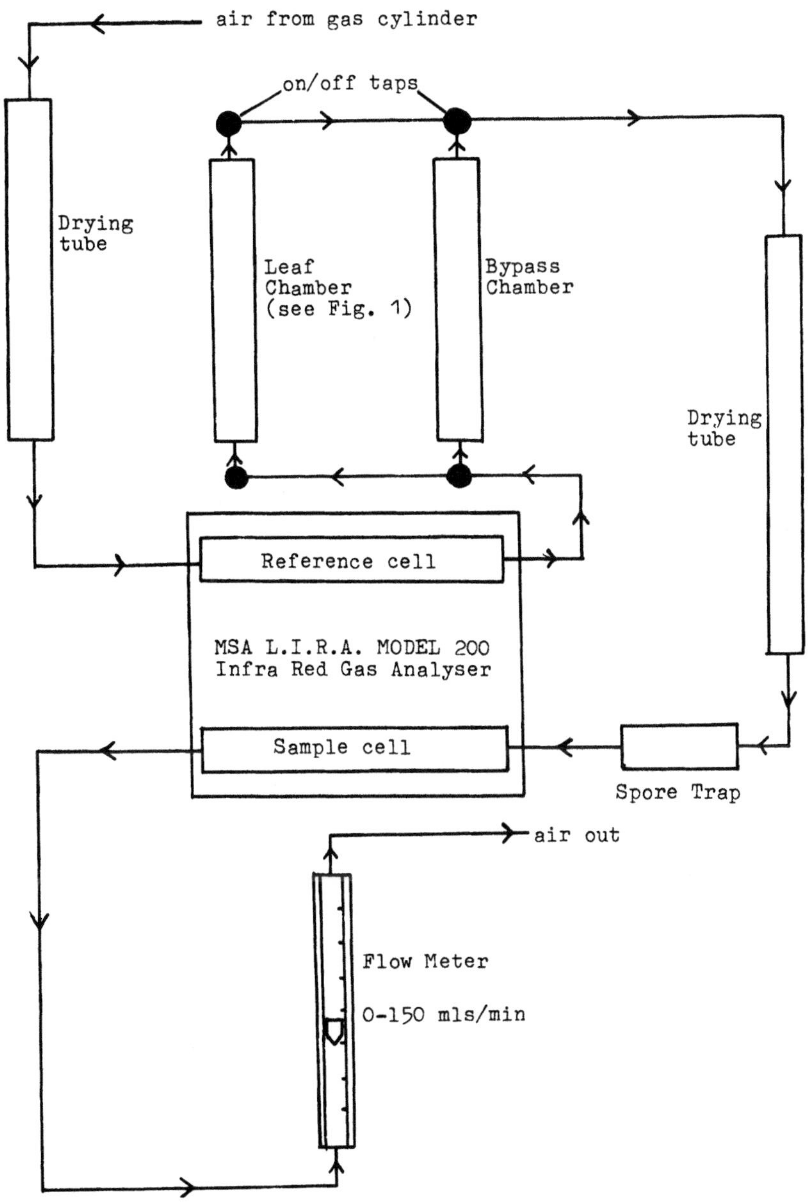

Fig. 2. Diagrammatic representation of the apparatus used to measure carbon dioxide uptake.

was switched back to the leaf chamber and left until the recorder showed a steady value for carbon dioxide depletion for 10 minutes. The light level was monitored continuously throughout, as was the temperature.

Determination of sugar levels

Frozen chopped leaves stored frozen after photosynthesis measurements were thawed by being covered with 2 ml of boiling distilled water. The tubes were then heated for 30 minutes in a boiling water bath and cooled to room temperature. The volume of extract was then adjusted to 2 ml and a 1 ml aliquot was taken for analysis using the method of Somogyi (1953). The results were calibrated using a standard glucose dilution series carried out for each batch of determinations.

RESULTS

Fig. 3 shows the percentage changes in net photosynthetic carbon dioxide uptake for six experiments. In general, the progress of both diseases was faster in Golden Promise than Maris Mink but there was a considerable difference between the rate of progress of the same disease in the same plant at different times. Fig. 4 relates the time from inoculation to the progress of the disease as measured by the area of leaf visibly affected by the disease organism. These two figures show that the spread of infection was paralleled by a reduction in the net photosynthetic uptake of CO_2 by the leaf. The usefulness of presenting the results in this way was limited since comparison between different experimental runs was not possible because of the different rate of disease development at different times. However, if the percentage reductions in net photosynthetic carbon dioxide uptake were expressed on the basis of the percentage of leaf affected by the organism, then the results of experiments carried out at different times could be compared directly, providing the experiments were similar in other respects.

The inclusion of the results of several experiments carried out over an extended period, allowed a statistical test of the hypothesis that the percentage reduction in the net photosynthetic carbon dioxide uptake was equal to the percentage of the leaf area colonized by the fungus. Perfect agreement with this hypothesis would give line A in Fig. 5. A necessary consequence of this hypothesis would be that the net photosynthetic carbon dioxide uptake by the uncolonized portions of the leaf would be the same as that of the control tissue. A recasting of the data calculating net carbon dioxide uptake rates on the basis of uncolonized leaf area, rather than total leaf area, should lie along X axis, line B on Fig. 5. The fit of the data to the line dictated by the hypothesis for the four disease/cultivar interactions are shown in Table 1 below. The response of Golden Promise to mildew infection fits the hypothesis line A in Fig. 5a surprisingly well, the best straight line is shown by the solid line. It was an excellent fit to the hypothetical line for both gradient and intercept on the Y axis. The other combinations were good fits for slope, but the intercepts on the Y axis were much too high for a good fit. The hypothesis cannot be readily accepted in these cases without modification. Since the line must go through or near the origin the easiest way of accomplishing this would be to replace a single line with two. A sharply rising line in the early stages of infection, line C in Fig. 5b and a portion parallel to, but higher than, that of the original hypothesis during the later stages of infection, line D in Fig. 5b. In the case of mildew infected Maris Mink there was sufficient data to do this if the data for levels of infection between 0 and 30% were analysed separately. The result is shown in Table 2. The slope of the line given by this data was significantly greater than that for the complete data presented in Table 1 and the intercept on the Y axis was somewhat closer to the zero value required. The reduction in photosynthetic capacity appeared to progress in two stages, an initial phase of fairly rapid loss of photosynthetic capacity followed by a further reduction from this level proportional to the loss of leaf to the disease in the later stages. There

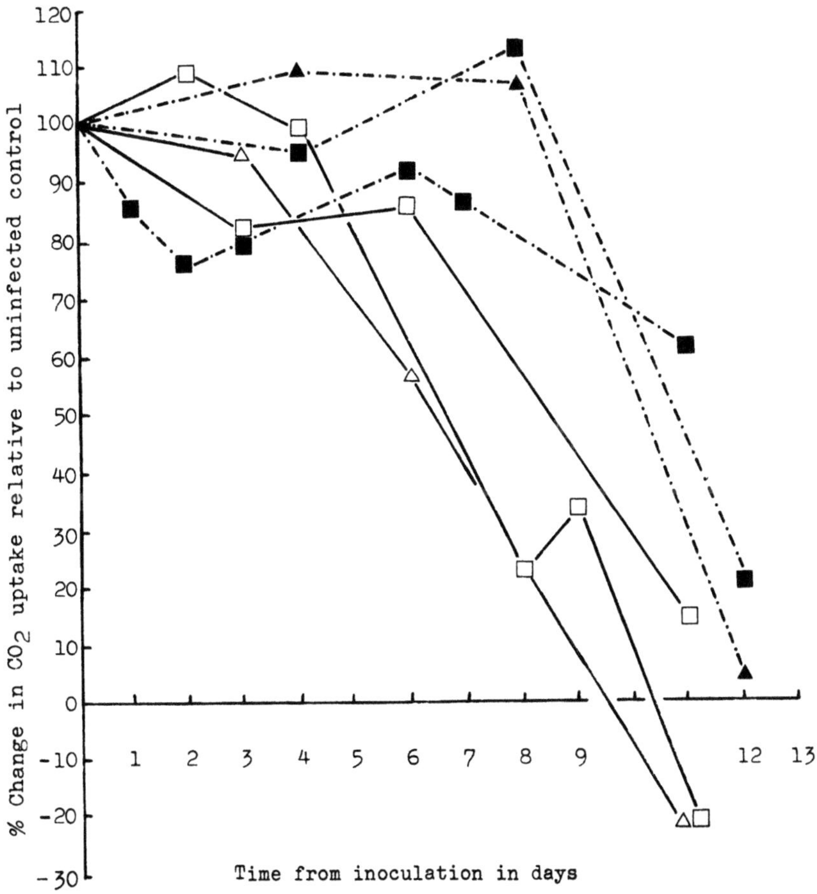

Fig. 3. Percentage changes of carbon dioxide uptake in infected plants relative to uninfected controls with time from inoculation. Open symbols refer to the Barley variety Golden Promise. Black symbols refer to the Barley variety Maris Mink. Triangles refer to *Rhynchosporium* infections. Squares refer to Mildew infections. Solid lines refer to Golden Promise infected with Mildew or *Rhynchosporium*. Dashed lines refer to Maris Mink infected with Mildew or *Rhynchosporium*.

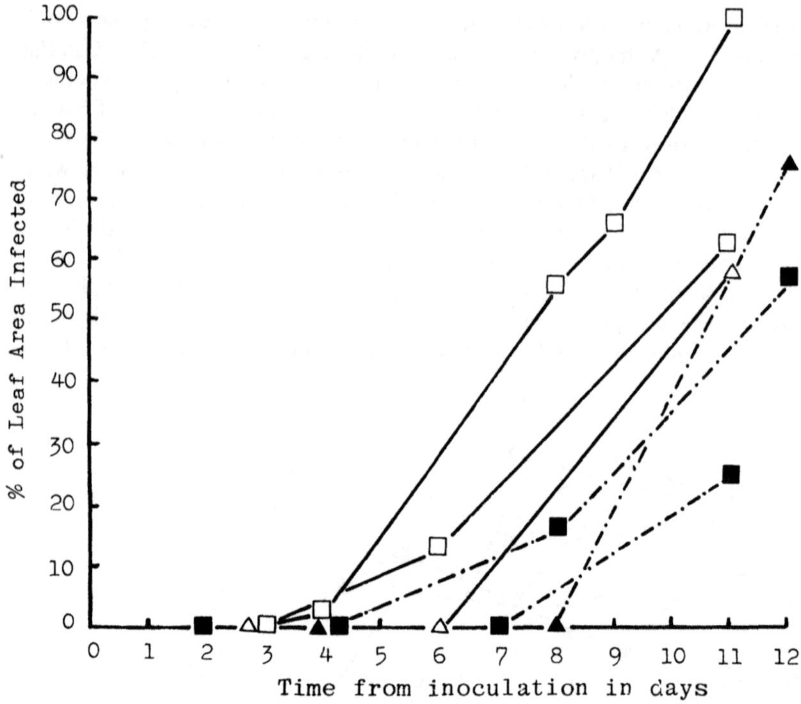

Fig. 4. Percentage of the leaf area colonized by the disease at different times after inoculation. Open symbols refer to the Barley variety Golden Promise. Black symbols refer to the Barley variety Maris Mink. Triangles refer to *Rhynchosporium* infections. Squares refer to Mildew infections. Solid lines refer to Golden Promise infected with Mildew or *Rhynchosporium*. Dashed lines refer to Maris Mink infected with Mildew or *Rhynchosporium*.

Table 1. Regression analysis of the percentage reduction in carbon dioxide uptake (Y) against the percentage of the leaf area affected by *E. graminis* or *R. secalis*.

		Disease	
Variety	Symbol	Mildew	Leaf Blotch
	r	0.862	0.515
	b	1.052	0.999
Golden Promise	a	2.25	34.34
	N	33	12
	t(b)*	0.57	0.004
	t(r)*	8.74	3.4
	r	0.824	0.903
	b	1.114	1.019
Maris Mink	a	11.18	17.96
	N	19	12
	t(b)*	0.83	0.14
	t(r)*	6.21	6.15

* t (p 0.05) for 11 degrees of freedom = 2.4.

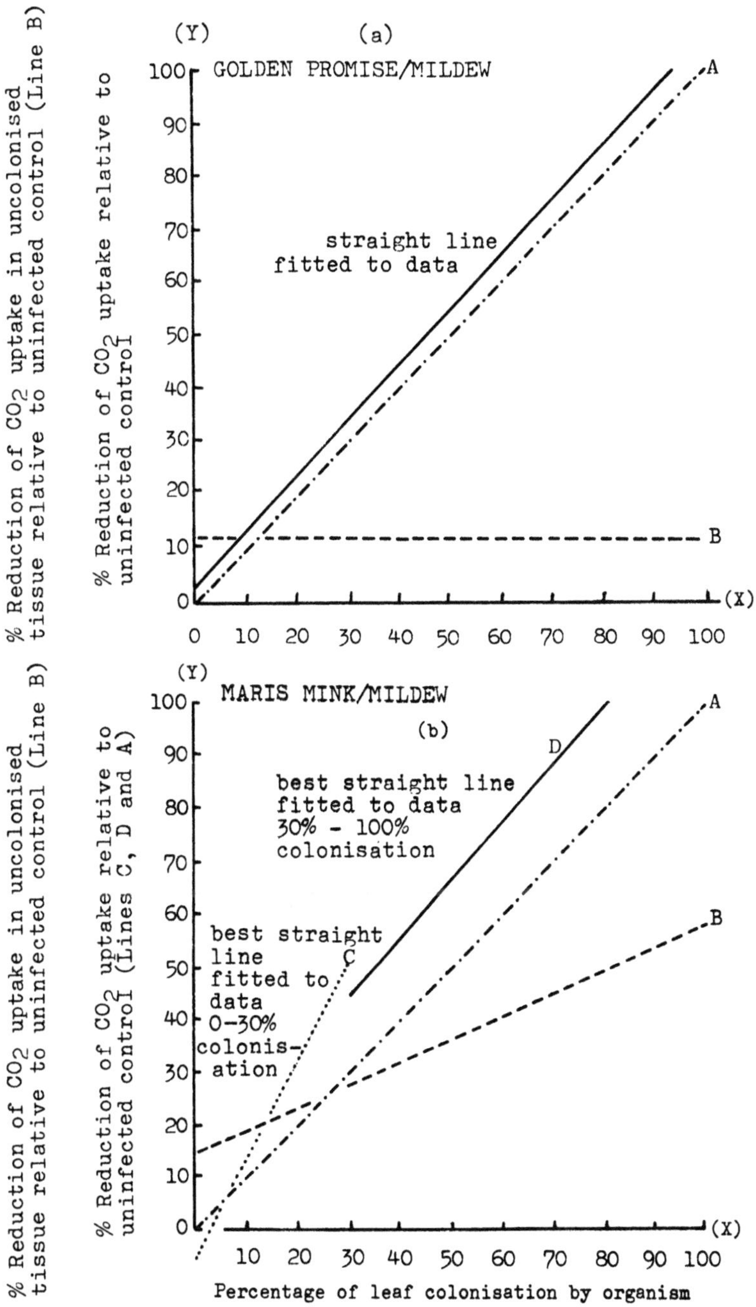

Fig. 5. Graphs showing the regression lines for the percentage reduction of carbon dioxide uptake relative to uninfected control and the percentage reduction of carbon dioxide uptake in uncolonized leaf tissue relative to uninfected control (YX) against the percentage of leaf colonized by the infective organism in the case of Golden Promise and Maris Mink infected with cereal mildew.

was insufficient data to repeat the test for *Rhynchosporium* infections, but the data in Table 1 seem to indicate that they may well follow the Maris Mink/mildew pattern.

Table 2. Regression analysis of the percentage reduction in carbon dioxide uptake (Y) against the percentage of leaf area of Maris Mink colonized by *E. graminis* during the initial period of infection. The initial period of infection being defined as up to 30% of the leaf area colonized by the infective organism.

Disease organism	Symbol	Mildew
Maris Mink	r	0.81
	b	1.889
	a	−6.19
	N	10
	t(b)*	4.29
	t(r)*	4.1

* t (p 0.05) for g degrees of freedom = 2.5.

As a check on the conclusions reached by this analysis of the data, the hypothesis that the loss in photosynthetic capacity was confined to infected tissue only was tested by calculating the regression analysis for the 4 combinations of host and pathogen, with data recalculated on the assumption that only uninfected tissue had any photosynthetic capacity. The results are given in Table 3. The mildew infected Golden Promise data satisfies this test supporting the results in Table 1, and showing no tendency for the photosynthetic capacity of uninfected leaf tissue to fall as the disease progresses. (See line b in Fig. 5a.) In the other 3 cases the loss of photosynthetic capacity in

Table 3. Regression analysis of the percentage reduction in carbon dioxide uptake in the portion of the leaf uncolonized by the infective organism (Y) against the percentage of the leaf area visibly affected by *E. graminis* or *R. secalis* (X).

		Disease	
Variety	Symbol	Mildew	Leaf Blotch
Golden Promise	r	0.001	0.903
	b	−0.0009	1.849
	a	11.66	21.59
	N	30	12
	t(b)*	0.00	13.6
	t(r)*	0.005	15.5
Maris Mink	r	0.501	0.904
	b	0.505	0.793
	a	12.49	21.19
	N	19	12
	t(b)*	2.4	5.87
	t(r)*	2.8	15.6

* t (p 0.05) for 11 degrees of freedom = 2.4.

regions of the leaf in which no fungus was present increases steadily as the disease progresses. (See line B in Fig. 5b.).

Leaf reducing sugar levels

The declining photosynthetic capacity of the leaf as disease progresses should have been reflected by a decline in leaf reducing sugar contents. In one series of experiments the sugar contents of leaves were determined after determinations of photosynthetic capacity had been made. The results were considered in two parts. The changes in the initial stages of infection, this being defined as before as up to 30% of the leaf infected, and the pattern of change over the whole period of infection. The changes in sugar concentration for leaves up to 30% infected are given in Table 4. This shows that in three of the four combinations there were significant increases in reducing sugar concentration over the control uninfected leaf. In the fourth case of Golden Promise infected with *R. secalis* there was a slight but not significant fall in reducing sugar levels.

Table 4. Concentrations of reducing sugar in leaves during the initial period of infection. The initial period of infection being defined as up to 30% of the leaf area colonized by the infective organism. Sugar concentrations expressed as mg per square centimeter of leaf.*

Treatments	Barley Varieties Maris Mink	Golden Promise
Control (uninfected leaf)	0.0230	0.0382
Mildew affected leaf	0.0415	0.0485
Leaf Blotch affected leaf	0.0415	0.0347

* SE ± 0.0073.

This result raises the possibility that it was the rise in leaf reducing sugar levels which provided the photosynthetic inhibitory mechanism acting in the uninfected leaf tissue. If this was true then the rise in leaf sugar contents should show a positive correlation with the reduction in photosynthetic rates. After expressing both these factors as percentages of the appropriate controls, the correlation coefficients for the four combinations of host and pathogen were calculated. The results are presented in Table 5. This shows that three of the correlations were in fact negative, so only one of the cases has the correct sign. In none of the cases does the correlation coefficient (r) approach the 5% level of significance. In this evidence there was no clear case for supposing that photosynthetic reductions can be associated with rising sugar levels in the initial stages of infection.

The changes in sugar contents over the whole period of infection are shown on a time from infection basis in Fig. 6. These show that while the mildew infections showed a rise in the early stages of infection followed by a rapid fall as the disease entered its final stages. In contrast the reducing sugar levels in *R. secalis* infected leaves tended to rise in the terminal stages of infection.

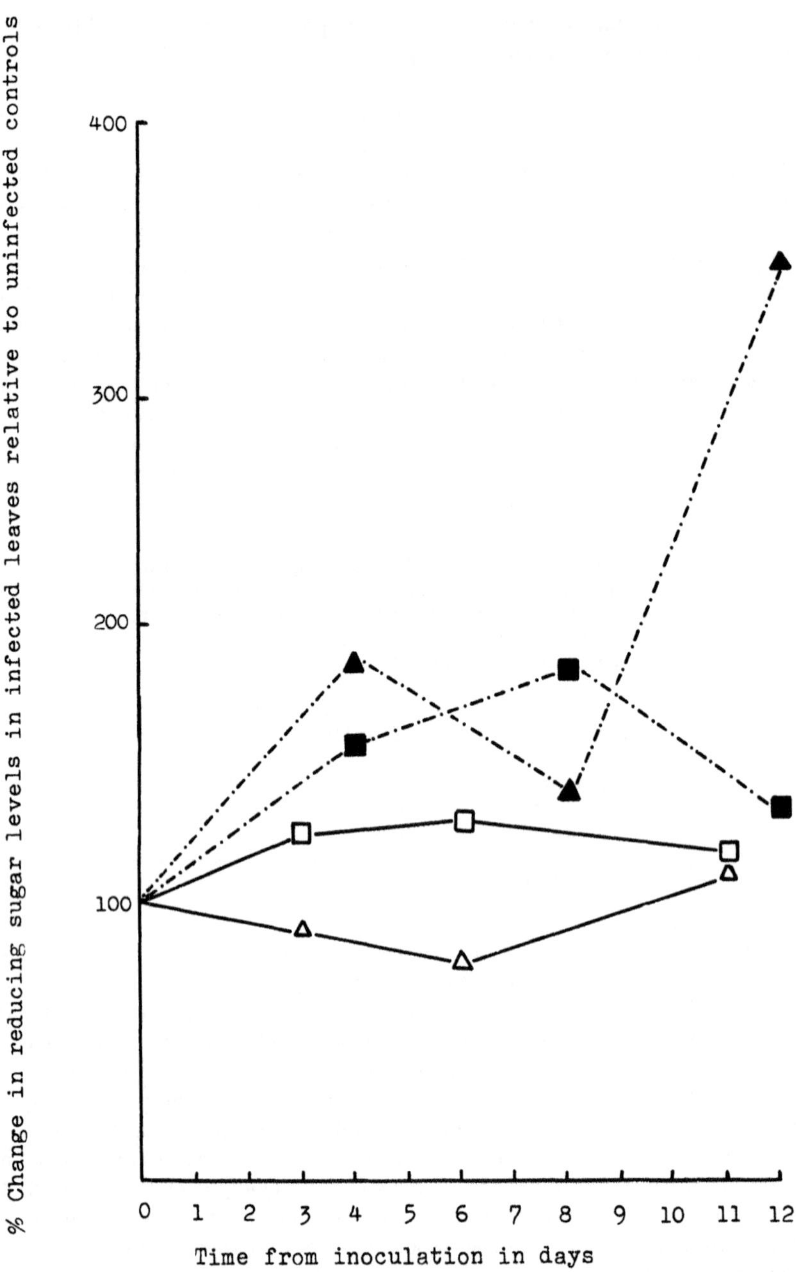

Fig. 6. Percentage changes in leaf reducing sugar content in infected plants relative to uninfected controls with time from inoculation. Open symbols refer to the Barley variety Golden Promise. Black symbols refer to the Barley variety Maris Mink. Triangles refer to *Rhynchosporium* infections. Squares refer to Mildew infections. Solid lines refer to Golden Promise infected with Mildew or *Rhynchosporium*. Dashed lines refer to Maris Mink infected with Mildew or *Rhynchosporium*.

Table 5. Correlation coefficients and levels of significance of the correlation coefficients for the percentage increases in leaf reducing sugar content (Y) against the percentage reductions in carbon dioxide uptake during the initial period of infection (X). The initial period of infection being defined as up to 30% of the leaf area visibly affected by *E. graminis* or *R. secalis.**

Variety	Symbol	Disease Mildew	Disease Leaf Blotch
	r	−0.719	−0.538
Maris Mink	p	0.1	0.2
	r	0.604	−0.170
Golden Promise	p	0.2	0.8

* N = 6.

Expressing the results as the percentage changes in sugar levels relative to the controls and the percentage of leaf occupied by the fungus relative to the controls and calculating the correlation coefficient in each of the four cases gave the result presented in Table 6.

Table 6. Correlation coefficients and levels of significance of the correlation coefficients for the percentage increases in leaf reducing sugar content (Y) against the percentage of leaf area visibly affected by *E. graminis* or *R. secalis* (X).*

Variety	Symbol	Mildew Disease	Leaf Blotch
	r	−0.323	0.52
Golden Promise	p	0.7	0.2
	r	−0.560	0.961
Maris Mink	p	0.1	0.01

* N = 9.

Both mildew infections were negative and the correlation coefficients were not significant at the 5% level. This was to be expected from the form of the sugar level changes with time in Fig. 6. Both *R. secalis* infection showed a positive correlation and in the case of Marris Mink the progressive rise in reducing sugar content with the rise in the percentage of leaf occupied by the fungus was significant at the 5% level.

DISCUSSION

The results show two different patterns of net photosynthetic carbon dioxide uptake change during the progress of disease in the four host/pathogen combinations. The two types of response can be defined. The 'minimum' response in which the reduction in net photosynthetic capacity was directly proportional to the reduction in uninfected leaf area and the 'wider' response where the reduction in net photosynthetic capacity was proportionately greater than the reduction in uninfected leaf area. The 'minimum' response of photosynthesis was shown by the Golden Promise plants infected with mildew. The 'wider' response by the other three combinations which were Golden Promise/leaf blotch, Maris Mink/mildew and Maris Mink/leaf blotch. There were thus two types of leaf symptom and two types of net photosynthetic capacity response to infection. While there seemed to be reasonable evidence to associate the necrotic/resistant characteristics with the 'wider' photosynthetic response compatible/susceptible characteristics were not clearly associated with the 'minimum' photosynthetic response. Susceptibility cannot therefore be taken as indicating a specific type of physiological response if different pathogens are involved.

There were thus, problems requiring interpretation and explanation. Firstly, why in the case of Maris Mink/leaf blotch, were compatible/susceptible leaf symptoms associated with the 'wider' photosynthetic response? Secondly, what explanations can be proffered for the mechanism of photosynthetic reduction and finally, what contributions can results of this sort make to an understanding of the disease in crops under field conditions? No special physiological explanations were necessary to explain the 'compatible/susceptible' response of the Golden Promise/mildew combination and its association with the minimum photosynthetic response. Cells penetrated by fungal haustoria rapidly lose their chlorophyll and presumably their capacity for photosynthesis (Dyer, 1972). The leaf, however, remains turgid till very late in the infection and tissue unaffected by the fungus stays green even if completely surrounded by infected tissue (Fig. 7a). The mycelium spreads only on the leaf surface so its extent was easily seen. Observation of leaves infected with mildew and showing a compatible/susceptible response showed that the extent of the infected area was easily determined, there was no collapse of the leaf until late in infection, uninfected tissue appeared normal, and there were no areas of necrotic tissue surrounding primary lesions (as exemplified in Fig. 7b). The effect of infection on the photosynthetic capacity of the leaf was almost completely accounted for by the loss of photosynthesising cells to the fungus. The slight deviation of the actual line from the hypothetical line in Fig. 5a may well be accounted for by the rise in respiratory rates observed by other workers (Scot, 1966). The technique was not sensitive enough to detect changes in carbon dioxide output in the dark, since in the leaves examined these were about an order of magnitude lower in photosynthetic rates in the initial stages of infection. However, tests of the sensitivity of the apparatus showed that respiration rate increases high enough to account for the observed photosynthetic reductions would have been detectable. The observation in these experiments of an increase reducing sugar content in the initial stages of Golden Promise/mildew infections may well be related to a higher respiration rate in this period.

The reason why 'susceptibility' leaf symptoms in Maris Mink/leaf blotch were associated with a 'wider' photosynthetic response may well be associated with the way in which this organism exploits its host. *R. secalis* after penetrating the leaf spreads its mycelium subcuticularly, the epidermal cells are killed by the secretion of enzymes (Olutiola, 1973) and the fungus grows by feeding on cell

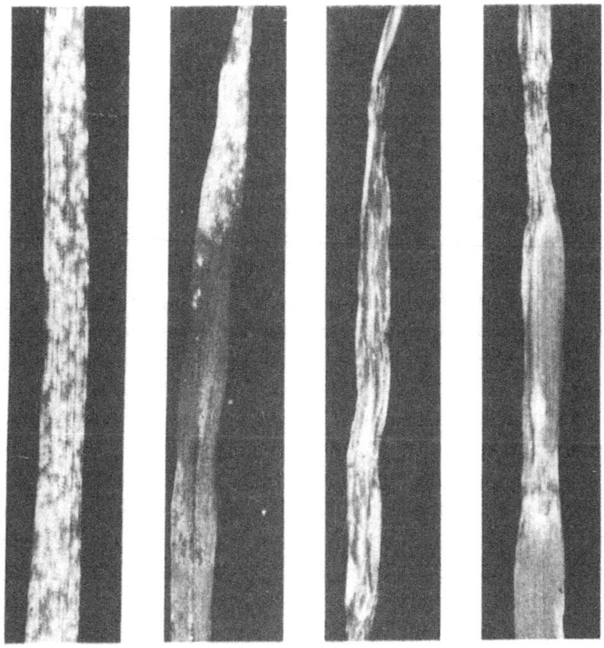

Fig. 7 (a) Mildew infected leaf; (b) Mildew infected leaf showing necrosis; (c) leaf blotch infected leaf showing extensive lesions; (d) leaf blotch infected leaf showing restricted lesions.

exudates. The area occupied by the fungus rapidly becomes water soaked, turns brown and eventually dries out producing a twisted, distorted leaf (Fig. 7c and 7d). The essential difference between susceptibility and resistance in this disease may rest, at least in part, with the rate of spread of the fungus, the rate of development of necrosis and the ability of the leaf to resist desiccation of infected areas. It is the rapid collapse of infected tissue which distinguishes the resistant from the susceptible reaction, not an ability to tolerate the presence of the fungal hyphae. So, whatever the symptomatic appearance of the leaf in *R. secalis* infections, for a given level of infection a similar number of cells were being destroyed in the same way, so there was no reason to suppose that a different type of physiological reaction would be present in cases of susceptible and resistant varieties. The differences, if present, would be of degree only. In the case of mildew however, a real difference exists between the type of response to infection associated with the two symptom types. In the compatible /susceptible reaction described previously the invaded cells remain alive till late in the infection process. The necrotic/resistant response results in the rapid death and desiccation of areas of the leaf penetrated by the fungal hyphae. This reaction has something in common with *R. secalis* infections and the physiological response would be expected to be similar to that of leaf blotch infections. A plausible explanation could be advanced for the association of the 'wider' type of photosynthetic response with a variety of leaf symptoms, but an explanation of the physiological basis of this response was required. Now the 'wider' photosynthesis response was characterized by a more rapid loss of photosynthetic activity in the initial stages of infection than could be accounted for by the loss of cells to the fungus. The response appears to be independent of the type of leaf symptom presenting and was not invariably associated with the appearance of localized 'necrotic' lesions on the leaf. However, it did appear to be associated with rapid cell death and the desiccation of areas of infected tissue within the leaf. It could be argued that with *R. secalis* infections the hypha spread subcuticularly and thus the extent of infection was harder to see and as a result the spread of the disease was underestimated, especially in the case of Maris Mink/leaf blotch where the absence of necrotic 'borders' to the lesions added to the difficulty. However, no such consideration applies in the case of Maris Mink/mildew which showed both necrotic lesions and a surface spread of mycelium. Underestimation of the area of leaf occupied by the fungus was unlikely to be the reason for the 'wider' photosynthetic effect. If the effect was not an artefact of the method used then three possible explanations remain. The respiration of the leaf could be raised as a whole or in the infected area. The respiration of the fungus itself could be responsible or the presence of the fungus could be inhibiting photosynthesis in the leaf as a whole by disrupting the working of the leaf physiological and biochemical processes. This latter state of affairs could be brought about by the presence of fungal metabolites in the leaf or by disruption of the water balance in the leaf or by interference with the distribution of photosynthesate within the leaf. There is no doubt that respiration rates do rise in response to infection but this is more marked in the later stages of infection where the existence of large areas of senescing leaf complicates the issue. Leaving aside the possibility of large increases in respiration rate in the terminal stages of infection, it was unlikely that increases in the respiration rate large enough to account for the observed reductions, would not be reliably observed even with the relatively insensitive technique used. The fact that the photosynthetic response of Golden Promise/mildew equates to the loss of cells to the fungus as it does, indicated that while rises in respiration could have been present they were small relative to the reduction in photosynthetic rate.

The argument that the respiratory rate of fungus itself was an important factor can be ruled out since only small respiration rises were consistent with the results in Golden Promise/mildew combination even though the extent of fungal growth and presumably bulk as well, was at least comparable with the other combinations tested. Moreover, in those combinations in which necrotic/resistant symptoms developed the growth of the fungus was restricted as a consequence, and as a result the development of fungal mycelium relative to the host tissue was likely to be smaller than in the case of Golden Promise infected with the *E. graminis*.

It appears likely, therefore, that the presence of the fungus in the leaf interferes in some manner with photosynthesis in addition to the reduction brought about by cell destruction. This could be brought about by fungal metabolites diffusing into uninfected tissues and this was a plausible alternative but there was nothing in the results of these experiments which could support or contradict this viewpoint. However, the results did allow some comment on the possibility that the reductions in photosynthesis could have been brought about by a disturbance in the water balance and assimilate distribution within the leaf.

The best case for water balance disturbances affecting photosynthesis was for *R. secalis* infections where there was a rapid desiccation of infected regions of the leaf under all but the most humid conditions. This was particularly marked during the periods when the leaf was in the dry air of the gas analyzer stream. With leaf blotch infections where one of the primary consequences of infection is the collapse of epidermal cells, it was not difficult to see that infected areas would very rapidly lose control over their capacity to regulate stomatal aperture. This could lead to a drop in leaf water content as a whole before mesophyll cells collapsed, and the infected area was sealed off from the rest of the leaf by necrotic reactions in the surrounding tissue. Controlled dehydration of Maris Mink and Golden Promise barley leaves with manitol solutions indicated that the leaf photosynthetic carbon dioxide uptake could be reduced by the necessary amount if leaf water contents were reduced by 3%. The initial rapid reduction in net photosynthetic rates in leaf blotch could thus be explained as a consequence of the regulation of stomatal aperture to compensate for higher rates of water loss from infected areas of the leaf.

This explanation was less satisfactory for the case of the Maris Mink/mildew combination where the degree of cell damage was relatively smaller, the necrotic reaction was relatively quicker in developing and the infected leaves showed less tendency to desiccate even in the dry air stream of the apparatus. An alternative explanation should therefore be considered in this case. High sugar levels in the leaf should be correlated with high photosynthetic rates. However, in certain circumstances an inverse relationship seems possible and a case of this type was discussed on a previous occasion for sugar beet (Habeshaw, 1973). The argument in this case was that sinks for the removal of sugar from the leaf were necessary to prevent a build up of sugar in leaf cells which would prevent the transport of sugars from the chloroplast. The inability of the chloroplast to dispose of the products of photosynthesis was thought to produce a reduction in photosynthetic rate. The presence of fungal lesions would create alternative sinks for photosynthate along the veins of the leaf effectively reducing the sugar flow out of the leaf to sinks in the rest of the plant. Cereal leaves like sugar beet leaves, do not normally store starch in their chloroplasts so the interruption of the loading sugar into the transport system could lead to a rise in sugar level which might inhibit photosynthesis in the areas of the leaf surrounding infected areas in which sugars accumulate. The results show that in the initial stages of infection leaf sugar levels do increase significantly in three out of four of the combinations examined, so there was some justification in considering this hypothesis. However, if such a relationship was present there should be a positive correlation of increased sugar levels with increasing reductions in net photosynthesis. No such correlation was observed at the 5% level of significance so that the hypothesis must be rejected. The 'wider' response of net photosynthesis in the case of Maris Mink/mildew might still have been due to a disturbance in water balance of the leaf due to the damage to leaf cells consequent to infection, or possibly it was due to the presence of some unknown inhibitory substance produced by the infecting organism or the leaf.

The immediate practical considerations of the work relate to the effect these physiological reactions have to the progress of the disease in the field. In Golden Promise/mildew infections the rate of photosynthesis in the uninfected portions of the leaf was maintained until a late stage in the progress of the disease, thus ensuring a good supply of food for the fungus enabling it to sporulate prolifically and spread rapidly. In Maris Mink/mildew infections the disease gains a foothold and eventually will destroy the leaf but at the expense of reducing the total leaf photosynthesis in the early stages of infection. The total food supply available to the fungus was thus less in this case reducing the capacity of the fungus to initiate a new cycle of infection. The response of Maris Mink and Golden Promise to *R. secalis* infections was similar for the reasons connected with its growth within the leaf and discussed earlier. Paradoxically, the 'compatible/susceptible' symptoms of the Maris Mink/leaf blotch combination seem to appear because the host seems able to tolerate the stress of infection better, allowing a higher level of infection to build up in the leaf before the photosynthetic output of the leaf fails completely (Table 3). The leaf of Golden Promise succumbed rapidly to *R. secalis* attack as far as photosynthetic output was concerned, so rapidly in fact that the leaf desiccates before the fungus can exploit the food reserve represented by the leaf. This can be confirmed by reference to Fig. 6 where the increases of reducing sugar resulting from digestion of leaf cells by fungal enzymes were much less marked for Golden Promise/leaf blotch than Maris Mink/leaf blotch. In the field therefore *R. secalis* will spread more slowly in Golden Promise than in Maris Mink and this difference will tend to increase as conditions get drier. These predictions agree well with the recorded behaviour of these varieties of barley infected in field conditions as can be seen by reference to the results of the Scottish Agricultural Colleges cereal trials (Recommended varieties of cereals 1978).

Physiological investigations of this type would therefore seem to be useful adjuncts to investigations of crop/disease interactions and I would like to record my thanks to Dr. Lennard who suggested the subject of this investigation, and the technical staff who supplied the material used in the experiment.

ABBREVIATIONS*

r = correlation coefficient
b = gradient of best straight line fitted to the data
a = intercept of best straight line on the Y axis
N = number of paired observations
t(b) = Student t value for the test of the hypothesis that b differs significantly from the value 1.
t(r) = Student t value for test of significance of r.
P = probability that a value of r greater than that observed could arise by chance.

* See tables.

REFERENCES

Dyer, T.A. & K.J. Scott. Decrease in chloroplast content of barley leaves infected with powdery mildew. *Nature Lond.* 236: *237–238* (1972).

Habeshaw, D. Translocation and the control of photosynthesis in Sugar Beet. *Planta* (Berl.), 110(3): *213–226* (1973).

Olutiola, P.O. & P.G. Ayres. A cellulase complex in culture filtrates of *Rhynchosporium secalis* (barley leaf blotch). *Transactions of the British Mycological Society* 60: *273–282* (1973).

Scott, J.J. & R.M. Smilies. Metabolic regulation in diseased leaves; I. The respiratory rise in barley leaves infected with powdery mildew. *Plant Physiol.* 41: *289–297* (1966).

Somogyi, M. Notes on sugar determination. *J. Biol. Chem.* 195: *19–23* (1953).

Author's address:
East of Scotland College of Agriculture
Department of Botany
Kings Buildings
West Mains Road
Edinburgh EH 9 3J6
United Kingdom.

LIST OF CONTRIBUTORS

SUBJECT INDEX

Numbers in italics refer to the first of a series of pages on which the subject is broadly covered.

GPSR Compliance

The European Union's (EU) General Product Safety Regulation (GPSR) is a set of rules that requires consumer products to be safe and our obligations to ensure this.

If you have any concerns about our products, you can contact us on ProductSafety@springernature.com

In case Publisher is established outside the EU, the EU authorized representative is:

Springer Nature Customer Service Center GmbH
Europaplatz 3
69115 Heidelberg, Germany

Batch number: 10583998

Printed by Printforce, the Netherlands